T0221437

CLASSIFICATION METHODS
FOR REMOTELY SENSED DATA

SECOND EDITION

CLASSIFICATION METHODS
FOR REMOTELY SENSED DATA

SECOND EDITION

BRANDT TSO • PAUL M. MATHER

CRC Press
Taylor & Francis Group
Boca Raton   London   New York

CRC Press is an imprint of the
Taylor & Francis Group, an **informa** business

CRC Press
Taylor & Francis Group
6000 Broken Sound Parkway NW, Suite 300
Boca Raton, FL 33487-2742

© 2009 by Taylor & Francis Group, LLC
CRC Press is an imprint of Taylor & Francis Group, an Informa business

No claim to original U.S. Government works

International Standard Book Number-13: 978-1-4200-9072-7 (Hardcover)

Library of Congress Cataloging-in-Publication Data

Tso, Brandt.
    Classification methods for remotely sensed data / Brandt Tso, Paul Mather.
-- 2nd ed.
      p. cm.
    Includes bibliographical references and index.
    ISBN 978-1-4200-9072-7 (hardcover : alk. paper)
    1. Remote sensing. 2. Pattern recognition systems. I. Mather, Paul M. II. Title.

G70.4.T784 2009
621.36'78--dc22
                                            2009004453

Visit the Taylor & Francis Web site at
http://www.taylorandfrancis.com

and the CRC Press Web site at
http://www.crcpress.com

Contents

Preface to the Second Edition

The first edition of this book was written between 1998 and 2000, and was published in 2001. In the ten years that have elapsed since the start of this project, developments in image classification technology have been considerable. In order to keep this book relevant and up to date, two new chapters have been added and the existing chapters have been updated. The new chapters cover the topics of support vector machines (SVMs) and decision trees, both of which are now the subject of research articles in the major journals. SVMs represent a recent development in the computational aspects of image classification, and one of the problems they face is in the allocation of data to one of several (rather than one of two) classes. A number of approaches to this problem are presented in Section 4.4, but further experimentation is required. The number of potential approaches to the use of decision trees is considerable, and these are covered in Chapter 6. Developments such as boosting and random forest generation are described in Section 6.8. Lopping branches that do not contribute to the effectiveness of the decision tree is an important aspect of the design of the tree, which is covered in Section 6.7.

Acknowledgment is due to our institutions, the Management College, National Defense University and to The University of Nottingham, for providing support while this book was being revised. Brandt Tso also recognizes the benefits of his year as a postdoctoral fellow in the Remote Sensing Laboratory, Naval Postgraduate School, Monterey, California.

Preface to the First Edition

We classify objects in order to make sense of our environment, by reducing a multiplicity of phenomena to a relatively small number of general classes. On a country walk, for example, you might point to cows, trees, tractors, or swans. What you are actually doing is identifying an observed object and allocating it to a preexisting class, or giving it a name. Before setting out on the walk, you knew that swans existed, and you could specify their characteristics. When you saw a large white bird, possibly swimming in a canal or river, with an orange and black beak, you compared those characteristics to those of a swan and thus identified the bird, giving it the name or label of *swan*. We must be careful, therefore, to distinguish between the definition of the classes to which objects may belong and the identification or labeling of individual objects, and to avoid confusion between the two meanings of the word *classification*—the definition of categories of objects and the assignment or allocation of individual objects to these classes.

The example of the swan can also help to define other concepts. First, you must already have a model (or idealized representation) of the key features of a swan before you can recognize one. You learned, presumably in your childhood, the names of categories and subcategories of animals, plants, and other objects. Now you use that knowledge to identify and name the things you see and hear. In the literature of classification, this approach is termed *supervised learning*, meaning that you have divided the phenomena of interest into a number of *a priori* groups. You have observed a number of examples from each group, and have characterized them in terms of a number of discriminating features. The sample set is called *training data*, and this approach is known as *supervised classification*. In fact, it is supervised identification because it is assumed that the classification (the definition of the groups and their characteristics) has been defined before any previously unknown objects were identified.

An alternative approach, known as *unsupervised classification* or *clustering*, is also widely used. In this approach it is assumed that (1) you have little knowledge of the characteristics of the data set, (2) that you wish to determine whether any natural groupings exist in those data, and if so, (3) whether they can be identified in terms of phenomena of interest. In a sense, this procedure is akin to exploring the data (and visualization methods can help considerably in the process) whereas the supervised approach is inductive.

This book is about pattern recognition for remotely sensed data. We prefer the term *pattern recognition* to *classification* because the latter term can be misleading, as noted above. However, the two terms are used in this book, partly for reasons of tradition. A *pattern* is a set of measurements made on an object. It can be described as a mathematical vector of measurements. For example, a person's height and weight can be represented by the vector [192, 50] (in cm and kg, respectively). If a supervised approach is used, then the pattern is compared in some way to members of the sets of patterns that define the categories of interest, and the given pattern is assigned to one of these categories (one of which may be "unknown"). This approach can be described as inductive. Alternatively, a clustering strategy may be used that is based on the similarity between patterns, in order to determine whether any distinct groups of patterns exist in the data.

In Earth observation by remote sensing, the objects to be labeled are normally the individual pixels forming a multispectral or hyperspectral image. Each pixel is represented by a pattern vector consisting of a set of measurements, one per image band plus, possibly, other measurements such as texture. If each spectral band is represented by one axis of a multidimensional space (the *feature space*), then the pixel can be represented as a point in that space. For simplicity, let the number of features be two, and let the x-axis represent the first feature and the y-axis represent the second. A pixel with a feature vector of [1, 5] can therefore be shown on a graph as a point with Cartesian coordinates [1, 5]. Now imagine that all the pixels in the two-band image have been plotted on the graph, and that they fall into clearly defined groups. We can separate these groups by lines or curves. These lines or curves are called *decision boundaries*, for they show the positions of the boundaries of individual categories. If a point lies on one side of the boundary, it is given a label such as "A," whereas if it lies at the other side of the boundary it is given the label "B." In higher-dimensional problems, the lines and curves become *hyperplanes* and *hypersurfaces*. So the labeling problem can be thought of as one that involves the positioning of hyperplanes or hypersurfaces, representing decision boundaries, in a multidimensional feature space. The algorithm that determines the position of the pixel with respect to the decision boundaries, and thus allocates a specific label to that pixel, is called a *decision rule*. The word *classifier* is widely used as a synonym for the term *decision rule*.

The use of pattern recognition methods in remote sensing has a long history. Air photo interpreters were perhaps the first to use intuitive methods to determine the information contained in reconnaissance photographs, and these methods continue to be of great importance, particularly in the gathering of military intelligence, as the human eye–brain combination can make decisions and judgments on complex problems in milliseconds, using experience as a guide. Automatic methods are more suitable to the routine processing of images that show predictable patterns. Such methods have been developed and applied in a number of disciplines, ranging from speech and handwriting recognition, industrial process management, and medical diagnosis, as well as in the collection of military intelligence. The main distinguishing characteristic of Earth observation data is its volume. Hence, methods that can be applied in other applications may not be suited to the analysis of remotely sensed data because of the computational requirements. A further point to note is that there is often a discrepancy between the dimensionality of remotely sensed data sets and the volume of training data that is available. Where training data is sparse relative to the dimensionality of the data, it becomes difficult to estimate the characteristics of each training class, and so error may become significant. This phenomenon of increasing error with increasing data dimensionality is sometimes known as the *Hughes effect*.

Advances in technology have led to rapid developments in methods of pattern recognition, leading to the formulation of new and more sophisticated decision rules.

Some of those new methods have been introduced into the field of remote sensing and have shown encouraging results. A further feature of remote sensing applications in recent years is the use of combinations of data derived from different sensors or from different time periods, plus terrain and other data extracted from geographic information system (GIS) databases. In addition, the spectral information contained in remotely sensed images is often augmented by derived measures such as values of texture and context. In dealing with multidata sources, a significant problem is the considerable increase in the computational cost. Other problems, such as data scale and data reliability, must also be considered. There is an increasing interest in seeking methods for efficiently manipulating multisource data in order to increase classification accuracy. It should always be remembered, though, that sophisticated algorithms cannot compensate for lack of training data or an inadequate definition of the problem (in terms of the number and nature of the classes to be recognized relative to the scale of the study).

Texture is the tonal variation within an area. A simple example that illustrates the concept is the pattern on a carpet. If we treat each pattern as a whole, then the carpet can easily be described. If the carpet is seen as a set of small rectangular units, then the problem of describing its properties is more difficult. In some cases, texture information seems to be more effective than tonal information to describe the objects, and one can develop *texture features* corresponding to different kinds of patterns to improve the performance of a classifier.

Contextual information describes how the object of interest may be affected by its neighbors. For instance, English words starting with the letter "q" are more likely to be immediately followed by the letter "u" than "z" or "c." In the case of classification of an agricultural area, a pixel labeled as "carrot" is more likely to be surrounded by pixels of the same class rather than by other classes such as "water" or "wheat." The ability to model such contextual behavior may reduce confusion in the classification process.

The decision to write this book was triggered by our experience in attempting to use new methods of describing and labeling pixels in a remotely sensed image. While a number of valuable but more general textbooks are available for undergraduate use, we know of no coherent source of advanced information and guidance in the area of pattern recognition for research scientists and postgraduate research students in remote sensing, together with students taking advanced remote sensing courses. We hope that this book will contribute to the increased understanding and adoption of recently developed techniques of pattern recognition, and that it will provide readers with a link between the remote sensing literature and that of statistics, artificial intelligence, and computing. We do suggest, however, that attention be paid to experimental design and definition of the problem, for an advanced pattern recognition procedure is no substitute for thinking about the problem and defining an appropriate set of features.

Chapter 1 introduces the basic concepts of remote sensing in the optical and microwave region of the electromagnetic spectrum. This chapter is intended to introduce the field of remote sensing to readers with little or no background in this area, and it can be omitted by readers with adequate background knowledge of remote sensing.

Chapter 2 introduces the principles of pattern recognition. Traditional decision rules, including the supervised minimal distance classifier, Gaussian maximum likelihood, and unsupervised clustering techniques are described, together with other methods such as fuzzy-based procedures and decision trees. The chapter also contains brief accounts of dimension reduction methods, including orthogonal transforms, the assessment of classification accuracy, and the principles underlying the choice of training data.

Chapter 3 describes widely used neural network models and architectures including the multilayer perceptron (also called the feed-forward neural network), Kohonen's self-organized feature map, counterpropagation, the Hopfield network, and networks based upon adaptive resonance theory (ART).

Chapter 4 deals with pattern recognition techniques based on fuzzy systems. The main topics of this chapter are the construction of fuzzy rules, fuzzy mapping functions, and the corresponding decision processes.

Chapter 5 presents a survey of methods of quantifying image texture, including fractal- and multifractal-based theory, the multiplicative autoregressive random field model, the grey level co-occurrence matrix, and frequency domain filtering.

Chapter 6 addresses the theory and the application of Markov random fields. The main application of Markov random fields is to model contextual relationships. Other related topics, including function formulation, image restoration, robust estimation in the presence of noise (outliers), and the derivation of Markov-based texture measures, are also presented.

Chapter 7 provides several approaches for dealing with multisource data. The methods described include the extension of Bayesian classification theory, evidential reasoning, and Markov random fields.

No one is more aware of a book's deficiencies and inadequacies than its authors. Even Socrates, after a lifetime of learning, is reported to have been impressed by the extent of his own ignorance. Had more time and space been available, the book would have contained a longer account of the use of wavelets in texture analysis, and of the applications of decision tree classifiers. Publishers, who live in the real world, impose constraints of time and space while authors naturally attempt to rewrite their manuscripts every month in order to include the latest developments. We hope that we have reached a happy compromise that should satisfy most readers. We have included references to further work in order to guide the more advanced reader toward the relevant literature.

Most of the research underlying the ideas presented in this book was carried out while Dr. Brandt Tso was a postgraduate student, and later a postdoctoral fellow, in the School of Geography, The University of Nottingham, under the supervision of Professor Paul M. Mather. The second author provided encouragement, support, contributions to the first three chapters, and numerous rewrites of the draft. We realize that any book written by human authors is necessarily flawed, and we accept responsibility for any errors that may be contained in these pages.

The School of Geography, The University of Nottingham, provided computing facilities as well as a stimulating and encouraging environment for research. The second author is grateful to his many postgraduate research students from different parts of the world who have, over the past decade or so, educated and trained him in many areas of remote sensing. In particular, he would like to thank Valdir Veronese, Taskin Kavzoglu, Carlos Vieira, and Mahesh Pal, who have carried out research projects in areas relevant to the subject matter of this book. The contribution of others, while not directly related to the topic of image classification, has helped by broadening the intellectual debate within my research group as well as helping in many other ways. We also thank Dr. M. Koch of Boston University for help and guidance with the Red Sea Hills data set, which is used in a number of examples in this book. The help, good humor, and patience of Tony Moore of Taylor & Francis is greatly appreciated. Finally, both authors recognize the contributions of their families, and dedicate this book to them.

Brandt Tso
Taipei, 2000

Paul M. Mather
Nottingham, 2000

Author Biographies

Dr. Brandt Tso has served as a scientific officer in the Taiwan military service, studying modeling and pattern recognition techniques since 1988. Dr. Tso completed his Master's degree in information science at the Management College, National Defense University, Taiwan. Between 1994 and 1998, Dr. Tso was awarded a scholarship by the Taiwan government to pursue a Ph.D. degree in the School of Geography, The University of Nottingham, U.K., under the supervision of Professor Paul M. Mather, concentrating on the field of remotely sensed data classification. In 2003, Dr. Tso was an invited postdoctoral fellow in the Remote Sensing Laboratory, Physics Department, Naval Postgraduate School, Monterey, California, U.S.A., to study more complex remotely sensed data classification skills. Currently, Dr. Tso is an associate professor in the information science department, Management College, National Defense University, Taiwan. His main research interests include remotely sensed data recognition, real scene image retrieval, and machine learning algorithms. He has published numerous research papers relating to these fields.

Professor Paul M. Mather graduated with a degree in geography from the University of Cambridge in 1966. He then moved to The University of Nottingham to conduct research for his Ph.D. in geomorphology, which was awarded in 1969. As lecturer, senior lecturer, and full professor (1988), his attention was focused on the use of multivariate analysis in physical geography, a subject on which he published a detailed and well-received monograph in 1976. By the 1980s his interest in multivariate analysis had branched out to include remote sensing. In 1987 the first edition of his book *Computer Processing of Remotely Sensed Data* was published. It is now in its third edition (2004) with a fourth edition in preparation. He has always been fascinated by the applications of computers in physical geography and the environmental sciences, and is a proficient Fortran programmer. He retired in 2006, and was made Emeritus Professor. He received the Back Award from the Royal Geographical Society for his work in remote sensing in 1992, and in 2002 was awarded the Order of the British Empire (OBE) by Her Majesty Queen Elizabeth II for services to remote sensing. He has lectured in a number of countries around the world.

1 Remote Sensing in the Optical and Microwave Regions

Remotely sensed image data is widely used in a range of oceanographic, terrestrial, and atmospheric applications, such as land-cover mapping, environmental modeling and monitoring, and the updating of geographical databases. Hence, the quantized pixel values making up an image may be converted to physical values of radiance and related to some property of the surface being sensed. An example of this approach is the calibration of thermal infrared imagery to produce maps of temperature fields, such as sea surface temperature. In other applications, thematic information is required. A thematic map is one that displays the spatial variation of a specified phenomenon, such as land surface elevation, soil type, geology, or vegetation. It is this second approach that is considered in this book. The term *pattern recognition* is used to describe the procedures involved in relating vectors of measurements that are spatially referenced to individual pixel locations to the types or categories into which the phenomenon of interest is subdivided. If, for example, the phenomenon of interest is agricultural crops, then the categories are the individual crop types. Each crop type is represented in the thematic image by a numerical label. Vectors of measurements that are spatially referenced to individual pixel locations include image pixel values plus derived values such as texture, coherence, or context, as well as other geographical data that can be related to the pixel location, such as terrain elevation and slope, geology, and soil type.

Digital thematic maps can be represented in two ways, using either the raster or the vector models. The *vector model* uses the classical cartographic representation of map objects in terms of points, lines, and areas. Using this model, a continuously varying spatial attribute such as terrain elevation is represented by contour lines and spot heights, while an attribute such as soil type or underlying geology is represented in terms of boundary lines that enclose areas that are homogenous with respect to the property of interest and at the chosen scale of observation. The *raster model* represents spatial attributes in terms of their values over a contiguous set of small individual (and usually square) areas. Thus, variations in land surface elevation over a region of interest are represented by numerical values stored in a rectangular grid or raster, each element of which is the average elevation of the ground area represented by that element or cell of the grid. This representation of terrain height variation is known as a *digital elevation model* or DEM.

In the same way, variations in geology in a study region are stored in raster format as a set of labels. Each cell in the data array is given a numeric label that is linked to a description. For example, pixels with the Label 1 may be described as Carboniferous Limestone, while pixels given the Label 2 might be described as Silurian Grit. In this book, we are concerned with spatial data (specifically, remotely sensed images) that are represented in terms of rectangular rasters.

In the examples given in the preceding paragraphs, each cell of the raster stores either a physical value (such as elevation in meters above a given datum) or a label (such as 1 or 2, indicating rock types such as granite and basalt). In practice, rasters (and particular raster data sets containing images for display) are generally stored in the form of integer rather than floating-point numbers, in order to conserve storage space. Thus, a DEM may be represented in terms of an array of integers in the range 0 to 255, with each integer value representing a range of land surface elevations. The idea is similar to the use of a key in a printed map in which elevation is generally shown in shades of green and brown, with the map key showing the relationship between these hues and specific elevation ranges. In the same way, the values 0 to 255 contained in the DEM are connected to a range of real elevation values by the use of a table rather than a key. The table, known as a *lookup table* or LUT, allows the user to determine the actual range of elevations denoted by a particular label, such as 215. In some applications, the physical elevation value is required (for instance, if slope angle is to be calculated). Other questions can be answered by using the counts or labels directly. If Label 18 is used to represent land with a surface elevation between 200 and 205 meters, then the locations of such areas can be achieved by searching the raster for all values of 18 rather than converting the raster labels back to physical values and searching for cells holding numbers in the range 200 to 205.

Remotely sensed images are stored and manipulated in raster form. Each element of the raster is known as a pixel, and the value contained in any pixel location is simply a quantized count or a label rather than a physical value. For some applications, such as pattern recognition, these counts can be used directly as we are interested in interpixel similarities and differences. In other applications, such as sea-surface temperature determination, the pixel counts must be converted to physical values of radiance or reflectance, with corrections applied for such factors as sensor calibration changes and atmospheric influences. The range of quantized counts used in a raster representation of an image ranges from 0 to 255 (8-bit representation, for images derived from sensors such as Landsat Enhanced Thematic Mapper + (ETM++ and SPOT High Resolution Visible [HRV]), to 0 to 1023 (10-bit representation, as used for Advanced Very High Resolution Radiometer [AVHRR] data to more complex formats. For example, raw synthetic aperture radar data are commonly represented in terms of two 16-bit integers per pixel, with the first integer representing the real part and the second integer representing the imaginary part of a complex number. Whatever the precision (8-, 10-, or 16-bit) these pixel values are stored as rectangular rasters, and can be held in a computer in the form of a two-dimensional array. The data set used in pattern recognition consists of a number of co-registered raster images representing, for example, the measurements in the individual bands of a multispectral or hyperspectral image, the ground elevation (DEM), or some other spatial property of interest. The number of features used to represent terrain conditions is known as the *dimensionality* of the data. Multispectral and radar data have a low dimensionality; for instance, SPOT HRV produces three bands of data, and Landsat-7 ETM+ generates seven bands in wavelengths ranging from the optical to the thermal infrared, plus a panchromatic band, while the radar satellite ERS-2 operates in a single band. Hyperspectral sensors such as Airborne Visible/Infrared Imaging Spectrometer (AVIRIS), Compact Airborne Spectrographic Imager (CASI) and the Digital Airborne Imaging Spectrometer (DAIS) have the ability to collect data in tens or hundreds of narrow spectral bands. One problem in the classification of high-dimensional remotely sensed data is the paucity of samples ("training data") relative to the dimensionality of the feature space. This problem leads to difficulties in estimating statistical parameters such as the mean and covariance matrix.

The aim of pattern recognition in the context of remote sensing is to link each object or pixel in the study area to one or more elements of a user-defined label set, so that the radiometric information contained in the image is converted to thematic information, such as vegetation type. The process can be regarded as a mapping function that constructs a linkage between the raw data and the user-defined label set. A simple example is shown in Figure 1.1. Normally, each object or pixel is linked to a single label. However, it is also possible to perform a *one-to-many* mapping, so that a given pixel can be associated with more than one label, with the differing degrees of association between the pixel and each label being expressed as probabilities of membership. Alternatively, a *many-to-one* scheme will link groups of pixels to a single label. This approach can be used, for example, to give the same label to all of the pixels in a single agricultural field.

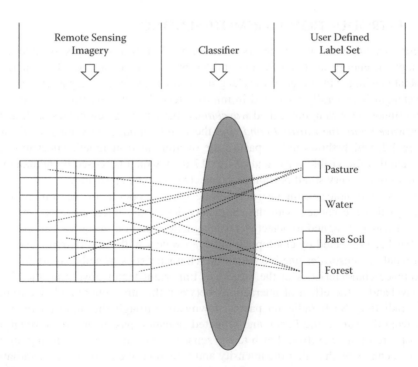

FIGURE 1.1 The concept of the classifier as a link between an image (left) and a set of labels.

Each application generally requires a different methodology, and each methodology is likely to generate different results. If reliable results are to be obtained, the analyst should understand the behavior of the method being used in order to achieve a satisfactory performance. For instance, the performance of a statistical procedure is strongly affected by the accuracy of the estimates of parameters such as the mean vector and the variance–covariance matrix for each class, which are obtained from samples of pixels called *training data sets*. Equally, the design of the architecture of a feed-forward neural net has an equally important impact on performance of the network.

The aim of this book is to provide a survey of pattern recognition methodology for use with remotely sensed imagery. Besides describing traditional approaches, more advanced techniques using artificial neural networks, support vector machines, fuzzy theory, and decision trees are introduced, and considerable space is devoted to the discussion of textural and contextual features. Some particular issues, such as pattern recognition using multiple data sources, change detection, and the analysis of mixed pixels, are also illustrated.

In the main part of this introductory chapter, the principles underlying remote sensing in the optical and microwave regions of the spectrum are described. Some important preprocessing techniques, such as corrections for atmospheric and topographic effects and noise filtering models, are also presented. These techniques are helpful in improving thematic accuracy in some kinds of applications, for example, in change detection.

1.1   INTRODUCTION TO REMOTE SENSING

Remote sensing is the use of sensors installed on aircraft or satellites to detect electromagnetic energy scattered from or emitted by the Earth's surface. This energy is associated with a wide range of wavelengths, forming the electromagnetic spectrum. Wavelength is generally measured in micrometers (1×10^{-6} m, μm). Discrete sets of continuous wavelengths (called *wavebands*) have been given names such as the *microwave band*, the *infrared band*, and the *visible band*. An example is given in Figure 1.2, which shows only a part of the overall electromagnetic spectrum. It is apparent that the visible waveband (0.4 to 0.7 μm), which is sensed by human eyes, occupies only a very small portion of the electromagnetic spectrum.

A specific remote sensing instrument is designed to operate in one or more wavebands, which are chosen with the characteristics of the intended target in mind. Thus, a sensor designed to detect electromagnetic radiation in the visible spectrum that has been reflected by chlorophyll in ocean waters will use different wavebands than would a sensor designed to detect the characteristics of soils. Apart from the reflectance characteristics of the target, an important factor governing the choice of waveband is the effect of interactions between the atmosphere and electromagnetic radiation. Such radiation passes downward through the atmosphere on its way from the sun to the Earth, and reflected radiation passes upward through the atmosphere on its way from Earth to the sensor. Absorption and scattering are the main mechanisms that alter the intensity and direction of electromagnetic radiation within the atmosphere. In some regions of the optical spectrum, these mechanisms (principally absorption) ensure that remote sensing is impossible. Spectral regions of

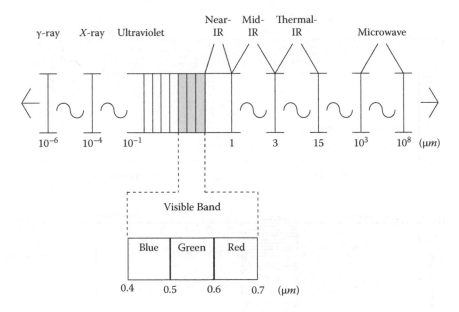

FIGURE 1.2 Regions of the electromagnetic spectrum that are of interest in remote sensing applications.

lower absorption are known as *atmosphere windows* (Figure 1.3), though it should be remembered that scattering and absorption affect all wavebands in the optical spectrum to a greater or lesser degree, and these effects are variable both in space and time.

1.1.1 ATMOSPHERIC INTERACTIONS

Electromagnetic radiation interacts with the Earth's atmosphere, the degree of interaction depending on the wavelength of the radiation and the local characteristics of the atmosphere. The basic interactions are known as *scattering* and *absorption*. Scattering is more likely to occur at shorter wavelengths. The most common scattering behavior is known as *Rayleigh scattering*, which is the main cause of haze in remotely sensed imagery. The atmosphere has different levels of absorption at different wavelengths. Regions of the spectrum that have a relatively high transmission are called atmospheric windows (Figure 1.3). The energy in some wavebands (e.g., from 15 to 10^3 μm) is almost completely absorbed by the atmosphere. These wavelengths cannot, therefore, be used for remote sensing. Wavebands with a high transmission could be potential candidates for remote sensing missions.

1.1.2 SURFACE MATERIAL REFLECTANCE

The choice of wavebands for remote sensing in the optical region is also affected by the characteristics of the surface material. Energy that is incident upon a target can be separated into three components, namely, energy that is transmitted,

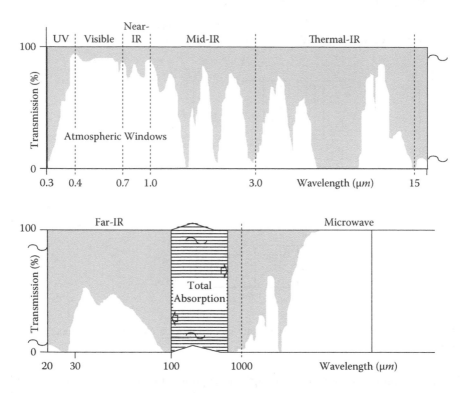

FIGURE 1.3 Atmospheric windows (unshaded). Vertical axis is atmospheric transmission (%). Horizontal axis is the logarithm of the wavelength in micrometers.

absorbed, and reflected. Surface material reflectance characteristics may be quantified by the spectral reflectance, which is a percentage measure obtained simply by dividing reflected energy in a given waveband by incident energy. The quantity of the reflected energy depends mainly on three factors: the magnitude of the incident energy, the roughness of the material, and the material type. Normally, the first two factors are regarded as constants. Therefore, only the third factor (i.e., the material type) is considered. However, it is worthwhile to first describe how roughness affects the reflected energy.

Surface roughness is a wavelength-dependent phenomenon. Given the same material, the longer the wavelength the smoother the material appears. For a perfectly smooth (specular) surface, reflected energy travels only in one direction such that the reflection angle is the same as incidence angle. In the case of a perfectly rough (Lambertian) surface, incident energy is reflected equally in all directions (Figure 1.4). However, in practical applications, most surface materials act neither as specular nor Lambertian reflectors; their roughness lies somewhere between these two extremes.

Figure 1.5 shows the average reflectance over the optical region of the spectrum for three ideal surface materials: dry bare soil, clear water, and green vegetation. This graph shows how these surface materials can be separated in terms of their reflectance spectra. It is apparent that vegetation reflectance varies considerably

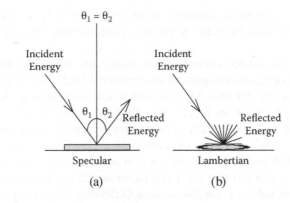

FIGURE 1.4 (a) Specular and (b) Lambertian reflectance.

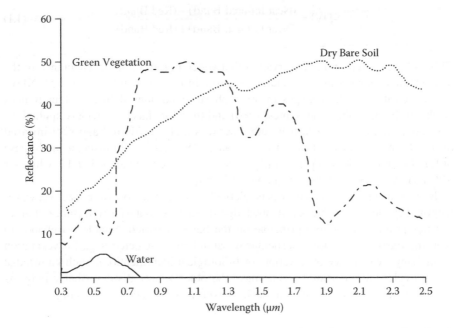

FIGURE 1.5 Typical spectral reflectance (%) of three materials: green vegetation, dry bare soil, and water.

across these wavebands. The lowest reflectance values occur at 0.4 μm (i.e., in the blue waveband), while the highest reflectance values occur around the near-infrared and part of the mid-infrared bands. The reflectance spectrum of bare soil, in contrast, shows reflectance increasing smoothly with wavelength. Its reflectance in the visible waveband is greater than that of vegetation, while in near-infrared and part of mid-infrared bands bare soil reflectance becomes less than that of vegetation, and eventually it dominates again beyond wavelengths of around 1.4 μm. The high near-infrared reflectance of vegetation, combined with its low reflectance in the red waveband, is used in the construction of vegetation indices. One such index, the NDVI (Normalized Difference Vegetation Index), is described in the following text.

Recent interest in global environmental change has led to the development of global climate models, which require inputs describing land cover type. The AVHRR sensor carried by the National Oceanographic and Atmospheric Administration (NOAA) satellite series (from NOAA-6 to NOAA-18) provides data that are useful for large-scale studies of terrestrial vegetation (Eidenshink and Faundeen, 1994; Lim and Kafatos, 2002). A variety of mathematical combinations (e.g., subtraction, ratioing) of AVHRR band 1 (red) and band 2 (near infrared) have been developed to characterize the spatial distribution of vegetation and its condition. The best known of these vegetation indices is the Normalized Difference Vegetation Index (NDVI), defined as

$$\mathrm{NDVI} = \frac{(\text{Near Infrared Band}) - (\text{Red Band})}{(\text{Near Infrared Band}) + (\text{Red Band})} \qquad (1.1)$$

This index is based on the observation that vegetation has a high reflectance in the near infrared band, while reflectance is lower in the red band (Figure 1.5). NDVI is less sensitive to the changes of atmospheric conditions than are other indices (Jackson, 1983; Holben and Kimes, 1986) and therefore has been widely applied for vegetation monitoring. Data from other sensors that provide red and near-infrared (near-IR) images can also be used to generate NDVI images. For instance, in the case of Landsat Thematic Mapper ™ images, NDVI is based on bands 5 and 7, while for SPOT HRV, NDVI is derived from bands 3 and 2.

Reflectance from water surfaces is relatively low, and is more or less zero at wavelengths beyond the visible red. Knowledge of surface material reflectance characteristics provides us with a principle on the basis of which suitable wavebands to scan the Earth surface for a particular mission can be selected (e.g., for vegetation monitoring, sea surface observation, or lithological identification). Such knowledge also provides an important basis to make the objects more distinguishable in terms of multiband image manipulations such as overlay, subtraction, or ratioing.

1.1.3 SPATIAL AND RADIOMETRIC RESOLUTION

The resolution of a remote sensing instrument can be expressed in terms of its spatial and radiometric resolution. The higher the spatial resolution the smaller the ground objects that can be distinguished. The spatial resolution is related to the *instantaneous field of view* (IFOV) of the sensor, which denotes the size of the area from which the sensor receives the energy at the given instant in time. In Figure 1.6 the energy transmission path to the sensor takes a conelike shape, and ground resolution is roughly equivalent to the diameter of the circle formed by the intersection of this cone and the ground surface. As the sensor scanning area moves away from the nadir, the larger the IFOV will be (Figure 1.6), and thus results in the distortion of the resulting image as the spatial scale decreases from the left and right edges of

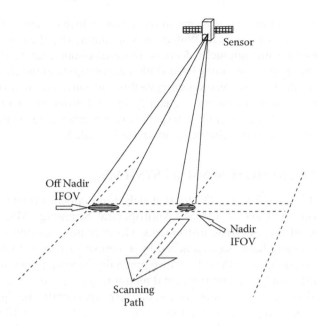

FIGURE 1.6 Showing the variation in the size of the instantaneous field of view (IFOV) of a sensor across a scan line.

the image toward the center. The correction for this scale distortion effect (and other effects due principally to the platform's orbit characteristics and to the eastward rotation of the Earth during scanning) is called *geometric correction*. This correction is usually performed by constructing a transform (using either an empirical procedure based on least squares methods or an analytical procedure using orbital information) in order to map ground coordinates to their corresponding image coordinates, and vice-versa. A review of local and global methods of geometric correction and image registration is provided by Brown (1992) and Zitová and Flusser (2003).

One's instinctive feeling might be that the finer the spatial resolution the better. However, in practice, this may not always be the case. The choice of spatial resolution should depend on what we want to see. For instance, in everyday life, we recognize a human face from the combination of features such as eyes, nose, lips, etc. Therefore, the appropriate spatial resolution should be set at a level to allow us to recognize each feature in the context of the whole face. It is arguable whether an increase in the spatial resolution of our eyes is likely to improve our identification of faces, once a certain limit has been reached. Thus, the choice of spatial resolution is case dependent.

Reducing the spatial resolution in terms of narrowing the instrument's IFOV also affects a related issue. The use of a smaller IFOV implies that the quantity of energy received by a detector is less (because area from which energy is collected is smaller, and the time available for the sensor to detect the upwelling energy is also shorter). It follows that the instrument's sensitivity to changes in the levels of energy will decrease, and thus the sensor may not be able to distinguish slight energy differences along a scan line. In other words, the radiometric resolution is degraded. A smaller

IFOV may thus result in a worse signal-to-noise ratio, which means that the signal is contaminated by more noise as IFOV decreases (assuming that the same radiometric resolution applies). Although such a decrease in signal-to-noise ratio can be compensated for by enlarging the scanning bandwidth (i.e., the region of the electromagnetic spectrum from which sensor receives energy), this will cause a reduction in the spectral resolution (i.e., sensor's ability to quantify spectral differences). Overall, it can be concluded that enhancement of both spatial and radiometric resolution cannot be achieved together, and some kind of compromise is needed.

1.2   OPTICAL REMOTE SENSING SYSTEMS

Over the past 20 to 25 years, the most widely used optical remote sensing systems have been the Landsat TM and Multispectral Scanning (MSS), the SPOT HRV, and the NOAA AVHRR instruments. Other remote sensing satellites carrying optical sensors have been launched by a number of national space agencies and private companies, and these have also gradually become the main sources for serving scientific study and environmental monitoring purposes. Examples are the Chinese-Brazilian remote sensing system, the Terra spacecraft, the Space Imaging Corporation's IKONOS satellite the European Envisat, the Indian IRS series, and several Japanese experimental projects. High-resolution sensing devices (4 m in multispectral mode, one meter or less in panchromatic mode) are now well established (e.g., the QuickBird satellite, and IKONOS). Radar systems are also becoming more numerous (Canada's Radarsat 1 and 2, the German TerraSAR-X, the European Advanced Synthetic Aperture Radar (ASAR) on Envisat, and the Italian COSMO-SkyMed X-band system). Considerable interest is also being shown in the application of hyperspectral imagery. Whereas multispectral sensors such as the Landsat ETM+ collect upwelling radiation in a small number of broad wavebands (seven in the case of the ETM+ instrument), a hyperspectral sensor collects data in a large number of very narrow wavebands. An example is the DAIS instrument, which collects data in 79 bands in the wavelength range 0.4 to 12.6 μm. The width of each spectral band varies from 15 to 20 nm in the visible to 2 μm in the middle infrared. The Compact Airborne Spectrographic Imager (CASI) instrument allows data to be collected for any 545-nm segment of the 0.4- to 1.0-μm region in 288 bands, with the bands spaced at intervals of approximately 1.9 nm. NASA's AVIRIS acquires data in 224 bands in the range 0.38 to 1.5 μm with a bandwidth of 10 nm. Up until the present time, hyperspectral data have been collected by aircraft-mounted sensors. NASA's Earth Observer I, launched in 2000, is the first orbiting spacecraft to carry a hyperspectral imager, Hyperion. The Hyperion instrument collects data over a narrow swath in 220 bands of 10 μm width. The large number of spectral bands produced by hyperspectral sensors (i.e., the high dimensionality of the data) poses significant problems in the pattern recognition process. These problems are discussed in Chapter 2.

Another class of satellites is the *small sat*, constructed using off-the-shelf components. The leader in this area is Surrey Satellite Technology Limited, which has developed the DMC or Disaster Monitoring Constellation of small satellites, which have been purchased by several governments, including Algeria, Nigeria, China, and the United Kingdom. The imaging sensor carried by these satellites is comparable to Landsat's TM in the visible and near-IR bands.

An exhaustive list of sensors and satellites is not supplied here; such lists tend to become outdated very quickly. The advent of the World Wide Web (WWW) means that researchers, teachers, and students can readily obtain up-to-date information via the Internet. Professional and learned societies, such as the Remote Sensing and Photogrammetry Society, maintain WWW pages that provide links to international and national space agencies and projects. The Committee for Earth Observation Satellites (CEOS) also maintains an information site.

Details of the operation of the main types of sensor carried by remote sensing satellites can be found in textbooks (e.g., Lillesand and Keifer, 2000; Mather, 2004) and readers who may be unfamiliar with these details are referred to one of these sources.

1.3 ATMOSPHERIC CORRECTION

Electromagnetic energy detected by remote sensing instruments (especially those that operate in the optical region of the spectrum) consists of a mixture of energy reflected from or emitted by the ground surface and energy that has been scattered within or emitted by the atmosphere. The magnitude of the electromagnetic energy in the visible and near-infrared region of the spectrum that is detected by a sensor above the atmosphere is dependent on the magnitude of incoming solar energy (irradiance), which is attenuated by the process of atmospheric absorption, and by the reflectance characteristics of the ground surface. Hence, energy received by the sensor is a function of incident energy (irradiance), target reflectance, atmospherically scattered energy (path radiance), and atmospheric absorption. Interpretation and analysis of remotely sensed images in the optical region of the spectrum is based on the assumption that the values associated with the image pixels accurately represent the spatial distribution of ground surface reflectance, and that the magnitude of such reflectance is related to the physical, chemical, or biological properties of the ground surface. Clearly, this is not the case unless corrections are applied to take account of variations in solar irradiance and in the magnitude of atmospheric absorption and scattering, as well as in the sensitivity of the detectors used in the remote sensing instrument. The response of these detectors to a uniform input tends to change over time. Correction for these effects is vital if thematic images of a given area are to be compared over time; for example, over a crop-growing season.

The necessity for atmospheric correction depends on the objectives of the analysis. In general, land cover identification exercises that are based on single-date images do not require atmospheric correction, as pixels are being compared to other pixels within the image in terms of similarity, for example. The validity of this statement depends also on the quality of the image (for example, an image displaying severe haze effects, spatially varying haze phenomena, or cloud–cloud shadow effects may be unsuitable for classification). Atmospheric correction and sensor calibration are necessary when multisensor or multidate images are being classified, or where the aim of pattern recognition is to identify land cover change over time, in order to ensure that pixel values are comparable from one image to the next in a temporal sequence. A more radical view is taken by Smith and Milton (1999, p. 2653), who emphasize that "to collect remotely sensed data of lasting quantitative value then data must be calibrated to physical units such as reflectance."

The value recorded for each pixel in a remotely sensed image is a function of the sensor-detected radiance. Owing to the atmospheric interaction, this *apparent* radiance is the combination of the contribution of the target object and the atmospheric effect. Their relationship can be approximated as

$$L_{app} = \rho TE/\pi + L_p. \qquad (1.2)$$

Here, L_{app} denotes the apparent radiance received by the sensor, L_p is the path radiance, ρ is the target reflectance (%), T is the atmospheric transmittance (%), and E is the solar irradiance on the target. Radiance is expressed in units of $Wm^{-2} sr^{-1} \mu m^{-1}$, and irradiance is expressed in the units of $Wm^{-2} \mu m^{-1}$. As these two terms are not expressed in equivalent units, solar irradiance is converted into *equivalent solar radiance* by introducing the term π into the denominator. This conversion is based on the assumption that the target behaves as a Lambertian reflector (as described in Figure 1.4) (Mackay et al., 1994).

In Equation (1.2), only the first term ρ contains information about the target. The atmosphere contributes the second term, the path radiance Lp, which varies inversely in magnitude with wavelength. In the case of multispectral images, the magnitude of the L_p term in the visible bands will be higher than that in the near- or mid-IR bands.

1.3.1 Dark Object Subtraction

Two kinds of methods are used for atmospheric effect correction. The first kind consists of the dark object subtraction techniques (Chavez, 1988; Ouaidrari and Vermote, 2001), which involve subtraction of a constant value (offset) from all pixels in a given spectral band. These methods are based on the assumption that some pixels in the image should have a reflectance of zero, and that the values recorded for these zero pixels result from atmospheric scattering. Thus, these pixel values represent the effects of atmospheric scattering. For example, the reflectance of deep clear water in the near-IR waveband is near zero. Dark object subtraction methods assume that nonzero pixel values over deep, clear, water areas in the near-IR band are contributed by the path radiance L_p, and that the path radiance is spatially constant (meaning that a single value of path radiance is subtracted from all pixel values in the image). In case the of the visible bands, one may use shadow areas due to topography as dark objects. In effect, one is using the histogram offset as a measure of atmospheric path radiance, as shown in Figure 1.7.

The empirical line method is described and evaluated by Smith and Milton (1999). The method requires that ground measurements of surface reflectance of dark and bright areas in the image are taken simultaneous with the overflight, though if the areas are spectrally stable, the measurements need not be simultaneous (though they must be taken under similar atmospheric and illumination conditions). The image data are converted to radiance using the standard calibration, and surface

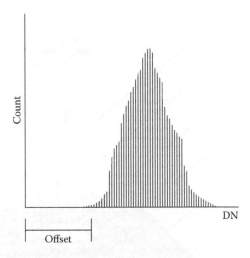

FIGURE 1.7   An estimate of path radiance is the image histogram offset from zero.

reflectance R is regressed against sensor radiance L to give a relationship of the form $R = (L - a) \times s$. The term a represents atmospheric radiance, which is subtracted from apparent radiance L before conversion to reflectance. Other descriptions of the empirical line methods are found in Moran et al. (2001) and Karpouzli and Malthus (2003).

1.3.2 MODELING TECHNIQUES

The methods described in Section 1.3.1 above involve the subtraction of the same pixel value from the whole image. They are easy to apply, but only provide an approximate correction. If the magnitude of the ground-leaving reflectance for each pixel is required, a more sophisticated method based on modeling techniques is necessary (e.g., Tanré et al., 1986, 1990; Vermote et al., 1997). These methods attempt to model atmospheric interactions, through which one can retrieve estimates of true target reflectance.

1.3.2.1 Modeling the Atmospheric Effect

The 5S model (Simulation of the Sensor Signal in the Solar Spectrum) developed by Tanré et al. (1986, 1990) is the best-known atmospheric calibration model. An improved version is the 6S (Second Simulation of the Sensor Signal in the Solar Spectrum) model (Vermote et al., 1997), which is capable of simulating a non-Lambertian surface to model the signal measured by the sensor. The 6S model also includes data for calculating atmospheric absorption using an increased number of atmospheric gases. In what follows, the core of the model is described in order to provide an introduction to atmospheric modeling.

Assume unit solar irradiance incident on top of the atmosphere. A fraction of the incident solar irradiance is scattered from the path between the sun and the ground target into the atmospheric volume, with the remainder of the radiation being

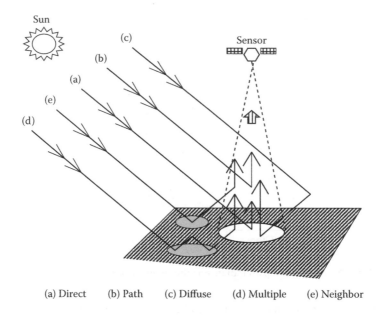

(a) Direct   (b) Path   (c) Diffuse   (d) Multiple   (e) Neighbor

FIGURE 1.8   Five kinds of radiative interaction with the atmosphere.

incident on the ground target as direct solar radiation (Figure 1.8a). This transmitted fraction, denoted by $T(\theta_s)$, is defined as

$$T(\theta_s) = \exp\left(\frac{-\tau}{\cos(\theta_s)}\right). \tag{1.3}$$

The term τ is known as the optical depth, and θ_s is the solar zenith angle, which is illustrated in Figure 1.9.

A fraction of the solar radiation that is scattered into the atmosphere will also appear to contribute to the illumination of the ground target (Figure 1.8c, marked "diffuse") and will compensate for some of the attenuation of the direct beam. If we

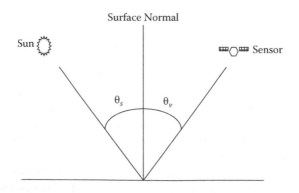

FIGURE 1.9   Solar and sensor view angles on a horizontal surface.

denote this diffuse skylight as $t_d(\theta_s)$, then the fraction of solar irradiance incident at the ground target becomes

$$T(\theta_s) = \exp\left(\frac{-\tau}{\cos(\theta_s)}\right) + t_d(\theta_s). \tag{1.4}$$

A second scattering flux that also should be taken into consideration is the trapping mechanism. The effect of this mechanism corresponds to the successive reflection and scattering of solar radiation between the ground target neighborhood and the atmosphere, so that the radiation then becomes incident upon the ground target, as shown in Figures 1.8d and 1.8e. The magnitude of this effect depends on the spherical albedo of the atmosphere, S, and the surface reflectance, ρ_s. The illumination at the ground target now becomes

$$\frac{1}{1-\rho_s S} \times T\left(\theta_s\right). \tag{1.5}$$

The proportion of the solar radiation reflected from the ground target is expressed as

$$\frac{\rho_s}{1-\rho_s S} \times T\left(\theta_s\right). \tag{1.6}$$

Consider a sensor that is receiving the reflectance from the ground target. The reflectance is generated by two main sources: one is the contribution of the total solar radiation reflected by the ground target and directly transmitted from the surface to the sensor, while the other is the contribution from the target neighborhood, which is scattered into the field of view of the sensor. The reflectance received by the sensor can thus expressed by

$$\frac{\rho_s}{1-\rho_s S} \times T\left(\theta_s\right) \times T\left(\theta_v\right). \tag{1.7}$$

In this equation, θ_v is the sensor view zenith angle, and Equation (1.4) can be again applied to express $T(\theta_v)$ as

$$T(\theta_v) = \exp\left(\frac{-\tau}{\cos(\theta_v)}\right) + t_d\left(\theta_v\right). \tag{1.8}$$

The sensor also receives a fraction of the solar radiation that has been scattered out of the downward solar beam into the sensor's field of view without interaction with the ground target, as shown in Figure 1.8b. This component is the atmospheric reflectance, denoted by the function $\rho_a(\theta_s, \theta_v, \phi)$, where ϕ is the relative azimuth between the sun and the sensor. Therefore, the apparent reflectance ρ^* at the sensor is

$$\rho^* = \frac{\rho_s}{1-\rho_s S} \times T(\theta_s) \times T(\theta_v) + \rho_a(\theta_s, \theta_v, \phi). \tag{1.9}$$

Equation (1.9) is a linear equation that specifies the relationship between the apparent reflectance ρ^* and the surface reflectance ρ_s. The term $T(\theta_s) \times T(\theta_v)$ is the total atmospheric transmittance along the sun–target–sensor path.

A second atmospheric interaction should be considered, namely, the process of absorption. In the solar (optical) spectrum, atmospheric gaseous absorption is principally due to the presence of ozone (O_3), oxygen (O_2), water vapor (H_2O), and carbon dioxide (CO_2). Both O_2 and CO_2 are uniformly mixed in the atmosphere and are constant in terms of their concentrations, whereas H_2O and O_3 concentrations vary with time and geographical location. If $T_g(\theta_s, \theta_v)$ denotes atmospheric gas transmittance after absorption, then Equation (1.9) can be modified to give

$$\rho^* = T_g(\theta_s, \theta_v) \times \left[\frac{\rho_s}{1-\rho_s S} \times T(\theta_s) \times T(\theta_v) + \rho_a(\theta_s, \theta_v, \phi) \right]. \tag{1.10}$$

In order to retrieve the surface target reflectance, ρ_s, Equation (1.10) is further expanded as

$$\frac{\rho^*}{T_g(\theta_s, \theta_v)} - \rho_a(\theta_s, \theta_v, \phi) = \frac{\rho_s}{1-\rho_s S} \times T(\theta_s) \times T(\theta_v)$$

$$\Rightarrow \frac{\rho^*}{T_g(\theta_s, \theta_v) \times T(\theta_s) \times T(\theta_v)} - \frac{\rho_a(\theta_s, \theta_v, \phi)}{T(\theta_s) \times T(\theta_v)} = \frac{\rho_s}{1-\rho_s S} \tag{1.11}$$

$$\Rightarrow \frac{\rho^*}{T_g(\theta_s, \theta_v) \times T(\theta_s) \times T(\theta_v)} - \frac{\rho_a(\theta_s, \theta_v, \phi)}{T(\theta_s) \times T(\theta_v)} = \frac{\rho_s}{1-\rho_s S}.$$

If we define

$$A = \frac{1}{T_g(\theta_s, \theta_v) \times T(\theta_s) \times T(\theta_v)}, \text{ and } B = -\frac{\rho_a(\theta_s, \theta_v, \phi)}{T(\theta_s) \times T(\theta_v)}$$

then Equation (1.11) eventually gives

$$\rho_s = \frac{A \times \rho^* + B}{[1 + S \times (A \times \rho^* + B)]}. \tag{1.12}$$

Thus, knowing the apparent reflectance ρ^*, the spherical albedo S, and the coefficients A and B, the surface target reflectance ρ_s can be obtained. Methods for obtaining these parameters are illustrated in Section 1.3.2.2.

1.3.2.2   Steps in Atmospheric Correction

The procedure to obtain estimates of the ground target reflectance involves three steps, as illustrated in Figure 1.10. The first step, converting the pixel value to radiance, is sensor-dependent. For instance, in the case of Landsat TM images, the relationship between the pixel values and apparent radiance L_{app} is expressed as

$$L_{app} = A_i \times DN + B_i \qquad (1.13)$$

while in the case of SPOT HRV data the relationship is

$$L_{app} = DN/A_i \qquad (1.14)$$

where A_i and B_i are the calibration *gain* and *offset* for band i, respectively. These calibration coefficients can be found either from the literature (e.g., Gellman et al., 1993; Thome et al., 1997; Teillet et al., 2001; Meygret, 2005) or from image header files. Note that header files may contain prelaunch calibrations that may differ significantly from the actual calibrations due to sensor degradation over time.

The second step, conversion from apparent radiance L_{app} to apparent reflectance ρ^*, is based on the observation that in case of 100% reflectance, the radiance measured by the sensor is the result of multiplication of equivalent solar radiance (E/π), the cosine of solar zenith angle ($\cos(\theta_s)$), and the Earth-to-sun distance multiplicative factor d (Kowalik and Marsh, 1982). The factor d is measured in astronomical units (au) and is described further below. One au is equal to the average Earth-to-sun distance. About January 3, at perihelion, the Earth-to-sun distance is approximately 0.983 au, and on July 5, at aphelion, the Earth-to-sun distance is about 1.0167 au.

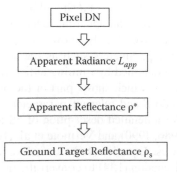

FIGURE 1.10   Procedure for retrieving ground radiance.

If the required coefficients are known, then the variation in sensor-detected apparent radiance L_{app} is caused by the difference in reflectance. One then obtains the following relation:

$$L_{app} = \frac{\rho^* \times \cos(\theta_s) \times E}{d\pi}$$

or, equivalently,

$$\rho^* = \frac{d \times L_{app}}{\cos(\theta_s) \times E/\pi}. \tag{1.15}$$

The value of the solar zenith angle can be retrieved from the image header file. The distance multiplicative factor d is used to compensate for the variation in solar irradiance E caused by the change in distance between the sun and the Earth. It is obtained from

$$d = au^2. \tag{1.16}$$

An alternative way to approximate Equation (1.16) is given by

$$d = \left(1 - 0.01673 \times \cos\left(0.9856 \times (JD - 4)\right)\right)^2 \tag{1.17}$$

where the term JD denotes the Julian day (e.g., in the case of 5th February, $JD = 36$).

The introduction of the factor d into Equation (1.15) is justified by the following observations. Since at perihelion, the magnitude of solar irradiance is greater than that at aphelion, if a sensor obtains the same radiance L_{app} at perihelion and aphelion, respectively, the apparent reflectance ρ^* at perihelion should be less than that at aphelion. Thus, if the sun-to-Earth distance is smaller than the average sun-to-Earth distance, one should use a lower (< 1) weighting factor d, and as the Earth is approaching aphelion, a higher value of d should be used.

The final step—that of converting apparent reflectance to ground target reflectance—uses Equation (1.12). We already know the apparent reflectance ρ^* from Equation (1.15), while other parameters such as the spherical albedo, S, and the coefficients A and B can be obtained by running either the 5S or the 6S model. An example is given in Table 1.1, which shows part of the output generated by the 5S model. The spherical albedo S and relevant parameters for deriving A and B are displayed in bold type. For a detailed description of 5S and 6S usage, readers are referred to Tanré et al. (1986, 1990) and Vermote et al. (1997). Once ground reflectance is retrieved, one may use Equation (1.15) to retrieve radiance followed by using either Equation (1.13) or Equation (1.14) to convert the radiance back to a corrected pixel value.

TABLE 1.1

Results from the 5S Model Used for Atmospheric Correction

	Downward	Upward	Total
Global gas transmission	0.961	0.965	$\mathbf{0.932} = T_g(\theta s, \theta_v)$
Water transmission	0.988	0.989	0.980
Ozone transmission	0.975	0.978	0.954
CO_2 transmission	1.000	1.000	1.000
O_2 transmission	0.999	0.999	0.998
NO_2 transmission	1.000	1.000	1.000
CH_4 transmission	1.000	1.000	1.000
CO transmission	1.000	1.000	1.000
Rayleigh scattering transmission	0.967	0.971	0.939
Aerosol scattering transmission	0.953	0.962	0.917
Total scattering transmission	0.922	0.934	$\mathbf{0.861} = T(\theta_s)\, T(\theta_v)$
	Rayleigh	**Aerosols**	**Total**
Spherical albedo	0.048	0.087	$\mathbf{0.123} = S$
Optical depth total	0.054	0.362	0.416
Optical depth plane	0.054	0.362	0.416
Atmospheric reflectance	0.019	0.017	$\mathbf{0.037} = \rho_a(\theta s, \theta v, \varphi)$
Phase function	0.993	0.099	0.215
Single scattering albedo	1.000	0.990	0.992

1.4 CORRECTION FOR TOPOGRAPHIC EFFECTS

Normally, the surface being measured by the remote sensor is assumed to be flat with a Lambertian reflectance behavior. Under this assumption, the magnitude of the radiance detected by the sensor is affected only by variations in the solar zenith angle, the wavelength, and the atmospheric interaction. The atmospheric correction model introduced in Section 1.3 is also based on such ideal assumptions, which may be invalid in the case of rugged terrain because the solar incidence angle will vary with topographic properties and will further contribute to differences in the level of radiance detected by the sensor. This is known as *topographic effect*. More specifically, the topographic effect can be defined as the variation in radiance exhibited by inclined surfaces compared to radiance from a horizontal surface as a function of orientation of the surface relative to the radiation source. Moreover, if we assume non-Lambertian reflectance for the surface being measured, the sensor position is another important variable that should be considered.

Calibration for topographic effects is intended to normalize the sensor-detected signal difference caused by the topographic variation. Various techniques (e.g., Smith et al., 1980; Colby, 1991; Dymond and Shepherd, 1999; Riano et al., 2003; Law and Nichol, 2004) have been published. Here, we present two approaches, using band ratioing and the Minnaert model, due to their better performance (Smith et al., 1980; Colby, 1991; Law and Nichol, 2004).

Band ratioing (Colby, 1991, Mather, 2004) is the most commonly used method for reducing the topographic effect. Colby (1991) uses a Landsat TM band 5/4 ratio image to study the effectiveness of topographic effect calibration by comparing several sample sites having the same vegetation cover but with differences in topography. In other words, the same vegetation cover should have a similar spectral response, irrespective of location, thus the differences between sample sites are assumed to be caused by topographic effects. Results showed that the variance of spectral response between sample sites in ratio image was lower than that obtained from original TM band 4 and 5 images. Colby (1991) concludes that band ratioing does partially compensate for topographic effects.

Smith et al. (1980) present two empirical photometric functions for studying the effect of topography on the radiance field. The first such function is based on a Lambertian reflectance assumption, while the second function assumes that reflectance is non-Lambertian. Although this is a relatively old model, it generates quite robust calibration results (Colby, 1991; Law and Nichol, 2004).

Figure 1.11 shows the geometrical relationships among the sun, the sensor, and an arbitrary surface element. The Lambertian model assumes that the surface reflects the incident radiation uniformly in all directions. If we treat wavelength as a constant and ignore atmospheric interactions, the variation in radiance detected by the sensor is mainly caused by the local incidence angle θ_i (i.e., the angle formed between the

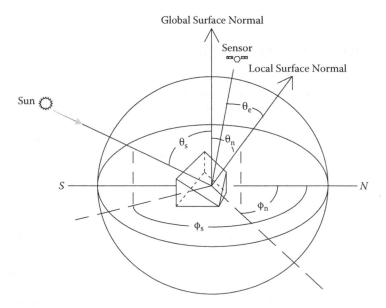

FIGURE 1.11 Geometrical relationships among the sun, the sensor, and the target position. Modified from Smith, J. A., T. L. Lin, and K. Ranson. "The Lambertian Assumption and Landsat Data," Figure 1. *Photogrammetric Engineering and Remote Sensing* 46 (1980): 1183–1189. (Reprinted with permission from the American Society for Photogrammetry and Remote Sensing).

solar radiation path and local surface normal). In this case, sensor-detected radiance L can be normalized in terms of

$$L_n = L/\cos(\theta_i). \tag{1.18}$$

where L_n denotes the normalized radiance. The $\cos(\theta_i)$ term can be derived from the spherical law as follows:

$$\cos(\theta_i) = \cos(\theta_s)\cos(\theta_n) + \sin(\theta_s)\sin(\theta_n)\cos(\phi_s - \phi_n) \tag{1.19}$$

where θ_s is solar zenith angle, θ_n is the angle of slope of terrain surface, ϕ_s is the solar azimuth angle, and ϕ_n is the surface aspect of the slope angle (see Figure 1.11). Slope is defined as a plane tangent to the surface containing two components: one is gradient, which specifies the rate of change in elevation, and the other is aspect, which measures the direction of the gradient. The values of several parameters are required to solve these equations, namely, solar zenith angle θ_s, solar azimuth angle ϕ_s, the surface slope θ_n, and aspect ϕ_n. Both θ_s and ϕ_s can be obtained from the image header file, while slope and aspect can be derived by co-registering the image with a digital elevation model (DEM). A variety of approaches can be employed to calculate the slope and aspect from a DEM (Skidmore, 1989; Jones, 1998).

In the case of the non-Lambertian reflectance assumption, Smith et al. (1980) suggest the following function to correct for the topographic effect:

$$L \times \cos(\theta_e) = L_n \times [\cos(\theta_i)\cos(\theta_e)]^k \tag{1.20}$$

where θ_e is effective view angle (Figure 1.11), $\cos(\theta_i)$ is defined in Equation (1.19), and k is known as the Minnaert constant (Minnaert, 1941) describing the bidirectional reflection distribution function of the surface, the type of scattering dependence, and surface roughness, respectively. A Lambertian surface is defined by a k value of 1.0, and Equation (1.20) then reduces to Equation (1.18).

In order to solve Equation (1.20), one needs to relate the effective view angle θ_e and k (the method for deriving $\cos(\theta_i)$ is described in Equation [1.19]). If Landsat TM or MSS imagery is used, θ_e can be regarded as the same as θ_n (i.e., the slope of terrain surface) since Landsat has a narrow view angle. In the case of SPOT HRV data, which can acquire imagery through an angle of ±27°, θ_e should be set to $\theta_e = \theta_n$, the satellite view angle. To estimate the Minnaert constant k, Equation (1.20) is converted into logarithmic form as

$$\log(L \times \cos(\theta_e)) = k \times \log(\cos(\theta_i)\cos(\theta_e)) + \log(L_n). \tag{1.21}$$

The term k is then equal to the slope of regression line of the plot made by the samples $log(cos(\theta_i)cos(\theta_e))$ plotted on the x-axis and $log(Lcos(\theta_e))$ plotted on the y-axis.

To calibrate for the topographic effect using a non-Lambertian assumption is more complicated than that based on a Lambertian reflectance assumption. As far as the computational cost and calibration accuracy is concerned, Smith et al. (1980)

suggest that when surface slopes are less than 25° and effective illumination angles are less than 45°, then the Lambertian assumption is more valid. Under such circumstances, one can use Equation (1.18) to carry out topographic effect correction, and the calibration accuracy should be preserved. If either of the above conditions is not satisfied, the use of Equation (1.20) is recommended.

It should also be appreciated that surface topographic variations will also cause distortions in the geometry of images. The map to which the image is referenced represents the relationship between features reduced to some datum such as sea level, while the image shows the actual terrain surface. If the terrain surface is significantly above sea level then the image pixel position will be displaced by an amount proportional to the pixel's elevation above sea level (or whatever datum is used).

1.5   REMOTE SENSING IN THE MICROWAVE REGION

The word *radar* is an acronym derived from the phrase "Radio Detection And Ranging." Imaging microwave sensors are known as *imaging radars*. These instruments transmit a signal in the wavelength range approximately 3 cm to 1 m, and receive reflection (backscatter) from the target. The level of backscatter for each pixel over the imaged area is recorded and the set of pixels forms the radar image. Remote sensing in the microwave region differs from optical remote sensing in a number of ways, the most important of which are

1. Radar backscatter is related to the roughness and electrical conductivity of the target. This information is complementary to that which is acquired by optical and thermal sensors.
2. Energy in the microwave region can penetrate clouds.
3. Microwave imaging radars are active, not passive instruments, and thus can operate independently of solar illumination.

An increasing number of space-borne radar systems is now in orbit, including the recently launched German TerraSAR-1 and the Italian COSMO-SkyMed, and it is probable that radar imagery will play an increasingly important role in supporting our understanding and monitoring of our environment. The main disadvantage of active microwave systems vis-à-vis optical systems is their power requirements, for the sensor transmits as well as receives energy. Optical sensors passively detect reflected solar radiation. Passive microwave sensors, which are not considered in this chapter, detect microwave radiation that is generated by the target.

This section introduces radar remote sensing. In Section 1.6, basic ideas underlying the use of radar images, including geometrical effects and the main factors affecting surface reflection or backscatter at radar wavelengths, are introduced. Section 1.7 considers the extraction of surface information from this backscattered signal. One of the main problems associated with the interpretation of radar imagery is the presence of noise, or *radar speckle*. The use of filters to reduce the noise effect is described in Section 1.8.

1.6 RADAR FUNDAMENTALS

An active radar system repetitively transmits short microwave energy pulses (normally of the order of microseconds, i.e., 10^{-6} second, denoted by μs) toward the area to be imaged. The energy pulse is likely to scatter in all directions when it reaches the surface. Part of the backscattered energy is reflected back to the radar antenna and recorded for later processing. Normally, each energy pulse has a duration of between 10 and 50 μs, and utilizes a small range of microwave wavelengths. A waveform can be characterized in terms of its wavelength and amplitude. Wavelength is defined as the distance between two adjacent crests or troughs of the waves (Figure 1.12). Amplitude measures the strength of an electromagnetic wave in terms of the maximum distance achieved by the waveform relative to the mean position (shown by the horizontal line in Figure 1.12). The amplitude may be a function of a complex signal including both magnitude and the phase; this point is discussed further below.

Frequency, rather than wavelength, can also be used to describe wavebands. *Frequency* is the number of oscillations per unit time or number of wavelengths that pass a point per unit time. One can obtain the frequency f (normally of the order of Gigahertz, GHz) corresponding to wavelength λ in terms of

$$f = \frac{c}{\lambda} \tag{1.22}$$

where c is the velocity of light ($3 \times 10^8\ ms^{-1}$).

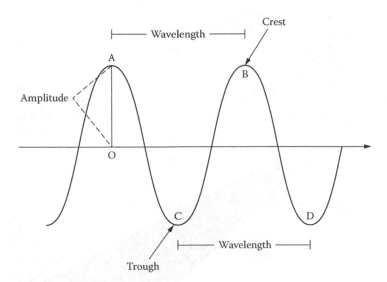

FIGURE 1.12 Wavelength is the distance between two adjacent crests (e.g., A and B) or troughs (e.g., C and D). The amplitude of a waveform (distance AO) measures the "power" or information carried by the wave.

The description of radar operation in the following pages is based on the most widely used radar system installed on aircraft platforms, the side-looking airborne radar (SLAR). Space-borne imaging radars operate on a similar basis, but use a synthetic rather than a real antenna. They are known as synthetic aperture radars (SAR).

1.6.1 SLAR Image Resolution

SLAR transmits and receives microwave energy using an antenna located to the side of the platform. The area imaged by the sensor is thus a strip of ground parallel to the flight track (known as the *azimuth direction*). SLAR image resolution is mainly dependent on pulse duration and antenna beam width, which is the ground area "illuminated" by the radar pulse at a given instant in time (Figure 1.13). Pulse duration affects the resolution in the range (cross-track) direction, while antenna beam width controls the azimuth (along-track) resolution. The ground range resolution and azimuth resolution are computed by

$$\text{Range Resolution} = \frac{c\tau}{2\cos\theta} \tag{1.23}$$

$$\text{Azimuth Resolution} = \beta \times d \tag{1.24}$$

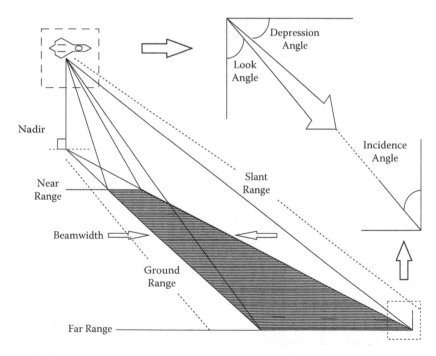

FIGURE 1.13 Some basic parameters of a SLAR system.

The term c is defined in Equation (1.22), while τ is the pulse duration, θ is the depression angle defined as the angle between the horizontal plane and the direction of emitted microwave energy (Figure 1.13), d is the ground range, and β is the antenna beam width. Range resolution can also be analyzed in terms of incidence angle or look angle (Figure 1.13). *Incidence angle* is defined as the angle between the radar beam and a line perpendicular to the illuminated surface. The *look angle* is complementary to the *depression angle*. If the illuminated surface is assumed to be flat, then one can also regard the incidence angle as the complement of the depression angle. The antenna beam width β, antenna length L, and wavelength λ are related as follows:

$$\beta = \frac{\lambda}{L}. \tag{1.25}$$

The combination of azimuth and range resolution determines the ground resolution of each pixel on a radar image.

It can be inferred from Equation (1.23) that the shorter the pulse duration τ, or the smaller the value of θ, the finer the range resolution. The depression angle θ varies across an image. The value of θ in the near range is relatively larger than that in the far range (Figure 1.13). Thus, the ground range resolution will also vary with respect to θ.

Equation (1.24) shows that the smaller the values of β and d, the finer the azimuth resolution will be. Thus, near ground range has a higher resolution than that in far ground range because both are smaller in near range than that in far range (Figure 1.13). According to Equation (1.25), one can use a long antenna length and short wavelength to obtain finer azimuth resolution. However, shorter wavelengths are more likely to be affected by the atmosphere and, furthermore, antenna length is constrained by physical limitations. For instance, to obtain an antenna beam width of 1×10^3 m at a wavelength of 20 cm, the required antenna length will be 200 m (Equation [1.25]). Clearly, if finer azimuth resolution is sought in terms of increasing antenna length, then serious practical difficulties will be encountered. An alternative strategy is to use a synthetic aperture radar (SAR), where the term *aperture* means the opening used to collect the reflected energy that is used to generate an image. In the case of radar, this opening is the antenna, while in the case of a camera, the opening is the shutter opening.

SAR increases the antenna length not in physical terms but by synthesizing a long antenna using the forward motion of a short antenna, a process that requires more complicated and expensive technology. SAR uses the Doppler principle in order to synthesize a longer antenna. The *Doppler effect* is the change in wave frequency as a function of the relative velocities of transmitter and reflector. A radar sensor can image a given target repeatedly from successive locations, as illustrated in Figure 1.14. Here, the frequency of the waveform reflected by the target will increase from location a to b because the distance between the sensor and the object is reducing. As the platform moves away from the target, from b to c, the frequency of the returned signal decreases. SAR uses the Doppler information to compute frequency shifts and thus determine the location and scattering properties of the target.

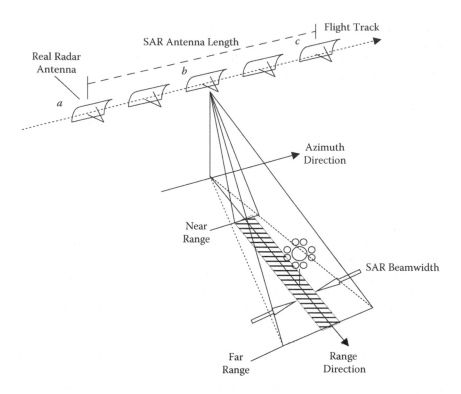

FIGURE 1.14 Concept of the synthetic aperture.

1.6.2 Geometric Effects on Radar Images

A radar image is generated from the timing data for transmitted energy to be returned to the radar antenna. This timing delay is dependent on the distance between the radar antenna and the target. This distance is the *slant range* (Figure 1.13), which is the path along which the microwave energy travels. Therefore, every target located on the terrain being observed by the radar will be mapped onto the slant range domain. Because of this slant range mapping, radar imagery is likely to be affected by geometric distortions. The most common distortions are those of layover, foreshortening, and shadow.

The *layover* effect results when the top of an illuminated target is seen by the radar as the bottom, and the bottom of the target is recorded by radar as the top. This phenomenon occurs when the time for the microwave energy to travel from the antenna to the top of an object is less than the time needed to travel to the bottom of the same object. Figure 1.15 shows two targets (a building and a mountain) that are illuminated by a radar sensor. The microwave energy transmitted by the radar will reach the tops of both objects (points *a* and *b* in Figure 1.15) before the bottoms (points *c* and *d*). The antenna will first receive the reflected energy from *a* and *b*, then some time later, the energy reflected from *c* and *d*. After projection onto the slant range domain, the result is called the *layover effect*.

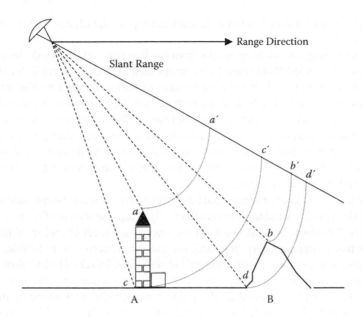

FIGURE 1.15 The layover effect in radar remote sensing.

It might be inferred from the preceding discussion that the higher the isolated target the greater the layover effect. However, layover is also controlled by another important factor: the angle, θ_f, between the front of the target and the energy path (Figure 1.16). Layover will occur only if θ_f exceeds 90° (see object A in Figure 1.16). If θ_f is smaller than 90°, as in the case of object B in Figure 1.16, then microwave

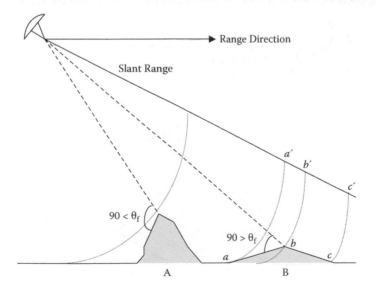

FIGURE 1.16 The angle θ_f controls layover effects (see text for discussion).

energy will first illuminate the bottom then the top of the object, and there will be no layover effect.

Foreshortening, like layover, results from the fact that radar is a side-looking sensor. The object labeled *B* in Figure 1.16 is symmetrical in cross-section, but the angle between its front slope *ab* and the microwave radiation emitted by the instrument is less than 90°. Hence, the front slope distance *ab* appears to be less than the back slope distance *bc* when projected onto slant range. Since the front slope also tends to reflect microwave energy more strongly than does the back slope, it will appear to be brighter, steeper, and shorter, while the back slope is shallower and darker. The darker back slope demonstrates another radar image geometry effect, that of shadow, as illustrated in the following text.

Radar shadow is due to the returned back energy from targets being affected by the nature of the terrain. A radar image is effectively a representation of returned energy levels plotted against the time taken for the energy to travel to and from the target. It follows that if, during a certain period, the antenna receives no reflection, then the image area corresponding to this time period will contain zero (dark) values.

The effect of radar shadow is controlled by the target height and angle θ_b (Figure 1.17), which is defined as the angle between the back slope of the target and the horizontal line parallel to the range direction. In Figure 1.17 the angle θ_b of object *A* is smaller than the corresponding depression angle θ_1. Thus, the back slope of object *A* is illuminated by the microwave energy. However, since the angle θ_b of object *B* is larger than the corresponding depression angle θ_2, the radar antenna will not receive any reflection from the back slope of object *B*, and this period of zero reflection is likely to continue until point *a* is reached. The resulting radar shadow after projection onto slant range is also illustrated in Figure 1.17.

The description of radar image distortions given above is a simplification. In real-world applications where terrains are often continuously sloping and rapidly varying,

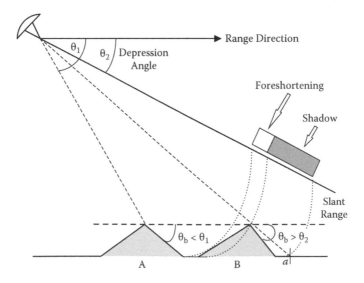

FIGURE 1.17   Showing the relationship between angle θ_j and radar shadow.

the resulting image will be a combination of different geometric effects. Hence, in order to compensate for these effects, one has to make careful case-by-case analyses (Kropatsch and Strobl, 1990; Goyal et al., 1999). That is, if one knows what effects are occurring at a given pixel, then one can use suitable algorithms to carry out calibration. As in the case of topographic calibration of optical imagery described above, geometric and radiometric correction of radar images requires a co-registration to a DEM. However, the calibration procedures are generally more complicated. Kwok et al. (1987), Riegler and Mauser (1998), and Hein (2003) provide descriptions.

1.6.3 Factors Affecting Radar Backscatter

It has already been noted that a radar image is a record of the strength of the backscatter from the targets making up the imaged area. The stronger the backscatter, the brighter the corresponding image element. The level of backscatter is determined by terrain conditions (such as roughness and electrical characteristics), and also by the parameters of the radar system. Understanding the factors affecting radar backscatter can help analyze landscape properties more knowledgeably.

1.6.3.1 Surface Roughness

Radar backscatter is stronger where the ground surface is rough relative to the radar wavelength. The roughness of a surface is dependent on both the wavelength of the incident energy and the angle of incidence. Rough surfaces act as Lambertian reflectors (Section 1.4) so that incident microwave energy is scattered in all directions and a portion is reflected back to the radar antenna. Smooth surfaces are specular, in that they act like a mirror and reflect the incident energy away from the sensor, resulting in extremely weak backscatter (Figure 1.18, point *a*). Normally, as the wavelength decreases, the surface appears rougher because smaller facets of the surface

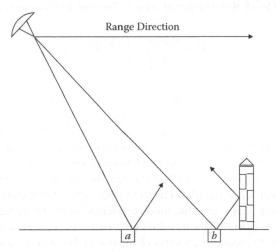

FIGURE 1.18 Microwave energy impinging on a smooth surface (point *a*) at which energy is reflected away from the sensor, and exhibiting a double bounce at point *b* when the reflected energy is reflected by a vertical wall.

contribute to the scattering process and thus stronger backscatter results. Likewise, as the wavelength increases, the surface tends to appear smoother. The strength of backscatter is also affected by the incidence angle. For a given wavelength, as the incidence angle increases, backscatter becomes weaker, and so the illuminated surface appears smoother. Some ground objects can behave like corner reflectors, which can reflect high energy back to the antenna and cause bright spots on the radar image. Such phenomena often occur in urban areas because energy can be returned by means of a *double bounce* from the corners of buildings (Figure 1.18, point *b*).

1.6.3.2   Surface Conductivity

Highly conductive ground surfaces tend to have higher reflectivity than surfaces with lower conductivities. Water and metal are good conductors. As a result, radar backscatter will be sensitive to metal objects and to the presence of moisture in the illuminated target area, even though the amount of moisture may be small. In a radar image, metal objects such as railway tracks and metal bridges generally result in bright spots. Moisture also affects the depth of microwave energy penetration of the soil surface. If soil contains a large amount of moisture, the signal does not penetrate the soil surface and is reflected back to the radar antenna. If the soil is dry, then the radar signal can penetrate more deeply into the soil surface layer. Wavelength is also another control on the depth of penetration. Lakes and other water bodies might be expected to exhibit high backscatter, but in fact the surfaces of rivers and lakes are generally smooth relative to radar wavelengths and act as specular reflectors. The ocean surface is generally rougher, and therefore the magnitude of backscatter depends on sea state as well as on wavelength and depression angle.

1.6.3.3   Parameters of the Radar Equation

The parameters of the radar equation (Van Zyl et al., 1993) are fundamental factors that influence the level of the returned signal. The radar equation is expressed as

$$P_r = \frac{P_t \lambda G_t(\gamma) G_r(\gamma)}{(4\pi)^3 R^4} \sigma^0 A. \tag{1.26}$$

P_t is the transmitted power from the antenna, λ is the transmitted wavelength, R is the distance to imaging area, γ is the radar look angle, A is the area on the ground responsible for scattering, G_t and G_r are the transmitted and received antenna gains (describing the system's ability to focus the transmitted microwave energy) at look angle γ, and σ^0 is the radar backscatter coefficient measured in decibels (dB). All of the parameters in Equation (1.26) affect the received power P_r. However, only σ^0 is related to the properties of the illuminated surface. Thus, the quantized pixel values (0 to 255) in a radar image are sometimes converted to σ^0 before being interpreted. The received power P_r in Equation (1.26) can also be characterized in terms of other parameters such as the scattering matrix (Equation [1.31]) and the coefficient of variation (Equation [1.35]), which also relate to the properties of surface objects, and can be used for classification purposes. A discussion of the scattering matrix is presented in the next section.

1.7 IMAGING RADAR POLARIMETRY

The polarimetry theory presented in this section is mainly derived from Evans et al. (1988), Zebker and Van Zyl (1991), Zebker et al. (1987, 1991), Van Zyl et al. (1987, 1993), Kim and Van Zyl (2000) and Hellmann (2002). Knowledge of radar polarimetry enables us to use a variety of features (such as complex format data, the elements of the scattering matrix, and the coefficient of variation of polarization signature) to perform image interpretation. Some basic concepts are described first.

An electromagnetic wave, besides being described in terms of wavelength and amplitude, can also be characterized using *complex number* format (a complex number consists of a two components, termed the *real* and *imaginary* parts). When the coordinate system is translated into both the real and imaginary axes (Figure 1.19), the wave is described by two parameters, namely, the *in-phase I* ($I = m \times \cos\varphi$) and *quadrature Q* ($Q = m \times \sin\varphi$) components. Both the I and Q parameters provide the wave's overall phase φ ($\varphi = \tan^{-1}(Q/I)$) and magnitude m ($m = (I^2 + Q^2)^{0.5}$) as illustrated in Figure 1.19. Radar phase represents the degree of coincidence in time between a repetitive radar signal and a reference signal having the same frequency. Over the complex domain, the amplitude a is expressed as $a = I + iQ$, where $I = (-1)^{0.5}$. The relationship between the magnitude m and a complex amplitude a, by definition, can be expressed as $m = |a|$, i.e., the absolute value of amplitude. The radar image can thus be formed by using any of the m, I, or Q components. As a result, this kind of radar image is said to be represented in *complex* format.

Complex format radar imagery can be used to generate interferometric information (Massonnet and Rabaute, 1993; Zebker et al., 1994; Gens and Van Genderen, 1996; Kim and Van Zyl, 2000), which is useful in producing digital elevation models (DEM) (Zebker et al., 1994; Lanari et al., 1996; Kim and Van Zyl, 2000), or in monitoring large-scale surface changes (Massonnet et al., 1993; Preiss et al., 2003). The potential of SAR interferometry (in the form of coherence maps) in land cover classification is the subject of current investigations (e.g., Ichoku et al., 1998).

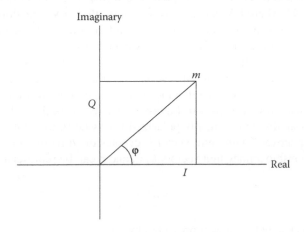

FIGURE 1.19 Wave described by complex (real and imaginary) coordinates.

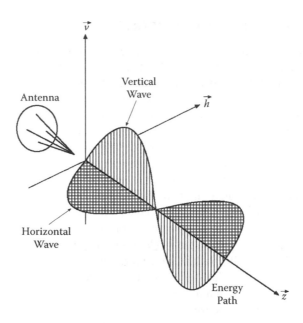

FIGURE 1.20 Polarized microwaves.

1.7.1 RADAR POLARIZATION STATE

Normally, microwave energy transmitted and received by a radar antenna can travel in all directions perpendicular to the direction of wave propagation. However, most radar systems polarize microwaves in such a way that the transmitted and received waves are restricted to a single plane perpendicular to the direction of wave propagation (Figure 1.20). The polarized wave is therefore transmitted and received in either the horizontal (H) or the vertical (V) plane. Consequently, there are four combinations of transmission and reception for the polarized waves. These combinations are HV, HH, VV, and VH, where HV denotes a wave transmitted in V direction and received in H direction. The other combinations can be inferred in a similar manner. Radar imagery generated in terms of HH or VV is called co- or like-polarized imagery, while imagery resulting from HV or VH polarization is called cross-polarized imagery. Cross-polarization detects multiple scattering from the target and thus generally results in weaker backscatter than that measured by a co-polarization configuration.

The coordinate system shown in Figure 1.20 determines the radar polarimetry, in which horizontally and vertically polarized waves lie in the unit vector $\vec{h}$ and $\vec{v}$ directions, respectively, while unit vector $\vec{z}$ denotes the direction of wave propagation. The relationship among unit vectors $\vec{h}$, $\vec{v}$, and $\vec{z}$ can be represented by

$$\vec{z} = \vec{h} \times \vec{v}. \tag{1.27}$$

The overall electric field can be represented by

$$\vec{E}(z,t) = \Re\left[(E_h\,\vec{h} + E_v\,\vec{v})e^{-i(\omega t - kz)}\right] \tag{1.28}$$

where E_h and E_v denote the electric field of vertical and horizontal plane, respectively, which are expressed by

$$E_h = a_h e^{-i\delta_h}$$
$$E_v = a_v e^{-i\delta_v}. \tag{1.29}$$

The terms a_h and a_v denote the positive amplitudes in the $\vec{h}$ and $\vec{v}$ directions (Figure 1.21), respectively, and the corresponding phases are δ_h and δ_v relative to the phase factor $\omega t - kz$, where ω is frequency, t is time, k is wave number, and z is distance traveled in the $\vec{z}$ direction (i.e., the direction of wave propagation).

In the most general case, the electric field vector of a plane monochromatic wave rotates in a plane perpendicular to the direction of microwave energy propagation, and in doing so traces out an ellipse, as shown in Figure 1.21. The wave is said to be *elliptically polarized*. If one refers to the relative amplitude and phase relationships of the components of a given wave as the elliptic polarization state, Equation (1.28) can be rewritten as

$$\vec{E}(z,t) = a\vec{p}\cos\chi\sin(\omega t - kz + \delta) + a\vec{q}\sin\chi\cos(\omega t - kz + \delta) \tag{1.30}$$

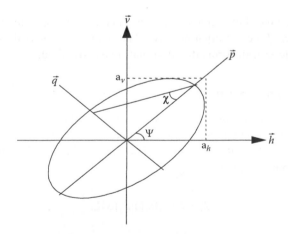

FIGURE 1.21 Elliptical polarization state for polarization synthesis. See text for explanation. Modified from Evans, D. L., T. G. Farr, J. J. Van Zyl, and H. A. Zebker. "Radar Polarimetry: Analysis Tools and Application." *IEEE Transactions on Geoscience and Remote Sensing* 26 (1988):774–789 © 1988 IEEE.

where $a^2 = a_h^2 + a_v^2$ is the intensity of the wave, $\delta = \delta_h - \delta_v$ is the phase angle, both $\vec{p}$ and $\vec{q}$ are unit vectors in a coordinate system rotated by angle ψ with respect to $\vec{h}$, and χ is the ellipticity angle. Note that the width of the ellipse is given by the parameter χ; so that $\chi = \pm 45°$ results in left- and right-handed circular polarizations, respectively. The orientation parameter ψ determines the orientation of the major axis of the ellipse; if $\chi = 0°$, then the values $\psi = 0°$ or $180°$ represent horizontal polarizations, while $\psi = 90°$ represents vertical polarization.

A polarimetric imaging radar measures the magnitude of the backscatter from a target as a vector quantity in such a way that the complex backscattered characteristics of any transmitting and receiving polarization configuration can be determined. Such backscattered characteristics are represented in terms of a scattering matrix [S] (Van de Hulst, 1981):

$$[S] = \begin{pmatrix} S_{hh} & S_{hv} \\ S_{vh} & S_{vv} \end{pmatrix} \tag{1.31}$$

where element S_{hv} is determined by measuring both the amplitude and phase of the electric field where a vertically polarized wave is transmitted and the scattered wave is received in horizontal polarization. The remaining elements are obtained in a similar fashion. The relationship among the electric field E_s of the scattered wave, the electric field E_t of the transmitting wave, and [S] is expressed as

$$E_s = \frac{e^{ikr}}{kr}[S]E_t \quad \Rightarrow \quad \begin{pmatrix} E_h \\ E_v \end{pmatrix}_s = \frac{e^{ikr}}{kr} \begin{pmatrix} S_{hh} & S_{hv} \\ S_{vh} & S_{vv} \end{pmatrix} \begin{pmatrix} E_h \\ E_v \end{pmatrix}_t \tag{1.32}$$

where r is the distance between the scatterer and the receiving antenna and k denotes the wave number of the illuminating wave. The scattering matrix thus describes how the ground scatterer transforms the illuminating electric field.

1.7.2 POLARIZATION SYNTHESIS

Knowledge of the scattering matrix [S] permits calculation of the received power P_r for any possible combination of transmit and receive antenna polarizations. This process is called *polarization synthesis*. The observed power P_r can be derived by evaluating the matrix equation:

$$P_r = K(\lambda, \theta, \phi) \left| E_s[S]E_t \right|^2 \tag{1.33}$$

where

$$K(\lambda, \theta, \phi) = \frac{1}{2} \frac{\lambda^2}{4\pi} \sqrt{\frac{\varepsilon_0}{\mu_0}} \frac{g(\theta, \phi)}{|E_s|^2} \tag{1.34}$$

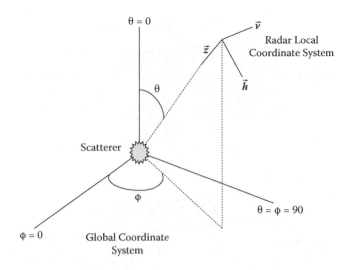

FIGURE 1.22   The local and global coordinate systems determine the angles θ and ϕ.

and $g(\theta, \phi)$ is the antenna gain function, θ and ϕ are the angles between the radar local coordinate system and the scatterer-centered global coordinate system shown in Figure 1.22, $[(\lambda^2/4\pi) \times g(\theta, \phi)]$ is the effective area of the antenna, and ε_0 and μ_0 are the permittivity and permeability of free space (i.e., transmission medium), respectively. Note that both permittivity ε_0 and permeability μ_0 determine the propagation velocity v (or phase velocity) of the wave; the relationship can be expressed as $v = 1/(\varepsilon_0 \times \mu_0)^{0.5}$.

Polarization synthesis can also be expressed in terms of either the Stokes matrix or the covariance matrix. Both of these representations consist of linear combinations of the cross-products of the four basic elements of the scattering matrix. The entries of these matrices can also provide a variety of features for classification purposes.

1.7.3 Polarization Signatures

A particular graphical representation of the variation of received power (or cross-section σ) as a function of polarization is called the *polarization signature* of an object. A polarization signature is a three-dimensional representation consisting of a plot of synthesized scattering power as a function of the ellipticity and orientation angles (i.e., χ and ψ in Figure 1.21) of the transmitted and received wave. Normally, analyses are based on only two types of polarization signatures, namely, co-polarization and cross-polarization.

Figure 1.23 illustrates the polarization signature based on a theoretical model of a large conducting sphere. Note that at the ellipticity angle $\chi = 0°$, the co-polarization signature reaches its highest value of cross-section, while the cross-polarization signature shows the lowest value at the same point. There is a measurement of surface roughness in accordance with polarization signature called the *coefficient of variation (CoV)*, which is defined as

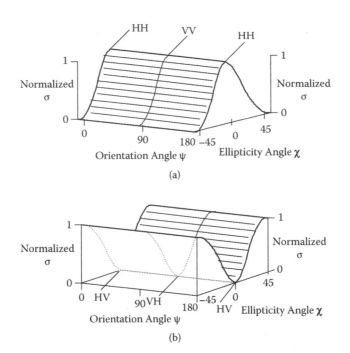

FIGURE 1.23 Polarization signature based on a theoretical model of a large conducting sphere. (a) At ellipticity angle 0 degrees the co-polarization signature reaches its highest cross-section value. (b) The cross-polarization value reaches its minimum point at the same ellipticity angle.

$$CoV = \frac{P_{r\min}}{P_{r\max}}. \qquad (1.35)$$

$P_{r\min}$ and $P_{r\max}$ each denote minimal power and maximal power occurring within a polarization signature, respectively. Since CoV relates to the surface roughness, it can be used as a discriminating feature in classification. As the value of CoV increases, the measured surface tends to be rougher. The concept of CoV is based on the following observations. The polarization signature for each resolution element represents the sum of the polarization signatures of many individual measurements. If the surface being measured is smooth, the scattering mechanisms from a group of scatterers should be identical. Therefore, the maxima (minima) of a scattering mechanism should coincide with the maxima (minima) of the other scattering mechanisms. When the composite polarization signature is derived, it will produce a composite signature in which there is a large difference in magnitude between maximal and minimal backscatter, and thus the polarization signature will result in more peak- and valley-like shapes. As a result, the value of the CoV will be small (i.e., closer to 0). Conversely, if the measured ground surface is rough, several different scattering mechanisms may result, the backscatter maxima and minima may occur together from different individual scatterers, and a relatively flat polarization signature shape will be produced (equivalently, CoV will be large, i.e., close to 1).

1.8   RADAR SPECKLE SUPPRESSION

Due to random fluctuations in the signal observed from a spatially extensive target represented by a pixel (or image resolution element), *speckle noise* is generally present on a radar image. Speckle has the characteristics of a random multiplicative noise (defined below) in the sense that as the average grey level of a local area increases, the noise level increases. In a SAR imaging system, speckle effects are more serious (Lopez-Martinez and Fabregas, 2003). SAR can achieve high resolution in the azimuth direction independent of range, but the presence of speckle decreases the interpretability of the SAR imagery. If such imagery is to be used in classification, then some form of preprocessing to reduce or suppress speckle is necessary.

There are two approaches to the suppression of radar image speckle. The first method is known as the multilook process, while the second method uses filtering techniques to suppress the speckle noise.

1.8.1   MULTILOOK PROCESSING

Radar speckle can be suppressed by averaging several looks (images) to reduce the noise variance. This procedure is called *multilook processing.* As the radar sensor moves past the target pixel, it obtains multiple looks (i.e., returned samples). If these looks are spaced sufficiently far apart they can be considered to represent individual observations. The relationships among the radar aperture length L, the resolution R, and the number of independent samples N_s is expressed by (Ulaby et al., 1982, 1986a)

$$N_s \approx \frac{R}{0.5L}. \qquad (1.36)$$

For instance, if radar aperture length is 10 m, and the desired spatial resolution is 25-m resolution, then the number of independent samples is 25/5 = 5. Figure 1.24 illustrates the relationship between the resolution and number of samples. Although the averaging of independent looks can reduce the noise variance, it also causes degradation in image resolution.

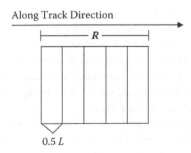

FIGURE 1.24   The relationship between resolution and number of looks in a SAR image. See text for explanation.

1.8.2 FILTERS FOR SPECKLE SUPPRESSION

The second method of speckle suppression uses filtering methods, which fall into two main categories, namely, adaptive and nonadaptive filters. *Adaptive filters* use weights that are dependent on the degree of speckle in the image, whereas *nonadaptive* filters use the same set of weights over the entire image. Adaptive filters are more likely to preserve details such as edges or high-texture areas (e.g., forest or urban areas) because the degree of smoothing is dependent on local image statistics.

The best-known nonadaptive filters are those based on the use of the mean or the median. The mean filter uses the same set of smoothing weights for the whole image without regard for differences in image texture, contrast, etc. The median filter does not use a weighting procedure, but is based on the ranking of image pixel values within a specified rectangular window (Mather, 2004). Both of these filters have a speckle-suppression capability, but they also smooth away other high-frequency information. The median is more effective than the mean in eliminating spike noise while retaining sharp edges. Both filters are easily implemented and require less computation than adaptive filters.

In comparison with nonadaptive speckle filters, adaptive speckle filters are more successful in preserving subtle image information. A number of adaptive speckle filters have been proposed, the best known being the Lee filter (Lee, 1980, 1981, 1986), the Kuan filter (Kuan et al., 1987), the Frost filter (Frost et al., 1982), and the Refined Gamma Maximum-A-Posteriori (RGMAP) filter (Touzi et al., 1988; Lopes et al., 1990; Baraldi and Parmiggiani, 1995). The effectiveness of these adaptive filters is dependent on the following three assumptions (Lee, 1980; Lopes et al., 1990; Baraldi and Parmiggiani, 1995a,b):

1. SAR speckle is modeled as a multiplicative noise (note that the visual effect of multiplicative noise is that the noise level is proportional to the image gray level).
2. The noise and signal are statistically independent.
3. The sample mean and variance of a pixel is equal to its local mean and local variance computed within a window centered on the pixel of interest.

All of the speckle filters described above rely strongly on a good estimate of local statistics (e.g., σ_z and μ_z) from a window. If the window center is located close to the boundary of an image segment (such as a boundary between agricultural fields), the resulting local statistics are likely to be biased and will thus degrade the filtering result. Nezry et al. (1991) notes this point, and proposes a refined GMAP filter called the RGMAP filter in which the local statistics extracted from a window do not cross image feature boundaries. Readers are referred to Sheng and Xia (1996) and Liu, Z. (2004) for comparisons of the above filters.

In recent years, the wavelet transform has been used for radar imagery denoising (Donoho, 1995; Fukuda and Hirosawa, 1998, 1999; Achim et al., 2003). The wavelet transform can decompose the multiplicative noise, and so simplify the speckle filtering process. Details of the wavelet transform are described in Chapter 7. Normally, using the wavelet transform for speckle suppression involves

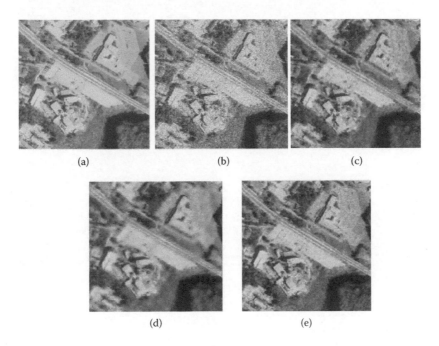

(a) (b) (c)

(d) (e)

FIGURE 1.25 Calculation of median and mean filters for 3×3 window.

three steps as follows. First, the SAR image is translated into logarithmic domain and is then decomposed by means of the wavelet transform in a multiscale sense. Note that translating the raw SAR image into logarithmic domain is to convert the multiplicative noise to an additive noise. Second, the empirical wavelet coefficients are shrunk using a thresholding mechanism. Finally, the denoised signal is synthesized from the processed wavelet coefficients through the inverse wavelet transform. It is noted that the quality of wavelet transform for speckle suppression is closely related to the thresholding method used (Donoho, 1995). Some good estimators and modeling techniques proposed by Simoncelli (1999), Pizurica et al. (2001), and Achim et al. (2003) can be applied for solving such thresholding issues and so as to make the speckle suppression quality well controlled. Gagnon and Jouan (1997) and Achim et al. (2003) conducted comparative studies between wavelet-based filters and several statistical adaptive speckle filters described in the previous paragraph and show that, from the perspectives of both qualitative and quantitative measures, the wavelet-based approaches outperform other kinds of filters for speckle removal. Figure 1.25 illustrates the results of speckle suppression in terms of various filters. In Figure 1.25a, the original remotely sensed imagery is shown. The image contaminated by the noise is displayed in Figure 1.25b. The noisy image is then subjected to a GMAP filter (Figure 1.25c), and wavelet based filters developed by Donoho (1995) (Figure 1.25d) and Achim et al. (2003) (Figure 1.25e), respectively.

Further reading on the material covered in this chapter is provided in Elachi (1987), Liang (2004), Slater (1980), and contributors to Asrar (1989).

FIGURE 2.25

2 Pattern Recognition Principles

In the context of pattern recognition, a pattern is a vector of features describing an object. This pattern is made up of measurements on a set of *features*, which can be thought of as the axes of a *k*-dimensional space, called the *feature space*. The aim of pattern recognition is to establish a relationship between a pattern and a class label. The relationship between the object and the class label may be one-to-one (producing a *hard classification*) or one-to-many (producing a *fuzzy classification*). The features describing the object may be spectral reflectance or emittence values from optical or infrared imagery, radar backscatter values, secondary measurements derived from the image (such as texture), or geographical features such as terrain elevation, slope, and aspect. The object may be a single pixel or a set of adjacent pixels forming a geographical entity, such as an agricultural field. Finally, the class labels may be known or unknown in the sense that the investigator may, in the case of a known label set, be able to list all of the categories present in the area of study. In other cases, the investigator may wish to determine the number of separable categories and their location and extent. Using this information, the separable classes are assigned labels or names based on the investigator's knowledge of the geographical characteristics of the area of study.

These two methods of labeling are known as the *supervised* and *unsupervised* approaches, though some approaches to pattern recognition use a combination of both. Supervised methods require the user to collect samples to "train" or teach the classifier to determine the decision boundaries in feature space, and such decision boundaries are significantly affected by the properties and the size of the samples used to train the classifier. For instance, if one decides to use the minimum distance between a pixel and the mean of each class as the classification criterion, one must first collect samples to construct estimates of the class means. The acceptability of the results will depend on how adequately these class means are estimated.

The label set selected for supervised classification experiments identify *information classes*. The investigator should have sufficient knowledge of the type and the number of information classes that are represented in the study area to allow him or her to collect training samples of pixels from the image that are representative of the information classes. In contrast, unsupervised pattern recognition methods are less dependent on user interaction. Normally, unsupervised classifiers "learn" the characteristics of each class (and possibly even the number of classes) directly from the input data. For instance, if the criterion used to label an object is the minimum distance between the object and the class means, this distance being measured in feature space, the unsupervised procedure will estimate the mean for each class and will refine this mean estimate iteratively (most unsupervised classifiers are iterative in operation). At each iteration, the previous set of estimates of the class means is refined until the process converges, usually when the means remain in the same place in feature space over successive iterations. The results output by unsupervised methods are called *clusters* or, sometimes, *data classes*. The pattern recognition process is complete when each cluster is identified, that is, linked to a specific information class by the user.

Although the unsupervised approach appears to be more elegant and automatic than the supervised procedures, the accuracy of unsupervised methods is generally lower than that achieved by supervised methods. In complex classification experiments, information classes often overlap. In the spectral domain, this implies that the reflectance, emittence, or backscatter characteristics of different classes may be similar. In the spatial domain, the implication is that any one object (a pixel or a field, for example) may contain areas representative of more than one information class. This is the *mixed pixel* problem. Spectral and spatial overlap of classes is the main barrier to the achievement of high classification accuracy. Even so, some interesting unsupervised algorithms are worthy of investigation as they may reveal useful information concerning the structure of the data set. Such methods can be thought of as *exploratory data analyses* or even *data mining* (Witten and Frank, 2005). A further problem with pixel-based classifiers is that radiance (carrying information) that apparently reaches the sensor from a given pixel actually includes contributions from neighboring pixels, due to atmospheric effects and the properties of the instrument optics (Chapter 1). Townshend et al. (2000) show that by considering this latter effect, improvements in accuracy can be achieved. They note that only where pixel size is small relative to the area of land cover units, will this effect be unimportant.

Until the mid-1990s, pattern recognition methods applied to remotely sensed imagery were mainly based on conventional statistical techniques, such as the maximum likelihood or minimum distance procedures, using a pixel-based approach. Although these traditional approaches can perform well, their general ability for resolving interclass confusion is limited. As a result, in recent years, and following advances in computer technology, alternative strategies have been proposed, particularly the use of artificial neural networks, decision trees, the support vector machine, methods derived from fuzzy set theory, and the incorporation of secondary information such as texture, context, and terrain features.

This chapter introduces the principles of pattern recognition, starting from the concept of feature space and its manipulation using feature selection and orthogonalizing techniques. Details of the statistical classifiers are then described. Algorithms based on artificial neural networks, decision trees, the support vector machine, the fuzzy rule base concept (including the mixed pixel problem), the incorporation of secondary information, and change detection are discussed in later chapters.

2.1   FEATURE SPACE MANIPULATION

Image coordinates give the relative location of a pixel in the spatial domain and, given the origin of the coordinate system and the pixel spacing (Δx and Δy), geometrical calculations, such as interpixel distance, can be performed. When we take into account the values associated with a pixel, which form a vector of measurements on a set of selected features, we can think of a space defined not by the x and y or row and column spatial coordinates, but by the features on which the pixel values are measured. These features may be image pixel values in separate wavebands, context or texture measurements, or geographical attributes of the area represented by the pixel, such as mean elevation, slope angle, or slope azimuth. Feature space is multi-dimensional and as such cannot be visualized. Nevertheless, standard geometrical measures such as the Euclidean distance as the shortest distance between two points are still valid (Alt, 1990, gives a nonmathematical description of *hyperspace*).

Figure 2.1a shows that the spatial domain coordinates of the shaded pixel (in row-column representation) are (5, 4). Figure 2.1b shows three co-registered images, perhaps representing reflectance in the green, red, and near-infrared (near-IR) wave-bands. The quantized pixel values in these wavebands are {35, 20, 46}, respectively. Figure 2.1c shows a plot of the position of the pixel in a feature space that has three axes defined by these three bands. The pattern recognition process involves the sub-division of feature space into homogeneous regions separated by decision boundar-ies. The various statistical, neural, and knowledge-based methods discussed in this book use different *decision rules* to define or specify these boundaries. In the fuzzy classification procedure (Chapter 5), decision boundaries can overlap. A number of techniques can be used to manipulate or transform the axes of the feature space in

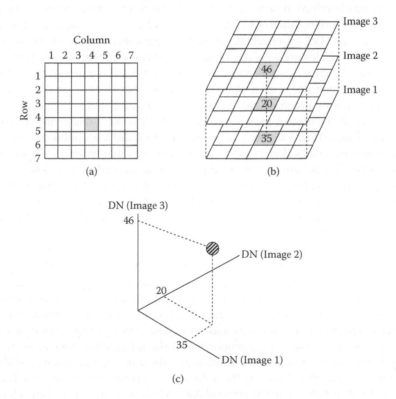

FIGURE 2.1 (a) Image spatial space. (b) Co-registration of three images. (c) Image feature space.

order to facilitate classification, for example, by determining a subspace that contains most of the information present in the original feature space. Orthogonal transforms, which accomplish this aim to a greater or lesser extent, are described later in this section.

In a survey of pattern recognition methodology in remote sensing, Landgrebe (1998) notes the need for accurate class statistics if supervised classifiers are to perform in a satisfactory way. The minimum size of a training data set depends to a considerable extent on the number of features used to characterize the objects to be classified. However, ground data are difficult and also expensive to procure. Landgrebe (1998) points out that the number of possible locations in feature space depends on the number of features and on the number of quantization levels used for each feature. For example, the Landsat-7 ETM+ produces 7 bands of data with 256 levels per band. It is theoretically possible, therefore, for a pixel to occupy any of $(28)7 = 7 \times 10^{16}$ discrete locations in feature space. This number is greater than the number of pixels making up a standard ETM+ image ($7 \times 3.6 \times 106$), so much of the feature space will be empty.

Given a fixed training set size, an increase in the dimensionality of the feature space (e.g., by adding additional spectral bands) means that the number of parameters to be estimated in a statistical classifier also increases. The maximum likelihood decision rule (described in Section 2.3.2.3) uses the values of the mean vector $\bar{x}$ and the variance-covariance matrix $\mathbf{C}$. The former has k elements to be estimated, while the latter has $k(k - 1)/2 + k$ elements, where k is the number of features, giving a total of $k(k - 1)/2 + 2k$ elements. As the value of k increases, the number of parameters to be estimated increases disproportionately. For $k = 6$, the number of parameters is 27. For $k = 12$ this value rises to 90. The fact that efficient estimation of statistical parameters requires a representative sample of a sufficient size is well known; consequently, as the number of parameters increases then, for a fixed sample size, the efficiency of the estimation decreases, which implies that the confidence limits for each estimate become wider. The effectiveness of the classifier will therefore begin to decrease once a certain number of dimensions is reached. This is known as the *Hughes phenomenon* (Hughes, 1968). It follows that if satisfactory results are to be obtained from the classification of remotely sensed data, the relationship between dimensionality and training sample size must be born in mind. The topic of sampling adequacy is considered further in Section 2.6.

One way of mitigating the effects of high dimensionality is to determine a subset of the k-dimensional feature space that contains most of the pixels. This is the aim of *orthogonal transforms*. If the measurements made by two instruments are correlated, then there is a degree of redundancy present in the data. When the correlation between two sets of measurements reaches ± 1.0, then one set of measurements is completely redundant. The features used in remote sensing image classification are generally correlated, and a high proportion of the information content of the data can be represented in terms of m dimensions, where $m < k$. Landgrebe (1998) also claims that the data distribution in the reduced-space representation is more likely to be normally distributed than the original data. This observation is an important one if statistical classifiers are used. Cortijo and Pérez de la Blanca (1999) suggest that where the sample size is small it may be advisable to compute a common covariance matrix so that all training data are used, rather than compute separate covariance matrices for each class.

An alternative approach involves the use of orthogonal transforms, which provide a means of reducing feature space dimensionality. Nielsen (1994), Park et al. (2004), and Tian et al. (2005) provide details and comparisons of these methods. In essence, the aim of an orthogonal transform is to replace the feature set with a derived or synthetic feature set. The covariances of the synthetic features are zero; hence there is no redundancy. The synthetic features are defined by linear combinations of the observed features, and much of the information content of the observed features is reproduced by m synthetic features, where $m < n$. In the following subsections, four orthogonalizing procedures are described. The tasseled cap transform differs from the remaining three methods as it uses three or four predefined linear combinations of Landsat multispectral scanner (MSS) or Thematic Mapper (TM) bands. Principal components analysis (PCA) is a well-known method of orthogonalizing a data set and of ordering the synthetic features (in this case, the principal components) in terms of their contribution to total variance. However, it is sensitive to the scale of measurement used for each feature, and there is no reason to believe that lower-order principal components represent noise or unwanted information. The MAF (minimum/maximum autocorrelation factors) transform and the MNF (maximum noise fractions) procedures aim specifically to separate information and noise and, unlike PCA, are independent of measurement scale.

These four transforms are described in the following subsections.

2.1.1 TASSELED CAP TRANSFORM

The tasseled cap transform was derived by Kauth and Thomas (1976) using four-band Landsat MSS images. The axes of this four-dimensional feature space are transformed into new four-dimensional coordinates defined by the concepts of *brightness*, *greenness*, *yellowness*, and *nonesuch*. The transformation involves rotation of the axes of feature space and translation of the origin of the coordinate system. For example, one of the axes could be moved to a position such that pixels with a low value of the ratio of infrared to red reflectance take low (near zero) values, while pixels that have a high infrared-to-red ratio take high values. Since the infrared-to-red ratio is correlated with vegetation vigor, then this "new" axis could be described by the term *greenness*.

The first transformed axis, brightness, is based on soil reflectance values (e.g., for dry and wet soil) to form a soil line. The second and the third axes are based on pixels of green vegetation and senescent vegetation, respectively. The fourth axis, nonesuch, has been interpreted as being related to atmospheric conditions.

To construct such a new coordinate system, the user selects at least two representative pixel values for each coordinate because the definition of a line requires at least two points. Each representative pixel value can be obtained by taking the average of pixels belonging to the same group. For instance, in the case of soil line construction, one can select several pixels belonging to wet soil class, then take the average as one end of the soil line. The other end can be obtained based on the average of dry soil pixels. Once these candidates have been selected, an orthogonalization process is carried out in order to make the coordinates perpendicular to each other. In the case of a k-band feature space, if the unit vector V_i, $V_i = (v_1, v_2, \ldots, v_k)^T$ (where ^{T} denotes the matrix transpose) of each transformed axis i is obtained, the projection p_i of the jth pixel with original vector (i.e., formed by the pixel values in each band) $\mathbf{x}_j$, $\mathbf{x}_j = (x_1, x_2, \ldots, x_k)^T$, on new coordinate i is calculated by

$$p_i = \mathbf{x}_j^T V_i + c \qquad (2.1)$$

where c is a constant required in order to ensure that the resulting p_i is always positive. Equation (2.1) is the basic relationship in coordinate transform theory for calculating the new mapping location of the points. There are studies (Crist and Cicone, 1984a; Huang et al., 2002) modifying the tasseled cap transform into three new coordinate axes called brightness, greenness, and wetness (sensitive to soil moisture) based on Landsat TM images. Readers should refer to Mather (2004) for further elaboration.

The advantages of the tasseled cap transform are

1. The dimensionality of the feature space is reduced, making the classification problem less complex.
2. The axes of the feature space represent specific concepts (brightness, greenness, and wetness) that can be considered to be defined externally to the specific data set under study.

The principal disadvantages of the transform are

1. The tasseled cap axes may not be well defined for a particular problem if the coefficients are not properly calculated (Jackson, 1983).
2. There can be no assurance that significant information is not omitted by the transformation of the six-band Landsat TM data set to a set of three tasseled cap axes.
3. The method has been widely used only for Landsat TM and MSS data.

2.1.2 PRINCIPAL COMPONENTS ANALYSIS

Principal components analysis (PCA) is another general tool for coordinate transformation and data reduction in remote sensing image processing. However, unlike the tasseled cap, the new axes formed by PCA are not specified by the user's prior definition of the transformation matrix, but are derived from the variance–covariance or correlation matrix computed from the data under analysis. The process of PCA can be divided into three steps:

1. Calculation of the variance–covariance (or correlation) matrix of multiband images (e.g., in the case of a four-band image, the covariance matrix has dimension 4×4),
2. Extraction of the eigenvalues and eigenvectors of the matrix, and
3. Transformation of the feature space coordinates using these eigenvectors.

Let $\mathbf{M}$ and $\mathbf{x}$ denote the multiband image mean and individual pixel value vectors, respectively. The first step, derivation of the covariance matrix $\mathbf{C}$, is expressed as

$$\mathbf{C} = \frac{\sum_{j=1}^{n}(\mathbf{x}_j - \mathbf{M})(\mathbf{x}_j - \mathbf{M})^T}{n-1} \tag{2.2}$$

where n is the number of pixels. If the correlation matrix is used, each entry in $\mathbf{C}$ should be further divided by the product of the standard deviations of the features represented by the corresponding row and column. For instance, let c_{12} denote the entry of 1st row and 2nd column (i.e., the covariance between image bands 1 and 2) in matrix $\mathbf{C}$; then the corresponding correlation r_{12} is obtained by

$$r_{12} = \frac{c_{12}}{\sigma_1 \sigma_2} \tag{2.3}$$

where σ_1 and σ_2 are the standard deviations of image bands 1 and 2, respectively. The correlations of other entries can be computed in a similar manner.

The second step, calculation of the eigenvectors of $\mathbf{C}$, is achieved by solving the following equation:

$$(\mathbf{C} - \lambda_i \mathbf{I})\mathbf{A}_i = 0 \tag{2.4}$$

where $\mathbf{A}_i = (a_1, a_2, \ldots, a_k)^T$ is the eigenvector corresponding to the eigenvalue λ_j, k is the total number of feature space dimension, and $\mathbf{I}$ is the identity matrix (i.e., a matrix with diagonal entries set to 1 and off-diagonal entries set to 0). All the eigenvalues λ can be determined by solving $(\mathbf{C} - \lambda \mathbf{I}) = 0$. The new coordinate system is formed by the normalized eigenvectors of the variance–covariance (or correlation) matrix. The mapping location f_i of each pixel $\mathbf{x} = (x_1, x_2, \ldots, x_k)$ on the ith principal component is given by

$$f_i = \mathbf{x}\mathbf{A}_i = x_1 a_1 + x_2 a_2 + \ldots + x_k a_k. \tag{2.5}$$

This is effectively a rotation of the axes of the feature space.

PCA has the property that the first principal component image (PC1, derived from the first eigenvector) represents the maximum amount of the total variance of the data set, and the variances of the remaining principal component images decrease in order, as denoted by the magnitudes of the corresponding eigenvalues. Normally, the variance contained in the last few principal components is small. Therefore, PCA is often used to condense the information in a multiband image set into fewer channels (represented by the higher-order components), and input them, rather than the raw data, into a classifier, thus reducing the computational demands and possibly improving performance. However, as noted already, there is no reason to assume that

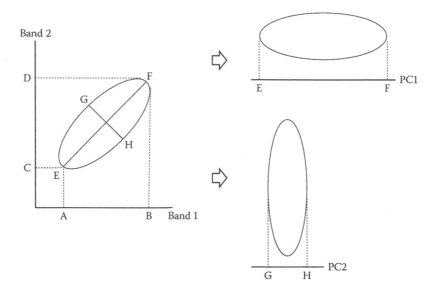

FIGURE 2.2   A two-band example for principal component analysis.

lower-order principal components do not contain any discriminating information. It may also be of interest to note that the locations of the principal component axes in the feature space are fixed by reference to two constraints. First, the axes are orthogonal. Second, each axis accounts for the maximum variance in the data set given that the influence of any higher-order components has been removed. Relaxation of the second constraint allows the use of orthogonal rotations that might make the resulting components more interpretable. It is also possible to relax the first constraint, but that does not seem to be helpful in the present context.

Figure 2.2 shows an example using a two-band image. Here, the original pixel values in the two images fall in the range [A, B] and [C, D], respectively. After the covariance (or correlation) matrix is formed, the resulting eigenvectors are orientated in the directions [E, F] and [G, H], respectively. The data range [E, F] in the first principal component (PC1) is larger than either [A, B] or [C, D], and thus a higher contrast image will result. The data range [G, H] of PC2 is much smaller, and the corresponding principal component image will show less variation. However, the information content of the two axes will differ in that PC1 will represent the information that is shared between all or most of the original spectral bands (and may therefore be thought to represent average brightness), whereas the second and subsequent components will contain information that is statistically uncorrelated with brightness. For example, the second principal component of a multispectral image set of a vegetated area may be related to variations in the nature, vigor, and spatial cover of the vegetation.

In the next example, a three-band SPOT HRV image of an agricultural area is used (Figure 2.3). The resulting principal component images are shown in Figure 2.4a, and a clear decrease in data variance (information) in terms of histograms is illustrated in Figure 2.4b.

FIGURE 2.3   Three-band SPOT HRV image of an agricultural area in Eastern England.
© CNES (Centre National d'Études Spatiales) 1994—SPOT Image distribution.

As noted earlier, PCA can be based either on the matrix of variances and covariances or the matrix of correlations among the spectral bands of an image set, which can be derived either from sample data or from all the pixels of the image set. PCA based on the covariance matrix is sometimes also called unstandardized PCA, while standardized PCA uses the correlation matrix. These matrices have different eigenvectors, and thus the resulting principal component images will also be different. In the case of land cover/use change detection, the standardized method is preferred by Fung and Le Drew (1987), Baronti et al. (1994), and Liu et al. (2004). If the images subjected to PCA are not measured in the same scale, the standardized method is also a better choice than the unstandardized method because the correlation matrix normalizes the data onto the same scale (i.e., each spectral band is given unit variance, hence interband variation is ignored). Users should be aware of the fact that the use of the correlation matrix implies the equalization of within-band variance, and should also note that the technique does not differentiate between information and noise. In some image sets (for example, hyperspectral data sets) there may be

(a)

FIGURE 2.4 (a) Principal component images derived from the SPOT image shown in Figure 2.3.

considerable noise present, and this noise may have a variance that is greater than some of the information sources.

2.1.3 MINIMUM/MAXIMUM AUTOCORRELATION FACTORS (MAF)

A somewhat different approach to the orthogonalization of specifically spatial data is presented by Switzer and Green (1984), who argue that PCA does not differentiate between signal and noise, that it is not explicitly spatial (in the sense that the image data could be rearranged randomly without affecting the results), and that it is scale-dependent. They propose a method of distinguishing between signal and noise, and suggest that their approach should be preferred to filtering because it does not blur the data.

Their method is based on the principle that information, or signal, is spatially autocorrelated (in the sense that pixels in a neighborhood will tend to have similar values because they represent some geographical object such as a lake or a forest). Conversely, noise will have a low spatial autocorrelation, if systematic noise such as

(b)

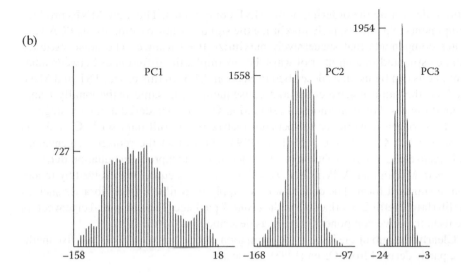

FIGURE 2.4   (continued) (b) Histograms of corresponding principal component images. The histograms show that the variance of the first principal component exceeds that of the second, which in turn is greater than the variance of the third component.

banding is excluded. The MAF procedure is based on the ordering of set of orthogonal functions in such a way that autocorrelation decreases from the low-order to the high-order functions, that is, autocorrelation is minimized rather than variance being maximized, as is the case with PCA. The lower-order functions may therefore be expected to contain mainly nonautocorrelated noise while the higher-order components would represent information. The procedure is scale-free in the sense that the same result is achieved irrespective of the scale of measurement used for each feature. This is because the procedure maximizes a ratio—that of signal to noise.

A detailed account of the computational procedures involved in the MAF transform is provided by Nielsen (1994) and Nielsen et al. (1998). The variance–covariance of the total data set, $\mathbf{C}$, is derived, together with the corresponding eigenvalues Λ and eigenvectors $\mathbf{P}$. Next, the variance–covariance of the noise component, $\mathbf{C}_\Delta$, is computed. Switzer and Green (1984) suggest that the original data set is converted to two difference images, one being shifted by one pixel horizontally and the other being shifted by the same amount vertically. The variance–covariance matrices of the two difference images are calculated and pooled to give $\mathbf{C}_\Delta$. Finally, the eigenvalues and eigenvectors of $\Lambda^{-\frac{1}{2}}\mathbf{P}^\mathsf{T}\mathbf{C}_\Delta\mathbf{P}\,\Lambda^{\frac{1}{2}}$ are obtained, and MAF images derived using procedures equivalent to those described above for the calculation of principal component images (Equation 2.5).

2.1.4   MAXIMUM NOISE FRACTION TRANSFORMATION

Maximum noise fraction transformation, which is sometimes known as *noise-adjusted principal components*, requires estimates of the variance–covariance matrices of the signal, $\mathbf{C}_S$, and noise, $\mathbf{C}_N$, components and it uses the signal-to-noise ratio to determine the ordering of the MNF components. Thus, the MNFs produce components that successively maximize the signal-to-noise ratio, just as PCA generates components that successively maximize the variance. The noise variance can be estimated in a number of ways. For example, the differences between adjacent pixels can be used, as described in Section 2.1.3. In this case, MNF and MAF produce the same eigenvectors, and consequently the same orthogonally transformed images. Other methods of estimating C_N are considered at a later stage.

The variance–covariance (dispersion) matrix of the full data set is C, and it is axiomatic that $C = C_S + C_N$. Nielsen (1994) shows that the required eigenvalues and eigenvectors are the solution of the generalized eigenproblem equation $\det(C_N - \lambda C_S) = 0$. Martin and Wilkinson (1971) show how this equation is solved by reduction to standard form. The code that they supply is available from libraries such as Netlib (http://netlib2.cs.utk.edu/) in the *eispack* package. The computed eigenvectors are used to derive component images, as described in Section 2.1.2.

Clearly, the nub of the issue is the estimation of C_N. Nielsen (1994) lists five methods, partly derived from Olsen (1993). These are

1. Simple differencing, using the method of Section 2.1.3.
2. Use of a simultaneous autoregressive synthetic aperture radar (SAR) model involving the pixel of interest and its neighbors to the W, NW, N, and NE.
3. Determine the difference between the value of the pixel of interest and the local mean, which is computed for a rectangular window.
4. Method (3), but the local median is used in place of the local mean.
5. Compute the residual from a local quadratic surface based on the pixel values neighboring the pixel of interest.

Neilsen (1994) also considers the problem (mentioned previously) of periodic noise, such as banding due to differences in sensor calibrations. Banding, which is seen as horizontal striping, is obviously autocorrelated, and so will be identified as part of the signal. Neilsen (1994) suggests that frequency-domain filtering (Mather, 2004) be used prior to the calculation of the noise covariance. The use of wavelets to estimate noise is another possible approach (Mather, 2004).

Other pertinent references to the derivation and use of MNFs are Green et al. (1988) and Lee et al. (1990), who discuss the mathematical details, while Van der Meer and De Jong (2000) show how the results of spectral unmixing are enhanced by the orthogonalization of end members. Neilsen et al. (1998) discuss the MAF transformation in the context of change detection. The use of the MNF and similar transforms is common in studies that utilize hyperspectral data, such as Compact Airborne Spectrographic Imager (CASI) (Jacobsen et al., 1998), AVIRIS (Harsanyi and Chang, 1994, Ustin et al., 2002) and DAIS (De Jong and Van der Werff, 1998; Kneubuehler et al., 1998).

2.2 FEATURE SELECTION

The computational requirements of classification are generally positively correlated with the number of features used as input to the classification algorithm. For instance, in the case of n input features, the computational requirements of the maximum likelihood classifier (Section 2.3.2) are proportional to n^2. As n increases in magnitude, the computational cost will rise nonlinearly. The use of large artificial neural networks also results in increased training times.

Two approaches can be used to reduce the number of input features without sacrificing accuracy. One is to project the original feature space onto a subspace (i.e., a space of smaller dimensionality). This can be done using either an orthogonal transformation, as described in Section 2.1, or a self-organized feature map (SOM) neural network (Chapter 3).

The second method is to use separability measurements in the input feature space, and then select the subfeature dimension in which separability is a maximum. The aim is to reduce the feature space dimension without prejudicing classification accuracy. Two separability indices, the divergence index (Singh, 1984), and the B-distance (Haralick and Fu, 1983) are widely used. The divergence index D_{ij} between two classes i and j is derived from the likelihood ratio of any pair of classes. For multivariate Gaussian distributions, D_{ij} for classes i, j based on a subfeature dimension m of the total p dimension is defined by

$$D_{ij} = \frac{tr\left((C_i - C_j) \times (C_j^{-1} - C_i^{-1})\right)}{2} + \frac{tr\left((C_i^{-1} + C_j^{-1}) \times (\mu_i - \mu_j) \times (\mu_i - \mu_j)^T\right)}{2} \quad (2.6)$$

where tr denotes the trace (sum of the diagonal elements) of the matrices, C is the class variance–covariance matrix of dimension $m \times m$, μ is the class mean vector, and ^{T} denotes the transpose of a matrix. In some cases, a transformation of D_{ij}, called the *transformed divergence* (TD_{ij}), is used. TD_{ij} is defined by

$$TD_{ij} = 2000 \times [1 - \exp(-D_{ij}/8)] \quad (2.7)$$

in which the value D_{ij} is projected into interval 0 to 2000. The greater the value of D_{ij} or TD_{ij}, the greater is the class separability based on selected m subfeature dimension. If expression (2.7) is used, Jensen (1986) points out that generally a poor separability is indicated by a value of TD_{ij} smaller than 1900 (corresponding to $D_{ij} < 23.97$).

The B-distance, B_{ij}, is defined by (Haralick and Fu, 1983)

$$B_{ij} = \frac{1}{8}(\mu_i - \mu_j)^T \frac{(C_i + C_j)}{2}(\mu_i + \mu_j) + \frac{1}{2}\ln\frac{\dfrac{(\mu_i - \mu_j)}{2}}{\sqrt{\mu_i}\sqrt{\mu_j}}. \quad (2.8)$$

The quantity B_{ij} is computed for every pair of classes given m subfeature dimension. The sum of B_{ij} for every pair of classes (except the class itself) is obtained and is a measure of the overall separability. Although the two separability measures defined by Equations (2.6) and (2.8) are based on different principles, Mausel et al. (1990) find that the divergence index and B-distance generally give similar results.

2.3   FUNDAMENTAL PATTERN RECOGNITION TECHNIQUES

2.3.1   UNSUPERVISED METHODS

As noted earlier, the main difference between the unsupervised and supervised approaches is that unsupervised methods do not require the user to select training data sets to characterize the targets or to train the classifier. Instead, the user specifies only the number of clusters to be generated, and the classifier automatically constructs the clusters by minimizing some predefined error function. Sometimes the number of clusters can be detected automatically by the classifier (Cheung, 2005). In theory, users do not need to interact with the classifier, which operates independently and automatically. However, in practice, it is more often the case that results are accepted or rejected on the basis of whether or not they meet the user's expectations.

2.3.1.1   The *k-means* Algorithm

The *k-means* algorithm has been the most popular unsupervised algorithm being widely used for automatic image segmentation in the field of remote sensing for many years (Ball and Hall, 1967; Mather, 1976, 2004). This algorithm is implemented by recursively migrating a set of cluster means (centers) using a *closest distance to mean* approach until the locations of the cluster means are unchanged, or until the change from one iteration to the next is less than some predefined threshold. Change may also be defined in terms of the number of pixels moving from one cluster to another between iterations, or by the value of a measure of cluster compactness, such as the sum of squares of the deviations of each pixel from the center of its cluster, summed for all classes.

The use of the *k-means* algorithm must estimate the initial number of clusters, say *n*, present in the data. The algorithm initially determines the location of *n* cluster means within the feature space, either by generating random feature vectors, or by selecting *n* pixels at random from the available data, or by using a predefined set of feature vectors. Each pixel is then associated with its nearest cluster center. *Nearest* is defined either by the Euclidean or Mahalanobis distance measure. At the next stage, the location of each cluster mean is recalculated based on the set of pixels allocated to that center. The process is repeated until the change between iterations becomes less than a user-specified threshold. Figure 2.5 illustrates the process.

The calculation of the distance between a pixel and a cluster center normally uses the Euclidean measure D_E represented by

$$D_E^2 = \left(\mathbf{x}_i - \boldsymbol{\mu}_j \right)^2 \tag{2.9}$$

where $\mathbf{x}_i$ is the observed vector of the ith pixel and $\boldsymbol{\mu}_j$ is the current mean vector of the jth cluster. The dimension of vector $\mathbf{x}_i$ is equal to the number of bands being used as input. For instance, if three band images are used, then $\mathbf{x}_i$ will be of dimension 3×1. Another choice for the calculation of distance between pixel and the cluster center is the Mahalanobis distance D_M, given by

$$D_M = \left(\mathbf{X}_i - \boldsymbol{\mu}_j \right)^{\mathbf{T}} \times \mathbf{C}_j^{-1} \times \left(\mathbf{X}_i - \boldsymbol{\mu}_j \right) \tag{2.10}$$

Begin
   Choose the number of clusters and mean change tolerance ε,
   Randomly assign the mean vector μ_j for each cluster j;
   Do while ($|\text{new } \mu_j - \text{old } \mu_j| < \varepsilon$) for al cluster j
      Do for each pixel i
         Calculate the distance D_{ij} between the pixel i and cluster
            mean μ_j for each cluster j;
         Assign pixel i to the cluster c according to
            $c = \min\{D_{ij}, \forall \text{ cluster } j\}$.
      EndDo
      Recalculate the cluster mean μj for each cluster j based on the
         pixels allocated to the cluster.
   EndDo
End

FIGURE 2.5   Algorithm for migrating mean clustering process.

where [T] denotes the matrix transpose, and C_j^{-1} is the inverse of the variance–covariance matrix for cluster j, respectively. Matrix C_j is obtained by Equation (2.2) (note that only pixels belonging to cluster j are used in the calculation of C_j).

The Mahalanobis distance takes into account the shape of the frequency distribution (assumed to be Gaussian) for a given cluster in feature space, resulting in ellipsoidal clusters, whereas the use of the Euclidean distance assumes equal variances and a correlation of 1.0 between the features, giving circular clusters (Figure 2.6). Here, although the distances between the pixel and two cluster centers are the same, the pixel will be assigned to cluster a when the Mahalanobis distance measure is used

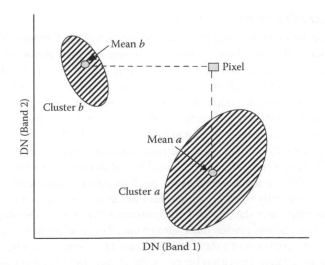

FIGURE 2.6a   General cluster shapes (ellipsoidal) using the Mahalanobis distance measure defined in Equation (2.10).

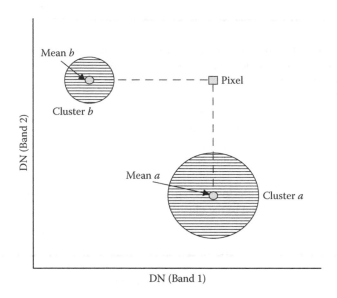

FIGURE 2.6b   Circular cluster shape using the Euclidean distance measure defined in Equation (2.9).

(Figure 2.6a) because cluster a is of higher value than cluster b in variance–covariance measure (thus the shorter the distance). If the Euclidean distance measure is used, then the decision to place the pixel in cluster a or b will be ambiguous. In addition to the above considerations, the initialization points of the clusters forming process may also play an important role for generating clustering outcomes. Pena et al. (1999) presents a comparative study for different initialization methods for the *k-means* algorithm.

2.3.1.2   Fuzzy Clustering

An alternative approach to unsupervised classification using *k-means* is fuzzy clustering. The main difference between fuzzy clustering and the *k-means* method is that the resulting clusters generated by fuzzy clustering are no longer "hard" or "crisp" but fuzzy. In other words, each pixel may simultaneously belong to two or more clusters and will have a "membership value" for each cluster. In general, the performance of fuzzy clustering methods is superior to that of the corresponding hard versions, and they are less likely to stick in a local minima (Bezdek, 1981).

   Most analytic fuzzy clustering algorithms are derived from Bezdek's (1981) fuzzy c-means (FCM) algorithm. The FCM algorithm and its derivations have been implemented successfully in many applications, such as pattern classification and image segmentation, especially those in which the final goal is to make a crisp decision. The FCM algorithm uses the probabilistic constraint that the membership probabilities of a data point across classes must sum to one. This constraint comes from generalizing a crisp c-partition of a data set, and is used to generate membership-update equations for an iterative algorithm based on the minimization of a least-square type of criterion function. The constraint on membership used in FCM is meant to avoid the trivial solution of all membership probabilities being equal to zero. FCM does provide meaningful results in applications where it is appropriate to interpret memberships as likelihood estimates (Goth and Geva, 1989), although some studies have provided a different perspective on the meaning of membership probabilities (see, for example, Krishnapuram and Keller, 1996). A more detailed description of FCM is provided in Chapter 5.

2.3.2 SUPERVISED METHODS

The supervised approach to pixel labeling requires the user to select representative training data for each of a predefined number of classes. Classification performance is highly dependent on how well the user is able to model the target class distribution. The user's experience can be very helpful in identifying and locating training areas. Ideally, the training areas should be sites where homogeneous examples of known cover types are found (Townshend, 1981). A supervised statistical classification can be carried out by the following three steps:

1. Define the number and nature of the information classes, and collect sufficient and representative training data for each class,
2. Estimate the required statistical parameters from the training data, and
3. Use an appropriate decision rule.

Although the selection of training data may be tedious, a supervised approach is preferred by most researchers because it generally gives more accurate class definitions and higher accuracy than do unsupervised approaches. Three statistical classifiers are in general use. These are the *parallelepiped method*, *minimum distance classifier*, and the *maximum likelihood algorithm*.

2.3.2.1 Parallelepiped Method

The parallelepiped method is implemented by defining a parallelepiped-like subspace (i.e., hyper-rectangle) for each class. The boundaries of the parallelepiped, for each feature, can be defined by the minimum and maximum pixel values in the given class, or alternatively, by a certain number of standard deviations on either side of the mean of the training data for the given class. The decision rule is simply to check whether the point representing a pixel in feature space lies inside any of the parallelepipeds. An example illustrating the specification of the topology of a parallelepiped classifier in the case of a two-dimensional feature space is shown in Figure 2.7.

The parallelepiped method is quick and easy to implement, but errors may arise, particularly when a pixel lies inside more than one parallelepiped or outside all parallelepipeds. These two situations are, in fact, likely to occur, because in the feature space the distribution of pattern vectors is often quite complex. Therefore, it is hard to provide a robust classification performance using this simple method.

2.3.2.2 Minimum Distance Classifier

Like the migrating mean clustering algorithm, the decision rule adopted by the minimum distance classifier to determine a pixel's label is the minimum distance

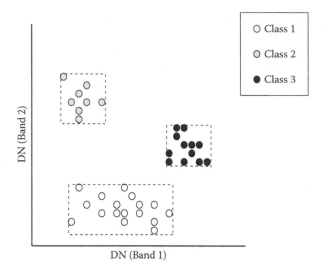

FIGURE 2.7 Each parallelepiped is bounded by minimal and maximal point values within the class.

between the pixel and the class centers, measured either by the Euclidean distance or the Mahalanobis generalized distance (Equations 2.9 and 2.10). Both of these distance measures can be described as *dissimilarity coefficients*, in that the similarity between objects i and j increases as distance becomes smaller. The mean spectral vector, or class centroid, for each class (plus the variance–covariance parameters if the Mahalanobis distance measure is used in place of the Euclidean distance) is determined from training data sets. A pixel of unknown identity is labeled by computing the distance between the value of the unknown pixel and each class centroid in turn. The label of the closest centroid is then assigned to the pixel.

An example is shown in Figure 2.8, in which pixel a is allocated to class 1 (i.e., given the label "1") according to the minimum distance between the pixel and class center. The shape of each class depends on which distance function (Mahalanobis or Euclidean) is used.

This type of classifier is mathematically simple and computationally efficient, but the theoretical basis may not be as robust as that of the maximum likelihood classifier, which is described in Section 2.3.2.3. A comparison reported by Benediktsson et al. (1990) shows that although the minimum distance classification using the Mahalanobis distance measure performs better than the same classifier using the Euclidean distance measure, its accuracy still falls considerably behind that achieved by the maximum likelihood algorithm.

2.3.2.3   Maximum Likelihood Classifier

The maximum likelihood (ML) procedure is a supervised statistical approach to pattern recognition. The probability of a pixel belonging to each of a predefined set of classes is calculated, and the pixel is then assigned to the class for which the probability is the highest. ML is based on the Bayesian probability formula:

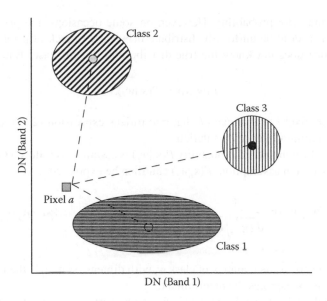

FIGURE 2.8   Example of minimum distance classification criteria. See text for details.

$$P(\mathbf{x}, w) = P(w|\mathbf{x})\, P(\mathbf{x}) = P(\mathbf{x}|w)\, P(w) \qquad (2.11)$$

where $\mathbf{x}$ and w are generally called *events*. $P(\mathbf{x}, w)$ is the probability of coexistence (or intersection) of events $\mathbf{x}$ and w, $P(\mathbf{x})$ and $P(w)$ are the prior probabilities of events x and w, and $P(w|\mathbf{x})$ is the conditional probability of event x given event w. $P(w|\mathbf{x})$ is interpreted in the same manner. If event $\mathbf{x}_i$ is the ith pattern vector and w_j is information class j then, according to Equation (2.11), the probability that $\mathbf{x}_i$ belongs to class w_j is given by

$$P(w_j|\mathbf{x}_i) = \frac{P(\mathbf{x}_i|w_j)P(w_j)}{P(\mathbf{x}_i)}. \qquad (2.12)$$

Since, in general, $P(\mathbf{x})$ is set to be uniformly distributed (i.e., the probability of occurrence is the same for all pixel features), Equation (2.12) can be rewritten as

$$P(w_j|\mathbf{x}_i) \propto P(\mathbf{x}_i|w_j)\, P(w_j). \qquad (2.13)$$

One can thus allocate pixel i to the class k, which has the largest value of the term $P(w_k|\mathbf{x}_i)$ in Equation (2.13). The classification criterion can be expressed as

$$w_k = \arg\max_{\forall w_j}\left\{ P(\mathbf{x}_i|w_j)P(w_j)\right\} \qquad (2.14)$$

where *arg* denotes *argument*. The criterion shown in Equation (2.14) is called the Maximum A Posteriori (MAP) solution, which maximizes the product of conditional probability and prior probability. However, on some occasions, the prior probability $P(w)$ is also set to be uniformly distributed (because of lack of prior knowledge or because one does not know the true distribution). In this case, Equation (2.13) reduces to

$$P(w_j|\mathbf{x}_i) \propto P(\mathbf{x}_i|w_j). \tag{2.15}$$

If we allocate pixel i to that class k that maximizes expression (2.15), the result is called the maximum likelihood solution.

Normally, the conditional probability $P(\mathbf{x}_i|w_j)$ is assumed to follow a Gaussian (or normal) distribution assumption. $P(\mathbf{x}_i|w_j)$ can then be expressed as

$$P(x_i|w_j) = \frac{1}{\sqrt{2\pi}^{\rho}\sqrt{|\mathbf{C}_j|}} \exp\left(-\tfrac{1}{2} \times (\mathbf{x}_i - \boldsymbol{\mu}_j)^T \times \mathbf{C}_j^{-1} \times (\mathbf{x}_i - \boldsymbol{\mu}_j)\right) \tag{2.16}$$

where $\mathbf{C}_j$ is the covariance matrix of class w_j with dimension ρ, $\boldsymbol{\mu}_j$ is the mean vector of class w_j, and $|\times|$ denotes the determinant.

In practical applications, Equation (2.16) can be further reduced to the following expression by taking the natural logarithm:

$$\ln\left[P(x_i|w_j)\right] = \frac{-1}{2}\rho \times \ln(2\pi) \times \frac{-1}{2}\ln|\mathbf{C}_j| - \frac{1}{2}(\mathbf{x}_i - \boldsymbol{\mu}_j)^T \times \mathbf{C}_j^{-1} \times (\mathbf{x}_i - \boldsymbol{\mu}_j). \tag{2.17}$$

Equation (2.17) avoids the computation of the exponential term. Since the term $\rho\ln(2\pi)$ is the same for all classes, it can be regarded as a constant and dropped from the equation without affecting the final ranking of the values of $\ln[P(\mathbf{x}_i|w_j)]$. Finally, the expression in Equation (2.17) is multiplied by the constant -2 to give

$$-2\ln\left[P(x_i|w_j)\right] = \ln|\mathbf{C}_j| + (\mathbf{x}_i - \boldsymbol{\mu}_j)^T \times \mathbf{C}_j^{-1} \times (\mathbf{x}_i - \boldsymbol{\mu}_j). \tag{2.18}$$

It is clear that maximizing Equation (2.16) is equivalent to minimizing Equation (2.18). Note that the second term of Equation (2.18) is the Mahalanobis distance, Equation (2.10). Thus, the geometrical shape of the cloud formed by a set of pixels belonging to a given class can be described by an ellipsoid. The shape of the ellipsoid depends on the covariance among the features belonging to a feature space. In a two-dimensional feature space the maximum likelihood function delineates ellipsoidal *equiprobability contours*, which can be viewed as decision boundaries.

Two parameters, the mean vector and the covariance matrix, are used to characterize each class. The importance of choosing a sample size that is adequate to provide for an unbiased and efficient estimate of these parameters is considerable (see the discussion above, Section 2.1, and below, Section 2.6).

The ML classifier assumes that the information class prior probability $P(w)$ is uniformly distributed. However, if one can model $P(w)$ in suitable way, the classification accuracy could be increased. For example, one may model prior probability as different weights associated with each class. A higher weight for a given class implies that there is a higher probability of a pixel receiving the label associated with that class. The effects of modeling $P(w)$ are discussed by Swain and Davis (1978), Strahler (1980), and Mather (1985, 2004). In recent years, there has been a trend toward modeling the prior probability using a smoothness assumption based on the concept of context. This method is called the smoothness prior. By using the smoothness prior together with the class conditional probability $P(\mathbf{x}|w)$, one can attempt to perform a MAP classification. The use of statistically robust methods is the main topic of Chapter 8.

As the performance of statistical maximum likelihood classifier is generally limited by frequency distribution assumptions, in recent years, more elegant classifiers such as artificial neural networks (Chapter 3), support vector machines (Chapter 4), fuzzy theory–based methods (Chapter 5), and decision trees (Chapter 6) have also been introduced into the field of remote sensing imagery classification. These state-of-the-art classifiers should draw our attention, because they normally are distribution-free, and are able to show a significant level of improvements over traditional methods introduced previously in this section.

2.4   COMBINING CLASSIFIERS

The different approaches to pattern recognition are often viewed as alternative methods, and many researchers have published comparisons among the various procedures in order to demonstrate that one is "better" than the other in some way. It appears that many of the methods are complementary; some are better in resolving one aspect of the labeling problem, while another method may be superior in another respect. Hence, some interest has been shown in combining the results obtained using different decision rules in order to improve the overall labeling. Brief details of combination methods known as *voting rules, Bayesian formulation, evidential reasoning,* and *multiple ANN* are given in the following paragraphs and later chapters. Vieira (2000) presents a more detailed survey.

The procedure using *voting rules* is quite simple. The labels output by a number of classifiers for a given pixel are collected, and the majority label is selected. This is known as the *majority vote rule.* A more stringent requirement is that all classifiers agree on a single label, and this approach uses the *conservative voting rule.* The latter rule is unlikely to produce a fully labeled thematic image if the number of classifiers used is more than two. On the other hand, the former rule may produce conflicts similar to the 2000 U.S. Presidential election, in that the winner may have only a few votes more than the loser. In order to avoid the possibility of choosing label A when the evidence supporting A is not much greater than the evidence supporting label B, the method of *comparative majority voting* can be used. This method requires that the "winning" vote should exceed the runner-up vote by a specified amount, thus ensuring a clear winner. Pixels that are not labeled because no clear winner emerges are labeled as unknown. See Hansen and Salomon (1990) for a discussion of combining artificial neural network classifiers. Brodley and Friedl (1999) consider the question of classifier combination and voting methods in the context of filtering training data (Section 2.6.3).

Bayesian formalism can involve a process as simple as averaging. It is used with multiple classifiers that output a probability (or probability-like) estimate of the likelihood of pixel A belonging to class j. Such classifiers include the maximum likelihood method and the artificial neural network. The probabilities (or pseudoprobabilities) for a pixel for each possible class are accumulated, and the winner is that pixel that has the greatest accumulated probability. As in the case of majority voting, a threshold may be set so that the winning margin is a clear one.

Evidential reasoning is described in Section 9.4. In essence, the method associates a degree of belief with each source of information, and a formal system of rules is used in order to manipulate the belief function. In the context of pattern recognition, this method is useful in handling multiple sources of data of different kinds and with different levels of accuracy. It can also be used to assess the plausibility of labels assigned to a given pixel by different classifiers.

Wilkinson et al. (1995) feed the output from several classifiers into an artificial neural net, which is trained to produce a single class label as output from the possibly several different labels output by the multiple classifiers. In a sense, these authors propose that the classifiers' output is itself classified.

If the results from several classifiers are to be amalgamated, then the classifiers should be independent. To be independent, these classifiers must each use an independent feature set or be trained on separate sets of training data. One possibility is to use subsets of the features for each separate classifier, or to use random sampling of the training data to generate p different subsets of training data, where p is the number of classifiers. Cross-validation is another possibility; the test data are subdivided into a number n of approximately equally sized subsets, and $(n - 1)$ of these subsets are used to train the classifier. The remaining n^{th} subset is used for estimating the error associated with the labeling. Each of the n subsets is used in turn for testing, with the remaining $(n - 1)$ subsets being used for training. Once this cycle of training and testing is completed, the error estimates are combined (Breiman, 1996, 2001; Schaffer, 1993).

Further elaboration of these multiple classifier approaches can be found in Tumer and Ghosh (1995), Kanellopoulos and Wilkinson (1997), Roli et al., (1997), Steele (2000), Smits (2002), Briem et al. (2002), and Bruzzone et al. (2004).

2.5   INCORPORATION OF ANCILLARY INFORMATION

The addition of ancillary (nonspectral) information, such as spatial relationships, may be a more powerful way of characterizing the classes of interest, and if added to the spectral information for image classification, the classification accuracy may be improved (Mather et al., 1998; Tso and Olsen, 2005a). Ancillary information can be extracted directly from an image or from other sources such as digital elevation models, and geological and soil maps. Some examples of ancillary features derived from an image are texture, context, and structural relationships.

2.5.1 Use of Texture and Context

Texture and context both measure aspects of the spatial structure of an image. They are frequently used in image interpretation and are generally extracted directly from an image. Texture refers to a description of the spatial variability of tones found within part of a scene. In visual terms, texture is expressed as the impression of roughness or smoothness created by the variation of tone or repetition of patterns across a surface. The application of tone and texture in digital image analysis required a quantitative characterization of these visual concepts.

Two texture descriptors for a defined area can be simply the mean and variance of the region. These are the simplest features to characterize textures, but our experience suggests that they are not sufficient to characterize texture in the best way possible. For example, it is possible for different texture patterns to have the same mean and variance. More comprehensive approaches are required. Different methods used to quantize texture patterns also have a profound effect on classification accuracy. Chapter 7 presents a review and comparative study of texture extraction techniques.

The context of a pixel or a group of pixels refers to its probability of occurrence based on the nature of the pixels in the remainder of the scene. Human photo interpreters have long exploited spatial information like texture and context, and attempts have been made to incorporate these attributes of the scene to computer-assisted image classification. There is clear evidence that data from future sensors, with their finer spatial resolving power, may not necessarily generate improved classifications when per-pixel classifiers are used (Townshend, 1981; Mather, 2004) because of the higher spectral variability of local areas of the image, which becomes apparent as the resolution becomes finer.

Contextual information can either be included in a statistical classifier or in procedures that amend some preliminary classifier output. A simple way to use context is in the form of a majority filter window (Gurney, 1980, 1981). For instance, let a window be centered on a pixel labeled i. If the majority of the pixels within the window belong to class j, the label of the central pixel is altered from i to j. In most cases, the majority filter is used for classification refinement. Although there is some improvement, the increase in classification accuracy is not impressive. More complex statistical models of spatial context are discussed by Haslett (1985), Kim and Swain (1995), Solaiman et al. (1996), and Nishii and Eguchi (2005).

Chapter 8 also provides details of methods of modeling context as prior probability using a Markov random field (MRF) model, and this prior probability is further combined with class-conditional probability to perform a MAP classification. The increase in classification accuracy compared with traditional methods demonstrates the practical value of this approach (Tso and Mather, 1999; Tso and Olsen, 2005b; Nishii and Eguchi, 2005).

2.5.2 Using Ancillary Multisource Data

A significant problem arises when one attempts to combine images and ancillary data, namely, the different nature of these data types. Data may be categorical and, even when continuous in nature, may violate the normality assumption required by most statistical classification techniques. Ancillary information can be continuous or categorical (Strahler et al., 1980; Schepers, 2004). Examples of continuous ancillary data are elevation, slope, and aspect derived from a digital elevation model. Examples of categorical data are land use, soil, and geological types. The different approaches proposed for the combination of spectral and ancillary information include: the logical channel approach, stratification, the use of prior probabilities, and file- or object-based classification.

The logical channel approach consists of adding each ancillary data set as an additional feature, so that the pixel vector is extended by the addition of this external information. This technique was called the logical channel approach (Strahler and Bryant, 1978; Ricchetti, 2000; Gercek, 2004). The advantages and disadvantages of this approach are discussed in Chapter 9.

Stratification involves the subdivision or segmentation of the study area into smaller areas or strata, based on rules derived from external knowledge, so that each stratum can be processed independently (Gorte and Stein, 1998; De Bruin and Corte, 2000). This process is performed before classification with the purpose of increasing the homogeneity of the individual data set to be classified, or to separate different objects that are spectrally similar. Some advantages relating to this kind of approach might be considered: one advantage is the convenience of dealing with smaller data sets; another benefit is the reduction of variability within individual strata. Stratification is effective and easy to implement, but is deterministic in the sense that it does not handle uncertainty about the occurrence of certain classes or the gradation between strata. An incorrect stratification can invalidate the entire classification process. Ricchetti (2000) compares the logical channel and stratification methods, and is in favor of the logical channel approach.

In most classification processes, the probability of class membership of a given pixel is assumed equal for all classes. However, if the information about an area shows the preference of certain classes for particular locations in the terrain, then this information can be expressed in terms of prior probabilities of occurrence for each class and this information can be incorporated into the classification process. The major difficulty involved in this approach is to define a suitable function of prior probability relating to each class in terms of achieving optimal result. Most experiments rely on the analyst's *ad hoc* decision.

A further application of spatial information is the use of boundary information in order to define objects prior to classification. Such a classification strategy is suitable for patchy areas (e.g., areas made up of agricultural fields), and might be less affected by boundary pixels, which generally result in classification error. In agricultural crop applications, the objects are fields, the boundaries of which can be derived either by digitizing paper maps of an appropriate scale, or by applying an edge detection algorithm to the image. Each object (field) is characterized by global statistical parameters, and is represented by a unique vector in feature space (Mégier et al., 1984; Erickson and Lickens, 1984; Mason et al., 1988; Jansen et al., 1990; Wang and Wang, 2004; Tso and Olsen, 2005b).

In contrast to the mainly *ad hoc* procedures described above, two specific multisource data fusion mechanisms known as extension of Bayesian theory and evidential reasoning have their particularly practical value because both approaches have a robust theory basis for dealing with multisource data sets. Details of these two methods for managing multisource data are provided in Chapter 9.

2.6   SAMPLING SCHEME AND SAMPLE SIZE

A sampling scheme describes the way in which sample pixels are selected from the image in order to characterize the thematic classes of interest. Sample size is important in terms of the accuracy with which estimates of statistical parameters describing these classes are obtained. Both sampling scheme and sample size are also of importance in assessing the accuracy of the thematic map derived from remotely sensed data. Samples may be derived from field observation, from farm records (in the case of agricultural crops), or from maps or air photographs. If the target classes are of a temporally changing nature, care must also be taken to ensure that the sample data adequately represent the temporal state of the phenomena being observed. There are certain restrictions on sampling, including cost, availability of source information such as maps and air photographs, and accessibility. If the area of interest is large, then it may not be possible to conduct a thorough and statistically valid sampling procedure close to the time of the satellite overpass. When clouds are a problem, the logistics of sampling are made more difficult by the need to be on call for a number of overpass times. Nevertheless, attention must be paid to the questions of sampling scheme and sample size, particularly if statistical methods of pattern recognition are employed.

A distinction is made between two kinds of sample data. *Training data* are used in supervised methods of pattern recognition to "teach" a classifier the main characteristics of each class. Campbell (1987) points out that

- The number of sample observations has a direct relationship with the confidence interval of the estimate of the accuracy of a classification, and on the estimates of statistical parameters used in the particular, chosen, classifier. For example, estimates of the mean vector and variance–covariance matrices of the individual classes are required by the maximum likelihood classifier, and
- There should be some minimum number of samples that are assigned to every class, to ensure that each class is properly represented.

Test data are used in the assessment of classification accuracy. These data are put to one side until the thematic map has been produced. The test data are then categorized, using the same procedure as that which generated the thematic map. A comparison of the label allocated to each sample element of the test data by the classifier and the label provided by the human observer allow an assessment of the accuracy achieved by the classifier. This topic is discussed in Section 2.6.1. Where it is not possible to acquire test data, then the use of cross-validation is a possible solution.

Some authors describe their test data as ground truth. This is a misleading and inaccurate description. It is human to err, and one presumes that this aspect of human behavior extends to the collection of test and training data in remote sensing. Some of the errors in a thematic map may not be due to the fallibility of the classifier but may be the result of mislocation of some elements of the test data, or to faulty identification of the attributes of the test data.

2.6.1 SAMPLING SCHEME

A number of sampling schemes are described in the literature (see, for example, Berry and Baker, 1968; Borak and Strahler, 1999; Ginevan, 1979; Fitzpatrick-Lins, 1981; Stehman, 1992; Congalton and Green, 1999; Wang et al., 2005). Congalton (1988) suggests that both *random sampling* without replacement and *stratified unaligned random sampling* generally provide satisfactory results. Stehman (1992) suggests, however, that the usual formulae for deriving the value of the Kappa coefficient, which is used in accuracy assessment (Section 2.7), may perform poorly when stratified sampling methods are used. Congalton and Plourde (2000) have a more relaxed view, but make the point that the placement of samples is as important as the sampling scheme. Wang et al. (2005) use the semivariogram technique (Chapter 7) to determine the optimal sampling space. Further discussion is contained in Brogaard and Ólafsdóttir (1997).

Atkinson (1991, 1996) notes that the standard statistical rules of sampling, as outlined in conventional statistical texts such as Cochran (1977), do not hold for spatial data because locations in space are in fixed positions and their attributes are therefore autocorrelated. He proposes the use of geostatistical methods, which take into account the spatial relationships between the pixels. Van der Meer et al. (1998) present the results of an investigation of mapping accuracy using geostatistical methods. They show that the optimum sampling distance varies with the nature of the target, being of the order of 100 m for vegetated areas. For soil, however, the optimum sampling distance is the highest achievable spatial resolution of the instrument. Guidance on the use of geostatistical methods in remote sensing is provided by Curran (1988) and Woodcock et al. (1988a, b).

In a random sampling scheme, the sampled pixels are randomly chosen from the image. The term without replacement means that no pixel is selected twice. Although this approach is quite easy to apply, it might not be always suitable in practice because some small information classes may be omitted. Moreover, if the number of samples selected is not sufficiently large, some information classes may be under- (or over-) sampled. Hence, stratified unaligned sampling may be preferred. One may choose several unaligned sub-areas for each class from the image, and perform random sampling within each sub-area. However, Congalton and Green (1999) note that the assumptions on which the computation of the Kappa coefficient is based are satisfied only by random sampling. The Kappa coefficient is discussed in Section 2.7.

A third sampling scheme, cluster sampling, can also be applied. In cluster sampling, groups of patchlike pixels, representing, for example, an agricultural field, are selected. This approach allows the collection of a large number of samples relatively quickly. However, the use of large cluster samples is not recommended, since pixels within a group are not mutually independent, which means that a sample of, for example, 40 contiguous pixels does not represent 40 independent samples, due to the autocorrelation effect that was noted previously in this section. Congalton (1988) suggested that no cluster should be larger than 10 pixels.

2.6.2 SAMPLE SIZE, SCALE, AND SPATIAL VARIABILITY

Several authors (e.g., Hord and Brooner, 1976; van Genderen and Lock, 1977; Hay, 1979; Fitzpatrick-Lins, 1981; Rosenfield et al. 1982; Mather, 2004) consider the question of sample size. Mather (2004) suggests as a rule of thumb that the number of training data pixels per class should be at least 30 times the number of features. This rule is based on the notion that in univariate studies, a sample size of 30 or more is considered "large." This rule could be modified to state that the sample size per class should be at least 30 times the number of parameters to be estimated. More soundly based advice uses some theoretical model of the data distribution in order to predict the sample size required in order to achieve a specified level of accuracy (Fitzpatrick-Lins, 1981; Congalton, 1991).

Congalton and Green (1999) present a method for estimating sample size that is based on the multinomial distribution. The sample size n is derived from the relationship $n = B\Pi_i(1 - \Pi_i)/b_i^2$, where b_i is the required precision (expressed as a proportion, so that 0.05 is equivalent to 5% precision), B is the upper $(\alpha/k) \times$ 100th percentile of the chi-square distribution with one degree of freedom, k is the number of classes, Π_i is the proportion of the area covered by class i, and α is the required confidence level. Vieira (2000) gives the following example: at a confidence level of 95% and a desired precision of 0.05, find the sample size necessary for the derivation of a valid confusion matrix. There are $k = 7$ classes, and class *wheat* occupies approximately 46% of the map area ($\Pi_{wheat} = 0.46$). The value of chi-square for the probability level $(0.05/7) = 0.00714$ with one degree of freedom is 7.348. The minimum training data sample size is therefore $7.348(0.46)(1 - 0.46)/(0.05)2 = 730$. Vieira (2000) notes that this value is quite close to the value of $30 \times 3 = 90$ samples per class (or $7 \times 90 = 630$ total samples) suggested by Mather (2004).

The size of the sample required for statistically valid measures of classification accuracy to be computed is not the only criterion that should be considered. Statistically based classifiers require that certain parameters be estimated accurately. In the case of the ML classifier, these parameters are the mean vector and the variance–covariance matrix for each class. The sampler size is related to the number of features (dimensionality of feature space) and, as noted earlier, as the dimensionality of the data increases for a fixed sample size so the precision of the estimates of these parameters becomes lower, leading to loss of classifier efficiency. This is the *Hughes phenomenon* (Section 2.1). For low dimensional data, the suggestion that there should be a minimum size of 30 × number of wavebands will give satisfactory results in most cases. However, as the dimensionality of the data increases (e.g., if hyperspectral data sets are used) then the required sample size will be unfeasibly large, and some method of dimensionality reduction, such as an orthogonal transform (Section 2.1) or the use of feature selection methods (Section 2.2), will be required.

These remarks are directed toward the use of statistical classifiers, which (in the case of the ML classifier) operate by defining a model of the data distribution, such as the multivariate normal distribution, and then estimating the parameters of the model from the training data. El-Sheik and Wacker (1980), Hseih and Landgrebe (1998), and Raudys and Pikelis (1980) discuss these matters in depth.

Other classifiers, such as decision trees, artificial neural networks, or support vector machines, are nonparametric and require that training data sets are large enough to represent the characteristics of each class. These methods are called *nonparametric* because they do not involve the estimation of statistical parameters. Evans (1998) notes that decision tree classifiers are susceptible to large changes in accuracy when only small changes are made in the composition of the training samples, indicating that both the size and the nature of the training samples assume considerable importance when such classifiers are used. There is some evidence that the artificial neural networks and support vector machines perform better than statistical classifiers even with small training data sets (Foody et al., 1995; Foody and Mathur, 2004, 2006), while the selection of training data sets from interclass boundary areas is said by Foody (1999a) and Foody and Mathur (2006) to improve classifier performance. Foody et al. (2006) adopt four principles, namely, intelligent selection of the most informative training samples, selective class exclusion, acceptance of imprecise descriptions for spectrally distinct classes, and the adoption of a one-class classifier to reducing training sample size. All four approaches were able to reduce the training set size required considerably below that suggested by conventional widely used heuristics without significant impact on accuracy.

Sample size may be related to the scale of observation (which determines the objects in the landscape that are deemed significant), and it also has an influence on observed variability both within and between classes. If a sample is to be representative, then its size should be related to the variance of the class. Scale of observation is generally proportional to variability; for example, consider the difference between two images with spatial resolutions of 1 km and 1 m, respectively. The latter shows more detail, in the sense that what appear to be homogeneous regions at 1-km resolution become fragmented and more variable as resolution increases (Woodcock and Strahler, 1987). Other pertinent references on the importance (and influence) of scale are Benson and Mackenzie (1995), Cushnie (1987), Foody and Curran (1994), Marceau et al. (1994), Marceau and Hay (1999), and contributions to Quattrochi and Goodchild (1997).

There is also an obvious link between the concept of spatial variability, mentioned in the preceding paragraph, and image texture. One can, in fact, consider the dependence between spatial scale/resolution and spatial variance from a number of aspects. For example, the multifractal dimension is based on the concept of sets each with its own fractal dimension (Pecknold et al., 1997). The use of fractal dimension as measures of texture are described in Chapter 7. The discrete wavelet transform is a second example of a hierarchical or multiscale procedure. Unlike the Fourier transform, which decomposes an image into its frequency components, the wavelet transform takes into account both the spatial and the frequency characteristics of an image. In other words, the wavelet transform can deal with nonstationary series. It is described as multiscale because it operates by decomposing the image into scale components in a recursive fashion. The applications of wavelets in remote sensing are described in detail in Starck et al. (1998). Zhu and Yang (1998) use wavelets to characterize image texture and demonstrate the utility of such texture measures in classification. Tso and Olsen (2005b) use wavelets to detect multiscale edge information to improve classification accuracy. See Chapter 7 for an extended discussion of the derivation of texture measures from remotely sensed images.

2.6.3 ADEQUACY OF TRAINING DATA

Since training data sets are used to teach a supervised classifier, it is important to ensure that the training data are adequate in the sense that no erroneous or unrepresentative samples are included. The presence of such outliers will have a lesser impact on supervised classifiers that are based on mean values alone, as one outlier will not have a particularly large influence on the value of the mean of a large sample (*large* means greater than 30 measurements on the given feature or variable). However, the variances and covariances are likely to be more badly affected because their computation includes the square of the difference between a given observation and the mean of its class. Artificial neural networks train directly on the sample data themselves, and are thus likely to be significantly influenced by the presence of training sample data that are not representative.

The presence of outliers can be accommodated by the use of robust statistical estimators, which are not as severely affected by outliers as are conventional estimators. Mather (2004) provides details of a method of weighting the observations used in the calculation of means and variance–covariance matrices. The weights are proportional to the Mahalanobis distance of the observation from its class mean. Since the position of the class mean vector in feature space will move as the weights change, the procedure iterates until the weight estimates converge. The use of the Mahalanobis distance implies that this weighting algorithm is appropriately applied to training data sets that are being used in statistical classifiers such as maximum likelihood. An alternative approach could use cross-validation procedures, in which the training data set is subdivided into a number of mutually exclusive groups; for example, into 10 groups each containing 10% of the training data. The classifier is trained on 90% of the available training data (i.e., nine subgroups combined) and is applied to the remaining 10%. If the label given to an observation by the classifier does not correspond to the class allocated by the analyst, then that observation is eliminated. The procedure is repeated until all 10 subsets have been labeled. An added sophistication is to use several classifiers and to eliminate only those training data observations that are mislabeled by all classifiers, or by a majority of the classifiers (Brodley and Friedl, 1996, 1999).

2.7 ESTIMATION OF CLASSIFICATION ACCURACY

No classification is complete until its accuracy has been assessed. In this context, the term *accuracy* means the level of agreement between labels assigned by the classifier and class allocations based on ground data collected by the user, known as *test data*. As noted already, ground data do not necessarily represent reality, due to observation and recoding errors, mislocation of test data sites, differences caused by changes in land cover between the time of observation and the date of imaging, and so forth. Where a separate set of test data is not available, accuracy can be assessed relative to the training data set, but the degree of accuracy will inevitably be overstated. The use of cross-validation methods is preferable in these circumstances.

Appropriate measures of classification accuracy can provide us with a measure of classification performance. The methods considered in this section are based on analysis of the confusion matrix. Questions relating to the impact of sample size on accuracy assessment are considered in Section 2.6.2.

The most common tool used for the classification accuracy assessment is in terms of a confusion (or error) matrix. A confusion matrix is a square array of dimension $n \times n$, where n is the number of classes. The matrix shows the relationship between two samples of measurements taken from the area that has been classified. The first set represents test data that have been collected via field observation, inspection of agricultural records, air photo interpretation, or other similar means. The second sample is composed of the labels of the pixels, allocated by the classifier, that correspond to the test data points. The columns in a confusion matrix represent test data, while rows represent the labels assigned by the classifier. An example, shown in Figure 2.9, involves four information classes, namely, grass, wheat, deciduous woodland, and coniferous forest. The numbers of sample pixels for each class in the test data set (column sums) are 76, 80, 102, and 79, respectively, giving a total of 337 test pixels. The row sums show that 55 pixels are classified as grass, 118 as wheat, 89 as deciduous forest, and 75 as coniferous forest. The main diagonal entries of the confusion matrix represent the number of pixels that are given the same identification by the test data and the classifier, and these are the number of pixels that are considered to be correctly classified. In the example, the numbers of correctly classified pixels are 48, 70, 65, and 59, for the four classes respectively.

Several indices of classification accuracy can be derived from the confusion matrix. The overall accuracy is obtained by dividing the sum of the main diagonal

	Grass	Wheat	Deciduous	Conifer	**Row Sum**
Grass	48	3	2	2	55
Wheat	18	70	24	6	118
Deciduous	7	5	65	12	89
Conifer	3	2	11	59	75
Column Sum	76	80	102	79	337

FIGURE 2.9   A confusion matrix composed of four information classes.

entries of the confusion matrix by the total number of samples. In the example, the overall classification accuracy is calculated as

$$(48 + 70 + 65 + 59)/337 = 242/337 = 71.8\%. \tag{2.19}$$

This result may be interpreted to mean that 71.8% of the image area is correctly classified. Alternatively, the interpretation may be that the probability that a pixel is classified correctly is 71.8%. This overall view treats the classes as a whole and does not provide specific information about the accuracy of each individual class, nor does it relate to the spatial pattern of errors. Moreover, the overall result can be misleading. For instance, if an image is classified into two classes, and the first class covers 90% of the image area, then one can simply classify the whole image as class one, and obtain an overall classification accuracy of 90%. Although the above example is an extreme case, it points out the potential drawback of using overall classification accuracy as a measure for assessing classification performance.

In order to assess the accuracy of each information class separately, the concepts of producer's accuracy and user's accuracy can be used. For each information class i in a confusion matrix, the producer's accuracy is calculated by dividing the entry (i, i) by the sum of column i, while the user's accuracy is obtained by dividing the entry (i, i) by the sum of row i. Thus, the producer's accuracy tells us the proportion of pixels in the test data set that are correctly recognized by the classifier. The user's accuracy measures the proportion of pixels identified by the classifier as belonging to class i that agree with the test data. Using the data shown in Figure 2.9, the producer's accuracy for each information class is calculated as

Grass: $48/76 = 63.1\%$ Wheat: $70/80 = 87.5\%$

Deciduous: $65/102 = 63.7\%$ Conifer: $59/79 = 74.7\%$

while the user's accuracy is determined as

Grass: $48/55 = 87.3\%$ Wheat: $70/118 = 59.3\%$

Deciduous: $65/89 = 73\%$ Conifer: $59/75 = 78.6\%$
$$(2.20)$$

The class wheat has the highest producer's accuracy of 87.5%, and thus the producer of the classification can estimate that this proportion of wheat pixels has been correctly classified. Classes grass and deciduous achieve only around 63% producer's accuracy, which indicates that a considerable number of pixels belonging to these classes have been classified erroneously. Producer's accuracy is really a measure of omission error.

User's accuracy denotes the probability that a classified pixel actually represents that information class on the ground. For instance, if an information class called potato has a user's accuracy of 100%, the user may infer that all the pixels classified as potato are actually covered by potatoes, assuming that the sample of test data is adequate (Section 2.6). Therefore, from the example results given for grass, wheat, deciduous, and conifer, we know that most of the pixels labeled as grass on the classified image are actually grass. However, although the wheat class shows the highest producer's accuracy, only 59.3% of the area labeled wheat is actually covered by wheat. In other words, 40.7% of pixels classified as wheat are actually other information classes. Thus, the user's accuracy is really a measure of commission error.

The accuracy measurements showing above, namely, the overall accuracy, producer's accuracy, and user's accuracy, though quite simple to use, are based on either the principal diagonal, columns, or rows of the confusion matrix only, which does not use the information from the whole confusion matrix. A multivariate index called the *Kappa coefficient* (Cohen, 1960) has found favor. The Kappa coefficient uses all of the information in the confusion matrix in order for the chance allocation of labels to be taken into consideration (though Foody [2000] suggests that chance agreement is overestimated and accuracy is underestimated. He presents an alternative formulation in Foody [1992]).

The Kappa coefficient is defined by

$$\hat{k} = \frac{N\sum_{i=1}^{r} x_{ii} - \sum_{i=1}^{r}(x_{i+} \times x_{+i})}{N^2 - \sum_{i=1}^{r}(x_{i+} \times x_{+i})}. \tag{2.21}$$

In this equation, $\hat{k}$ is the Kappa coefficient, r is the number of columns (and rows) in a confusion matrix, x_{ii} is entry (i, i) of the confusion matrix, x_{i+} and x_{+1} are the marginal totals of row i and column j, respectively, and N is the total number of observations (Congalton et al., 1983). For computational purposes, the following form is often used:

$$\hat{k} = \frac{\theta_1 - \theta_2}{1 - \theta_2}$$

where

$$\theta_1 = \sum_{i=1}^{r} x_{ii}/N,$$

and

$$\theta_2 = \sum_{i=1}^{r} x_{i+}x_{+i}/N^2. \tag{2.22}$$

The large-sample variance of Kappa is (Congalton and Green, 1999)

$$Var(\hat{k}) = \frac{N(x_{i.} - x_{ii})}{\left[x_{i.}(N - x_{.i})\right]^3} \left[(x_{i.} - x_{ii})(x_{i.}x_{.i} - Nx_{ii}) + Nx_{ii}(N - x_{i.} - x_{.i} + x_{ii})\right]. \quad (2.23)$$

The distribution of the ratio $\hat{k}/\text{var}(\hat{k})$ is approximately Gaussian, and so it can be used as a test statistic for the null hypothesis that the observed accuracy differs from zero only as a result of sampling from a large population. Individual class values of Kappa (conditional Kappa values) can be derived as follows:

$$\hat{k}_{cond} = \frac{Nx_{ii} - x_{i.}x_{.i}}{Nx_{i.} - x_{i.}x_{.i}}. \quad (2.24)$$

The Kappa coefficient $\hat{k}$ takes not just the principal diagonal entries but also the off-diagonal entries into consideration. The higher the value of Kappa, the better the classification performance. If all information classes are correctly identified, Kappa takes the value 1. As the values of the off-diagonal entries increase, the value of Kappa decreases. For the data shown in Table 2.1, the value of the Kappa statistic is calculated as

$$\theta_1 = (48 + 70 + 65 + 59)/337 = 0.718$$

$$\theta_2 = (76 \times 55 + 80 \times 118 + 102 \times 89 + 79 \times 75)/337^2 = 0.252$$

$$\hat{k} = \frac{\theta_1 - \theta_2}{1 - \theta_2} = \frac{0.718 - 0.252}{1 - 0.252} = 0.623. \quad (2.25)$$

It should be noted that the interpretation of the Kappa statistic is based on the assumption of a multinormal sampling model. If the test data are not chosen properly, the above assessments become less reliable. Another consideration relates to sample size and sampling scheme; the consensus view appears to be that simple random sampling is required for the use of the Kappa coefficient, and that a minimum sample size is needed in order to ensure a specific, predefined level of accuracy. Issues concerning sample size and sampling procedures are considered in Section 2.6. Recent surveys of accuracy measures are Congalton and Green (1998), Foody (2000a, 2002), Jansen and van der Wel (1994), and Stehman and Czaplewinski (1998).

None of the methods of assessing the accuracy of a classification derived from remotely sensed data considers the spatial distribution of error. Without considering any other factors, one might expect that erroneously classified pixels should be randomly distributed over the study area. Observations of the actual distribution of such erroneously classified pixels may show that this is not the case. For instance, in a per-pixel classification it might be seen that erroneously classified pixels are distributed around field boundaries, or along spatial features such as roads and railways, which are not included in the classification. Vieira and Mather (1999) present some methods of visualizing the spatial pattern of classification error, and of analyzing these patterns using simple methods of spatial statistics. The same authors show how cartographic methods can be used to depict the reliability of classifications derived from remotely sensed data. The spatial distribution of errors in agricultural crop classifications was found by Vieira and Mather (1999) to be closely related to the positions of field boundaries. The use of a buffering operation within a geographic information system (GIS)-improved classification accuracy without compromising the integrity of the results, and removed autocorrelated errors.

2.8   EPILOGUE

In this chapter we attempted to provide an overview of the methods of pattern recognition, plus a summary of related issues such as sampling and accuracy assessment methods. The range of material covered may appear to the reader to be daunting; however, the following chapters provide considerably more detail and descriptions of the principal methods of pattern recognition that are summarized in this chapter. The consensus view, in recent years, has emphasized the superiority of artificial neural network methods over statistical methods, largely because of their nonparametric nature (that is, the fact that artificial neural networks and support vector machines do not assume any particular statistical frequency distribution of the data). More recently, the use of decision trees has assumed a higher profile; like artificial neural networks and support vector machines they are nonparametric, but they do not need extensive design and training. However, their use of hyperplane decision boundaries parallel to the feature axes may restrict their use to cases in which classes are clearly distinguishable.

The use of multiple classifiers may present a way out of the spiral of increasing complexity. Rather than develop increasingly refined decision rules, it may be sensible to make best use of what is available. Users of these methods should be aware of the need to ensure that the component classifications are independent.

Hyperspectral data present difficult problems due to their high dimensionality. It is impracticable to consider the collection of large volumes of test and training data when the intrinsic dimensionality of the hyperspectral data set is considerably less than the number of spectral bands. A proportion of the variance exhibited by hyperspectral data is noise, either random or coherent. Hence, it makes sense to consider the use of orthogonal transforms, particularly those that explicitly discriminate between signal and noise. The two methods described in this chapter—the MNF and MAF procedures—offer the possibility of reducing the size of the dataset without compromising accuracy, while avoiding the problems associated with high dimensionality.

The need to estimate the accuracy of a thematic map, or validate the methodology, is an issue that requires considerable attention; in fact, no pattern recognition exercise is complete unless it includes validation and accuracy estimation. The simple overall accuracy measure deriving from the confusion matrix gives only a very rough estimate of the true accuracy. The use of Kappa is statistically more acceptable, provided that the user is aware of the need to ensure a sample of test data of sufficient size that has been collected using a random sampling scheme. None of these measures takes the spatial distribution of error into consideration; indeed, it can be argued that conventional sampling schemes applied to spatial data also fail to take into account the specific properties of spatial data. Considerably more research could be profitably directed into this area.

3 Artificial Neural Networks

The efficiency of the human eye–brain combination in solving pattern recognition problems led researchers in this field to consider whether computer systems based on a simplified model of the brain can be more effective than standard statistical classification methods. Such research led to the development of artificial neural networks (ANN), which have been increasingly used in remote sensing over the past ten years, mainly for image classification. An advantage of neural networks lies in the high computation rate achieved by their massive parallelism, resulting from a dense arrangement of interconnections (weights) and simple processors (neurones), which permits real-time processing of very large data sets.

Artificial neural networks are generally described as nonparametric; that is, the use of a neural network does not require any assumptions about the statistical distribution of the data. The performance of a neural network depends to a significant extent on how well it has been trained, and not on the adequacy of assumptions concerning the statistical distribution of the data, as is the case with the maximum likelihood classifier. During the training phase, the neural network "learns" about regularities present in the training data and, based on these regularities, constructs rules that can be extended to the unknown data. This is a special ability of neural networks. However, the user must determine the architecture of the network, and also define parameters such as the learning rate, which affect the training time, performance, and the rate of convergence of a neural network. There are no clear rules to assist with the design of the network, and only rules of thumb (or heuristics) exist to guide users in their choice of network parameters.

Five kinds of fundamental neural network architecture, including the multilayer perceptron with back-error propagation, the self-organized feature map (SOM), counter-propagation networks, Hopfield networks, and ART systems, are introduced in this chapter. All of these different types of network have been, or can be, used for classifying remotely sensed images. See Bishop (1996), Garson (1998), Haykin (1999), Hewitson and Crane (1994), and Ripley (1996) for additional reading on the topics covered by this chapter.

3.1 MULTILAYER PERCEPTRON

The multilayer perceptron using the back-propagation learning algorithm (Rumelhart et al., 1986b) is one of the most widely used neural network models. A typical three-layer multilayer perceptron neural network is shown in Figure 3.1a. It can be seen that the leftmost layer of neurones in Figure 3.1 (a) is the input layer, which contains the set of neurones that receive external inputs (in the form of pixel values in the different bands of a multispectral image or other feature values). The input layer performs no computations, unlike the elements of the other layers. The central layer

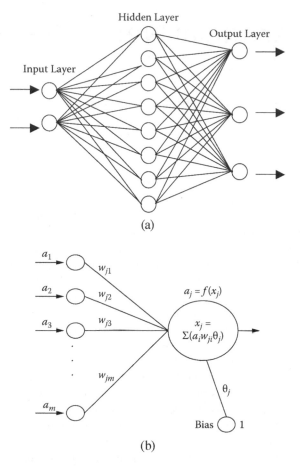

(a)

(b)

FIGURE 3.1 (a) A typical three-layer multilayer perceptron neural network. (b) Example of forward propagation procedure on neurone j.

is the hidden layer (there may be more than one hidden layer in complex networks). The rightmost layer of neurones is the output layer, which produces the results of the classification. There are no interconnections between neurones in the same layer, but all of the neurones in a given layer are fully connected to the neurones in the adjacent layers. These interconnections have associated numerical weights w_{jm}, which are adjusted during the learning phase. The value held by each neurone is called its *activity* a_m (Figure 3.1b).

3.1.1 BACK-PROPAGATION

The most popular algorithm used for updating the neuronal activities and the interconnection weights in a multilayer perceptron is the back-propagation algorithm. Back-propagation involves two major steps, termed *forward* and *backward* propagation, to accomplish its modification of the neural state. During training, each sample (for example, a feature vector associated with a single training pixel) is fed into the input layer, and the activities of the neurones are sequentially updated through the input layer to the output layer in terms of some mapping function. Once the forward pass is finished, the activities of the output neurones are compared with their expected activities. For example, if there are five classes and training pixel i is thought to belong to class 1, then the five output neurones (one per class) would be expected to output a pattern such as "1 0 0 0 0." Except in very unusual circumstances, the actual output will differ from the expected outcome, and the differences are the network error. This error is distributed through the network by a backward pass, starting at the output layer, which updates the weights. The process is similar to the distribution of closure error in a chain survey. The forward and backward passes are continued until the network has learned the characteristics of all the classes. This procedure is called *training the network*.

During the forward propagation process, the activities of neurones are updated sequentially layer by layer, from the input to the output layer, in order to generate output in the form of the activations of the output layer neurones. Let x_j denote the total input received by neurone j, which is represented by

$$x_j = \sum_i a_i w_{ji} \tag{3.1}$$

where a_i is the activity of neurone i, and w_{ji} is the weight of connection from the ith neurone to the jth neurone. Once the value of x_j is computed from Equation (3.1), it is converted to an output value (for transmission to the next layer if the neurone lies on an intermediate layer) using a mapping function. The sigmoid function is a common choice of mapping function (Figure 3.2). It is a monotonically increasing nonlinear function defined by

$$a_j = f(x_j) = \cfrac{1}{1 + \cfrac{1}{\exp\left(\cfrac{x_j}{T}\right)}} . \tag{3.2}$$

T is a parameter called temperature (normally $T = 1$; a more extensive discussion of the effect of T is given in Section 8.4), which renders the sigmoidal curve more abrupt ($T < 1$) or gradual ($T > 1$). The corresponding pattern is illustrated in Figure 3.2a. After the activity of each neurone within the same layer has been computed, a similar process is carried out on the next adjacent layer. Note that the input layer is a special case because the neurones in the input layer take the values provided by the training sample. For nodes in the input layer, the activity for neurone j is directly set as the jth component of the input pattern vector.

Sometimes an extra neurone, called the bias, is added to the neural network. It links to all the network layers except the input layer. This bias unit has a constant activity

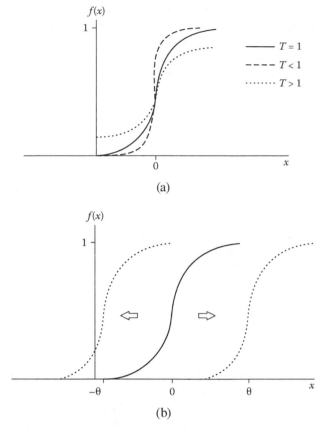

(a)

(b)

FIGURE 3.2   (a) The shape of the sigmoidal curve varies with parameter T. (b) The location of the sigmoid function has shifted after bias θ is added or subtracted.

of 1, but affects each neurone j via different weight values (e.g., θ_j in Figure 3.1b). If the bias unit is introduced, Equation (3.1) is modified as follows:

$$x_j = \sum_i a_i w_{ji} - \theta_j \qquad (3.3)$$

where θ_j is the bias associated with the neurone j. The effect of the bias term θ is to contribute to a left- or rightward shifting of the sigmoid mapping function, as shown in Figure 3.2b, depending on whether the value θ is negative or positive. The introduction of a bias unit to the network is believed to improve the convergence property of a multilayer perceptron neural network.

In backward propagation, the interneurone weights are modified starting from the output layer until the input layer is reached; that is, moving left from the rightmost layer in terms of the layout shown in Figure 3.1a. The aim of weight updating is to reduce the identification error of the network. Generally, the least mean square error criterion is applied. This is defined as

$$E(w) = \frac{1}{2} \sum_{j,k} (a_{j,k} - o_{j,k})^2 \qquad (3.4)$$

where w is a set of weights in a network, $a_{j,k}$ is the jth neurone in the output layer obtained from the kth training sample, and $o_{j,k}$ is the target output at neurone j in the output layer for the kth training sample. A number of search methods can be applied to find the minimum of the function $E(w)$. One widely used method is that of steepest descent, in which each weight is updated in terms of the expression

$$w_{ji}^{n+1} = w_{ji}^n - \eta \frac{\partial E}{\partial w_{ji}^n} \qquad (3.5)$$

where w_{ji}^n is the weight associated with the connection from neurone i to j at time n, η (the learning rate) is a small positive constant, in the range $0 \le \eta \le 1$, controlling the size of the step, and $\partial E/\partial w_{ji}^n$ is the gradient at point w_{ji}^n, which can be interpreted as the change in error E that results from a change in weight w_{ji}^n. The initial values of the weights, w_{ji}^0, are generally small random values. The gradient term in Equation (3.5) can be further expanded to

$$\frac{\partial E}{\partial w_{ji}} = \frac{\partial E}{\partial a_j} \frac{\partial a_j}{\partial x_j} \frac{\partial x_j}{\partial w_{ji}} = \frac{\partial E}{\partial a_j} f'(x_j) a_i = \delta_j a_i \qquad (3.6)$$

in terms of the chain rule of calculus, where $f'(x_j)$ is the first derivative of the sigmoid function, a_i is the activity in neurone i, and δ_j is the error value occurring on neurone j. If neurone j is in the output layer, the values of δ_j in Equation (3.6) are simply defined as

$$\delta_j = (a_j - o_j) f'(x_j) \qquad (3.7)$$

where o_j is the desired output for neurone j. Combining Equations (3.5) to (3.7), the weights w_{ji} for j = output layer can therefore be updated. For the other layers, the δ_j are computed by

$$\delta_j = \left(\sum_k \delta_k w_{kj} \right) f'(x_j) \qquad (3.8)$$

where k denotes the neurone that received output from neurone j. In other words, if neurone j is in the lth layer, then k denotes all neurones in the $(l + 1)^{th}$ layer. The combination of Equations (3.5), (3.6), and (3.8) determines the adjustment for each weight. This rule for the adjustment of the weights associated with network connections is also known as the generalized δ (delta) rule. Sometimes a momentum term is added to Equation (3.5) to give

$$w_{ji}^{n+1} = w_{ji}^n - \eta \frac{\partial E}{\partial w_{ji}^n} + \xi \left(w_{ji}^n - w_{ji}^{n-1} \right) \qquad (3.9)$$

where $0 \leq \xi \leq 1$. The momentum term is applied to avoid oscillation problems during the search for the minimum value on the error surface, and can therefore speed up the convergence procedure.

The performance of a multilayer perceptron is controlled by several factors: the model-associated parameters, the network structure, and the nature of the training samples. It is very difficult to choose an optimum combination of those factors to construct an ideal network for a given classification task.

3.1.2 PARAMETER CHOICE, NETWORK ARCHITECTURE, AND INPUT/OUTPUT CODING

In the back-propagation algorithm, one or more parameters need to be defined by the user. The choice of values for these parameters can have a very significant effect on the performance of the network. In addition, the design (architecture) of the network is important in ensuring that the network is able to generalize from the sample (training) data. Finally, encoding of the data can make a substantial difference in the accuracy achieved by the network.

Kavzoglu (2001) reports on an extensive survey of the impact of these choices on a network's ability to classify unknown patterns. He notes that a multilayer perceptron requires at least one hidden layer, in addition to the input and output layers, and that the number of neurones in the hidden layer(s) will significantly influence the network's ability to generalize from the training data to unknown examples (i.e., pixels in the image to be classified). Small networks cannot fully identify the structures present in the training data (known as *underfitting*), while large networks may determine decision boundaries in feature space that are unduly influenced by the specific properties of the training data. The latter phenomenon is known as *overfitting*. A single hidden layer is thought to be adequate for most classification problems, but where there are large numbers of output classes, two hidden layers may produce a more accurate result. Kanellopoulos and Wilkinson (1997) suggest that where there are 20 or more output classes, two hidden layers should be used, and that the number of neurones in the second hidden layer should be equal to two or three times the number of output classes.

Sarle (2000) lists a number of reasons to illustrate the factors involved in determining an appropriate number of hidden-layer neurones. These factors include the number of input and output neurones, the size of the training data set, the complexity of the classification being performed, the amount of noise present in the data to be classified, the type of activation function used by the hidden layer neurones, and the nature of the training algorithm. Kavzoglu (2001) reviews a number of *ad hoc* rules and recommendations, and concludes that Garson's (1998) proposal produces good results in the cases tested. Garson (1998) proposes that the number of hidden-layer neurones should be set to the value $N_P(r(N_I + N_O))$, where N_P is the number of training samples, N_I is the number of input features, and N_O is the number of output classes. The parameter r is related to the noisiness of the data and the cleanness or simplicity of the classification. Typical values of r are in the range of 5 (clean data) to 10 (less clean data), but values as high as 100 or as low as 2 are possible. Pruning algorithms (Section 3.1.4) can be used once the network has been trained in order to remove links between neurones that are ineffective (Kavzoglu and Mather, 1999) and thus increase the generalization capabilities of the network. Note that sometimes the strategy of selecting training band combination may also provide ways to improve the generalization of the ANN (Kavzoglu and Mather, 2002).

The learning rate η in Equation (3.5) must be specified. If Equation (3.9) is used for updating the weights, then the user must define both learning rate η and momentum coefficient ξ. The value of the parameter η in the steepest descent minimization procedure should not be too large, or the result of the search will be poor. Conversely, if η is too small, the network training phase can be very time-consuming. The selection of the learning rate η may be case-dependent. On the basis of a large number of trials, Kavzoglu (2001) recommends that the value 0.2 is suitable for the learning rate when no momentum term is used. Where a momentum term is used, the learning rate should be set to a value in the range of 0.1 to 0.2, with the momentum term taking a value between 0.5 and 0.6. It is possible to vary the parameters during the training process, as suggested by Pal and Mitra (1992).

A further consideration involves the selection of the initial values to be assigned to the weights associated with interneurone links. These initial values are generally assigned randomly, with values in a specified range, and the choice of these initial values can have a significant effect on classifier performance, as demonstrated by Ardö et al. (1997), Blamire (1996), and Skidmore et al. (1997). Ardö et al. (1997) found that classification accuracy varied from 59% to 70% using different initial weight sets selected from the range [−1, +1]. Kavzoglu (2001) notes that classification performance improves for small initial weight ranges and recommends that the initial values of the weights should be in the range [−0.25, 0.25].

Training using the back-propagation algorithm is an iterative process. However, if the network is trained over too many iterations, it learns the specific characteristics of the data and fails to recognize some of the image data to be classified. If the network is insufficiently trained, the positions of the decision boundaries in feature space are not optimal. The practice of allowing the network to train until some predetermined level of error is reached is therefore unsatisfactory. A cross-validation approach can be used in which the available training data set is subdivided into *validation* and *training* subsets. The network learns from the training subset and is stopped at several points during learning. At each stopping point, the network is used to classify the samples contained in the validation subset. Training continues until the classification error of the validation subset begins to rise. This strategy is rather more involved than a simple error-threshold stopping rule, but it has been found to be effective, provided it is realized that the error of the validation subset may fluctuate in the early period of training. The training process should not be terminated too quickly in response to a fall in the error of the validation subset. The effects of overtraining are considered further in Section 3.1.4.

The coding scheme used for both the input and the output vectors requires some thought. For the input vector, each component can be normalized onto the interval [0, 1]. Normalization is believed to reduce noise effects (Lee et al., 1990). Several coding schemes, such as the binary code or the *spread* approach, can be applied to the network outputs. For example, in the case of four output classes, only two neurones are needed in the output layer if the binary coding method is applied as classes labeled 1 to 4 can be coded as 00, 01, 10, and 11, respectively. The spread approach, however, has the advantage that the result of classification is directly mapped onto the output vector, and so this approach is more widely used. For example, in the previous four-class case, if an input pattern belongs to class 3, the output vector will be coded as (0, 0, 1, 0). Such a data representation can enlarge the *Hamming distance* between individual vectors. For any two words x and y, the Hamming distance $d(x, y)$ is defined by

$$d(x,y) = \sum_i \#(x_i, y_i)$$ (3.10)

where x_i and y_i denote alphabets at location i and $1 \le i \le n$, and the function $\#(a, b) = 1$ if $a \ne b$, and 0 otherwise.

3.1.3 DECISION BOUNDARIES IN FEATURE SPACE

A classic example showing the relationship between the multilayer perceptron network structure and the location and shape of decision boundaries in feature space is the problem of classifying the output from a simple XOR (exclusive OR; Boolean operator) operator. Figure 3.3a shows the results of the XOR operation and the corresponding positions of the required decision boundaries are shown in Figure 3.3b.

It can be shown that a 2|0|2 network (i.e., two input neurones, no hidden neurones, and two output neurones in case of spread coding) cannot solve this classification problem because this network is only able to construct a single straight-line decision boundary. The XOR classification problem can only be solved if the network contains a hidden layer of neurones. Moreover, different sizes of neural networks (i.e., with different numbers of hidden neurones) will, in general, produce different decision boundaries. A larger network has the potential to form more complicated decision boundaries than a smaller network. To demonstrate this point, we use a three-class classification problem (modified from the XOR problem) over a two-dimensional domain within the data range of [0, 1]. Each dimension is further partitioned into 100 × 100 subsquares. Only five training patterns are used, and the corresponding locations are illustrated in Figure 3.3c.

Two networks are used for this experiment. One has a 2|3|3 structure, with 15 weights (2 × 3 = 6 weights connecting the input and hidden layers, and 3 × 3 = 9 weights connecting the hidden and output layers) and the other has a 2|40|60|3 structure, with 2660 weights. The output vector is coded using the spread approach. There are two neurones in the input layer because the feature space shown in Figure 3.3c

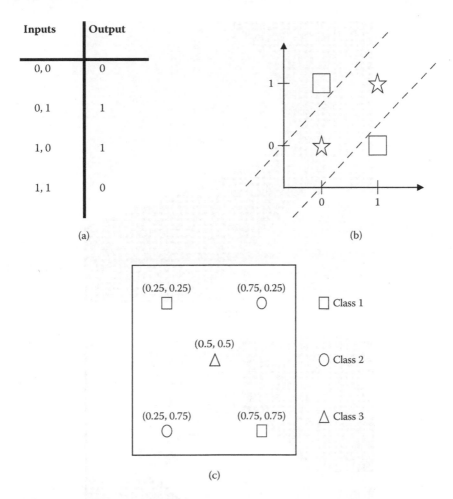

Inputs	Output
0, 0	0
0, 1	1
1, 0	1
1, 1	0

(a)

(b)

(c)

FIGURE 3.3 (a) Results of the XOR operation. (b) Spatial position of input data with two classes denoted by ☆ and □ showing that at least two decision boundaries (dashed line) should be generated in order to separate the classes. (c) Training pattern locations for three-class problem. (d) Decision boundary formed by a 2|3|3 network. (e) Decision boundaries formed by a 2|40|60|3 network.

is two-dimensional. There are three neurones in the output layers because there are three classes in the problem shown in Figure 3.3c. The results shown in Figures 3.3d and 3.3e are based on the *winner takes all* rule (i.e., a pixel is placed in class *i* if output neurone *i* has the largest activity of all neurones in the output layer). The decision boundaries formed by the larger network are more satisfactory than those produced by the smaller network. Does this indicate that larger networks always achieve better classification performance? Before answering this question, we consider a second example.

The second example is a two-class classification problem (Figure 3.4a) in which a two-dimensional domain within the range of [0, 1] has again been partitioned into 100 × 100 subsquares. Figure 3.4b illustrates the spatial position of the training

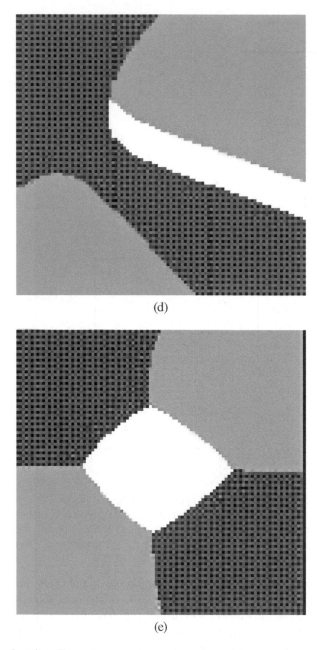

(d)

(e)

FIGURE 3.3 (continued).

patterns. The symbol "□" denotes class 1, and "☆" indicates class 2. In Figure 3.4c, one more training sample that should belong to class 1 is added (shown by the symbol "☆" inside a dashed circle), but is treated as if it were a member of class 2. This spurious sample is intended to test network performance in the presence of

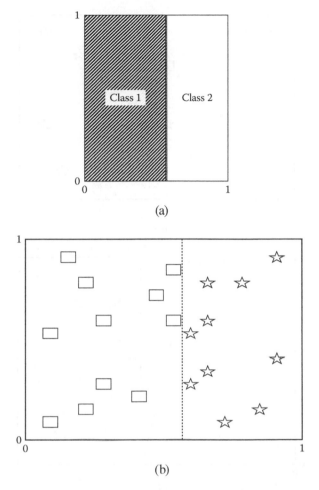

(a)

(b)

FIGURE 3.4 (a) A two-class classification problem. (b) Position of training patterns in a two-class case: □ denotes class 1 and ☆ denotes class 2.

training noise. The 22 samples shown in Figure 3.4c are used to train two networks. The first has a 2|60|10|2 structure with 740 weights, while the second is a 2|3|2|2 network with only 16 weights. Figures 3.4d and 3.4e illustrate the decision boundaries formed by the two networks. It is apparent that these decision boundaries are affected by the nature of the training data. The decision boundary shown in Figure 3.4e is clearly the better choice (i.e., decision boundary closer to Figure 3.4a).

Both examples in the preceding text indicate that the number of hidden layers and the number of nodes per hidden layer that are required to achieve satisfactory classification performance are case-dependent. Although there are studies analyzing the bounds on the number of hidden neurones and samples (e.g., Huang and Huang, 1991; Mehrotra et al., 1991), the assumptions are all quite strict. Methods that are more practical are described in the following sections.

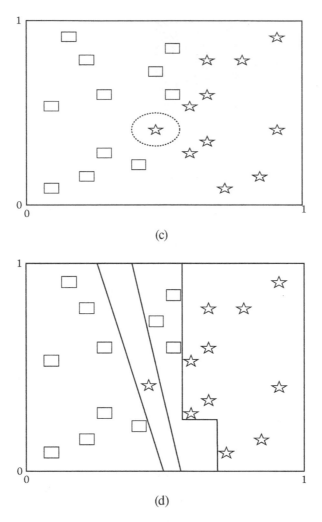

FIGURE 3.4   (continued) (c) One training pattern is added. (d) Resulting decision boundaries realized by a 2|60|10|2 network.

3.1.4   Overtraining and Network Pruning

During the training phase, the state of the network is iteratively updated in order to decrease the error between the actual and the desired output vectors. If the training data contain noise or are incomplete, then after a certain number of iterations, the network may be misled by these noisy training data elements. If a slightly different set of samples is used to test the network, the accuracy might be worse, even though the training error may be less. This issue is known as *overtraining*, and an example is given in Figure 3.5.

One way to avoid overtraining is to consider the error derived from a test data set. That is, the user constructs two data sets. One is the data set used for training the

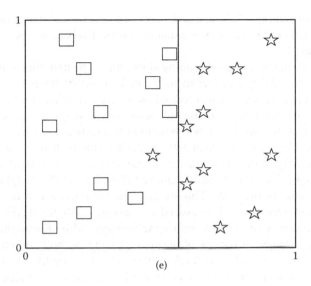

(e)

FIGURE 3.4 (continued) (e) Decision boundary formed by a 2I3I2I2 network.

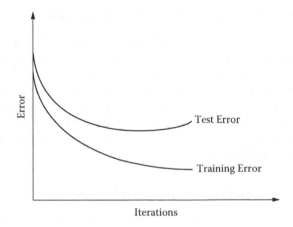

Iterations

FIGURE 3.5 During the early stage of training, the test error may follow the same decreasing trend as the training error. However, after a certain number of iterations, the test error may start to rise due to the overtraining effect.

neural network in the normal way, and the other is a test data set used for monitoring the *test error*. If the magnitude of the test error begins to increase, then network training should stop. Another approach to overcome the overtraining effect is to limit the ability of network to take advantage of noise (as illustrated in Figure 3.4) using one of the techniques of *network pruning* to reduce the number of connections between nodes. A common method of network pruning is to begin with a large network and then simplify the network structure by cutting some of the links between neurones. Network pruning leads to several advantages, specifically (1) it avoids the problem of overtraining, (2) it makes the network smaller and thus more capable of generalizing, and (3) it reduces the computational burden because fewer weights and neurones are used.

There are two major groups of pruning algorithms. The first kind of method eliminates weights by adding extra terms to the error function (refer to Equation 3.4). This can help the network to detect redundant weights. In other words, the objective function will force the network to use only some of the weights to achieve a solution and the redundant weights, which will have magnitudes near zero, can be eliminated (Ji et al., 1990). The second group of network pruning methods is based on the analysis of the sensitivity of the error function to the removal of a particular link and its associated weight. The weight can be removed if the lack of that weight only results in a slight increase in the error. This weight removal process can be iterated until some predefined error threshold is exceeded (Karnin, 1990). Reed (1993) provides a comprehensive survey of network pruning techniques, while Kavzoglu and Mather (1999) give an example of the use of network pruning methods in remote sensing. Other references are Hassibi and Stork (1993), who describe the algorithm known as *optimum brain surgeon*, and Le Cun et al. (1990), whose technique is known as *optimal brain damage*.

3.2 KOHONEN'S SELF-ORGANIZING FEATURE MAP

The self-organizing (feature) map, or SOM, was developed by Kohonen (1982, 1988b, 1989). Unlike the multilayer perceptron, this kind of network contains no hidden layer or layers, but is made up of one input and one output layer. The SOM network has an interesting property: it automatically detects (self-organizing) the relationships within the set of input patterns. This property can be applied to the problem of image mapping from higher dimensions to a two-dimensional feature space and, via the use of the learning vector quantization training algorithm (Kangas et al., 1990), to perform image classification.

3.2.1 SOM Network Construction and Training

The input and output neurones in a SOM are known as the *sensory cortex* and the *mapping cortex*, respectively, by analogy with their function in the biological neural system. Since the SOM training algorithm is based on the competitive concept, the output layer is also called the *competitive layer*.

The number of neurones in the input and output layers defines a SOM network. The number of input neurones is equal to the number of input features. However, there are no clear rules about the specification of the number of output neurones. Generally, the output layer of a SOM is a two-dimensional layer made up of $n \times m$ ($n, m > 1$) neurones, with each neurone relating to a fixed position in the two-dimensional output space. Although a one-dimensional output layer is possible, such an arrangement is seldom seen in remote sensing applications. It is assumed that adjacent neurones in the rows and columns of the output layer are spaced apart at a Euclidean distance of unity. The neurones in the input layer and the output layer are linked by synaptic weights w_{ji} where i and j are the identifiers of the input and

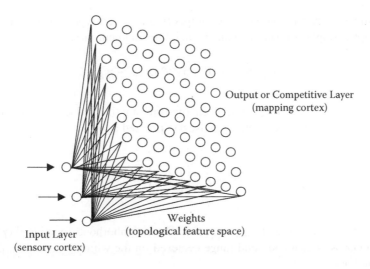

Output or Competitive Layer
(mapping cortex)

Weights
(topological feature space)

Input Layer
(sensory cortex)

FIGURE 3.6   Example of the topology of a self-organizing map (SOM) with three components: the input layer (sensory cortex) with three neurones, the linking weights (topological feature space), and the output layer (mapping cortex) made up by a grid of 8 × 8 neurones all equally spaced.

output neurones, respectively. The weights w_{ji} are initialized randomly and are then continually adjusted during training in order to organize the relationships among the input patterns. Once the training is complete, the final weights w_{ji} describe what is called a *topological feature space*, which is the characterization of input features in terms of the weights. An example of a SOM structure is shown in Figure 3.6.

The SOM training strategy is based on the concept of *competitive learning*. The neurones in the output layer have to compete with other neurones in that layer in order to "win" the opportunity of interaction with the input pattern. The result is that the weights connecting the input layer to the winning neurone and its neighbors are adjusted simultaneously, while other weights remain unchanged. Eventually, the neurones that are close together will have similar properties (in terms of weight magnitude).

3.2.1.1   Unsupervised Training

Training of a SOM begins with the initialization of the weights w_{ji} to random values. Then, for each input feature vector $x = \{x_1, x_2, \ldots, x_k\}$, where k is the input data dimension, the squared distances d_j^2 between an input neurone and each output neurone j are calculated using the Euclidean distance measure:

$$d_j^2 = \sum_i^k (x_i^n - w_{ji}^n)^2 \tag{3.11}$$

where x_i^n is the input to neurone i at iteration n, and w_{ji}^n is the weight from input neurone i to output neurone j at iteration n. The selected output neurone is determined from: $\min\{d_j^2\}$, $\forall j \in$ output layer. A competitive Hebbian-type learning law adjusts the synaptic weights of neurone j and its neighboring neurones:

$$w_{ji}^{n+1} = w_{ji}^n + \alpha^n \times \beta_{j'}(\gamma^n) \times (x_i^n - w_{ji}^n) \qquad (3.12)$$

where the learning rate α^n is a time-decaying function (i.e., reducing in magnitude as the number of training iterations increases) expressed as

$$\alpha^n = \alpha_{max} \left(\frac{\alpha_{min}}{\alpha_{max}} \right)^{\frac{n}{n_{max}}} \qquad (3.13)$$

with constraints $1 \geq \alpha$, and α_{min}, $\alpha_{max} \geq 0$. The neighborhood function $\beta_j(\gamma^n)$ determines a Gaussian neighborhood range centered on the winning neurone j. $\beta_{j'}(\gamma^n)$ is calculated by

$$\beta_{j'}(\gamma^n) = \exp \left(\frac{-(j-j')^2}{2(\gamma^n)^2} \right) \qquad (3.14)$$

where j' denotes the neighborhood of j including j itself, and γ^n is also a time-decaying function, and is defined by

$$\gamma^n = \gamma_{max} \left(\frac{\gamma_{min}}{\gamma_{max}} \right)^{\frac{n}{n_{max}}} \qquad (3.15)$$

Usually, a value of γ_{min} in the region of 1.0 is chosen, and γ_{max} can be set sufficiently large to make the neighborhood function $\beta_{j'}(\gamma^n)$ cover the whole output layer.

Equation (3.14) indicates that, for $j' = j$, that is, for the winning neurone itself, the updating magnitude for the associated weights is proportional to the learning rate α (because $\beta_{j'} = \beta_j = 1$). The updating magnitude for other j' (i.e., the magnitude of change in j') is sequentially decreasing with the increasing distance between the neighborhood j' and the winning neurone j. Figure 3.7 illustrates the update in neighborhood function as a time-decreasing function.

Different values of the functions γ and α produce different effects on the learning mechanism. Several studies have investigated the effects of varying the values of the input parameters for both γ and α (e.g., Ritter and Schulten, 1988; Bavarian and Lo, 1991; Erwin et al., 1992a,b). The general conclusion is that if a suitable value of γ_{max} is chosen, such that the neighborhood function covers the whole mapping cortex, then the SOM network will be likely to terminate in a well-ordered state. These authors also note that the learning rate α should be large (of the order of 0.1) at the beginning of training and should decrease during the training process. At the end of training, the value of α may be as small as, for example, 0.01.

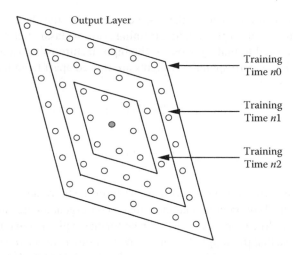

FIGURE 3.7 Modification of the weights of the neurones around a selected neurone begins with a wide neighborhood that decreases in size following training time $n2 > n1 > n0$.

Other extensions to the SOM approach have been proposed. The use of a hierarchical clustering scheme is described by Furrer et al. (1994). Another extension is the use of an additional layer (called the Grossberg layer) to achieve supervised training as proposed by Cappellini and Chiuderi (1994). This method is described in Section 3.3.

At the end of the unsupervised training phase, the SOM can characterize the distribution of input samples, and thus generate a two-dimensional view of the multidimensional input features (Schaale and Furrer, 1995). However, it should be noted that one should not use the results of this stage of the SOM to perform pattern recognition or other decision processes because there is a considerable difference between feature mapping and detecting clusters. It can be shown that the SOM can be further trained, for example, by the use of the learning vector quantization (LVQ) algorithm, to increase recognition accuracy (Kangas et al., 1990; Pal et al., 1993). Specifically, after the first stage of unsupervised training, each information class will generally activate several neurones on the SOM mapping cortex. The purpose of the subsequent supervised labeling phase is to tune the network weights and to define the boundaries of these information classes in order to reduce misclassification error.

3.2.1.2 Supervised Training

The supervised labeling algorithm introduced here is based on the concept of majority voting using the LVQ algorithm. The mapping cortex neurones are initially labeled using the training set. Each training pattern is input to the SOM and the selected (i.e., winning) neurone is determined by choosing the minimum Euclidean distance between the training pattern and the weights (Equation [3.11]). For instance, for a certain neurone a on the mapping cortex, if a was triggered r and s times by the training patterns for class 1 and 2, respectively, then neurone a will be labeled as class 1 if $r > s$, and class 2 otherwise. This is the concept of majority voting.

If the winning neurone matches the desired output (i.e., the class allocated to neu-rone a is the same as the corresponding training pattern class), then the correspond-ing weight is adjusted to make it close to the input feature using Equation (3.16). If it does not match, the corresponding weights are moved apart from the input feature using Equation (3.17):

$$w_{ji}^{n+1} = w_{ji}^n + \delta^n (x_i - w_{ji}^n) \tag{3.16}$$

$$w_{ji}^{n+1} = w_{ji}^n - \delta^n (x_i - w_{ji}^n) \tag{3.17}$$

where δ^n is a gain term, which may decrease with time. The value of δ^n has to be cho-sen in the range (0, 1) in order to guarantee that a convergence state can be obtained.

This weight modification mechanism can be further explained in terms of a simple example. Suppose that the weight linking to the winning neurone a has the value 30, and the input feature value is 25. According to the steps introduced above, if the win-ning neurone a matches the desired output for the current input value, then the weight should be adjusted in order to make it close to the input value. Let the gain term δ^n be 0.01; so, from Equation (3.16), the new weight will be $[30 + 0.01(25 - 30)] = 29.95$, which is closer to the input value 25 and thus will enhance the probability that in the next iteration for the same input value of 25, the desired neurone a is again likely to become the winning neurone. If neurone a is not our desired output, then according to Equation (3.17), the new weight will become $[30 - 0.01(25 - 30)] = 30.05$, which is further away from the input value 25, and thus the probability that the undesired neurone a wins again in the next iteration for the same input value of 25 is reduced.

As pointed out previously, the purpose of supervised training of the SOM is to label the clusters formed at the unsupervised training stage. Such a procedure should not be confused with the unsupervised classification methods described in Chapter 2. Although traditional unsupervised classification methods also embody a two-stage process comprising clustering and labeling, the labeling stage contributes nothing to the location of the clusters in feature space. It is an identification phase. However, the supervised training algorithm for the SOM updates the cluster location as defined by Equations (3.16) and (3.17).

3.2.2 EXAMPLES OF SELF-ORGANIZATION

Self-organization is an interesting characteristic of a SOM network. It implies that the network can detect relationships between the input patterns, and then arrange and express such relationships through the network weights. The process of self-organization consists of two steps: detecting relationships and spanning. During the early stages of training, the SOM focuses on finding the ordering of input pat-terns. Once the interpattern relationship is found, then the SOM will span its output neurones to make each neurone respond to approximately the same number of input patterns. The second step of this process requires more time than the first step. To help understanding of the behavior of the SOM, two examples are presented below.

In the first example, 65,536 samples, which are uniformly distributed on a two-dimensional space within the range [0, 1] on each coordinate axis are input to a SOM for unsupervised training. The SOM was constructed by using two neurones in the input layer, and 4 × 4 neurones for the output layer. Figure 3.8 shows the

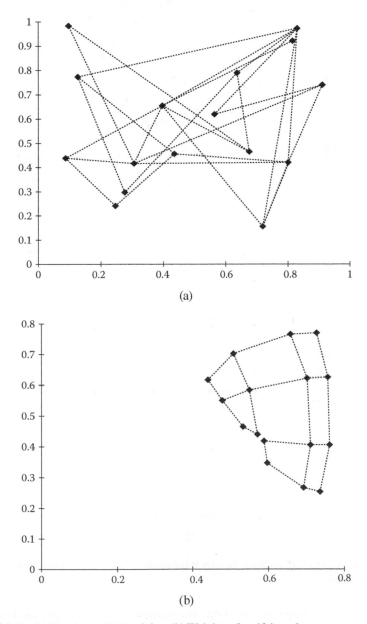

FIGURE 3.8   (a) Random initial weights. (b) Weights after 10 iterations.

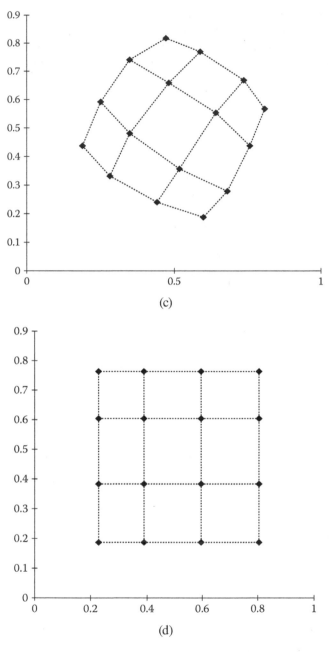

FIGURE 3.8   (continued) (c) Weights after 1000 iterations. (d) Weights after 6000 iterations.
See text for details.

weight distribution output by the SOM. The symbol "◊" indicates the location of the
output neurones in terms of associated weights (i.e., in feature space). The lines in
Figure 3.8 show the links between spatially adjacent neurones.

Before training begins, the weights are initialized to random values within the range [0, 1] (Figure 3.8a). After 10 iterations, the SOM weights start to expand (Figure 3.8b). After 1000 iterations, the ordering between input patterns can roughly be seen by the topology made up by output neurones and weights as shown in Figure 3.8c. After 6000 iterations, the weights are approximately uniformly spread as illustrated in Figure 3.8d.

Figure 3.9 shows a second example in which the number of test data points and the feature ranges are the same as in example 1. However, a slight change is made so that 75% of the random samples fall in the shaded area (Figure 3.9a). The resulting distribution of weights after 6000 iterations is shown in Figure 3.9b, which is clearly different from the results presented in Figure 3.8b. Most of the weights in Figure 3.9b are distributed within the shaded area. Only four of the 16 neurones are located in the remaining area. These examples indicate that a change in sample size will result in change in SOM performance. Therefore, if one intends to distinguish a slight difference within some specific feature domain, then more samples within that area should be presented in order to weight the SOM toward the area of interest.

One of the capabilities of SOM is to achieve a nonlinear mapping of the feature space (Kraaijveld et al., 1995), which can be used to reduce the input data dimension. Principal component analysis (PCA, Chapter 2) is often used for data dimension reduction. PCA is based on the eigenvalues and eigenvectors of the variance–covariance (or correlation) matrix of the data set. Such a method is linear and has its drawbacks. The simplest example involves just two dimensions. If the data distribution on a two-dimensional space has a correlation of near zero, one may experience

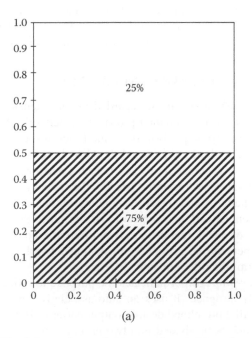

(a)

FIGURE 3.9 (a) 75% of the random training samples used to train a SOM fall within the shaded area; the remaining area contains 25% of the random samples.

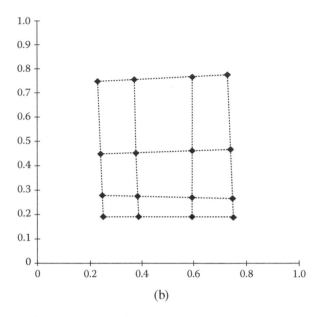

(b)

FIGURE 3.9 (continued) (b) Result of weight distribution after 6000 iterations. Most of the output neurones are located inside the shaded area.

difficulty in using PCA to perform a suitable data mapping. However, by using a SOM, the resulting weights will be located in such a way as to match the shape of data distribution. Such a mapping is nonlinear and may more clearly reflect the data distribution than linear mappings such as PCA. Kraaijveld et al. (1995) provide a detailed description of the use of SOM in nonlinear mappings.

3.3 COUNTER-PROPAGATION NETWORKS

Counter-propagation networks can be regarded as an extension of the SOM network. Three layers make up a counter-propagation network. The first two layers (input and competitive layers) perform the same functions as the two layers of the SOM (Section 3.2). The third layer (called the Grossberg layer) is fully connected to the competitive layer. Compared to a SOM, such a network structure provides an improved means of supervised training. Hence, the inputs to this kind of network contain both input patterns and the corresponding output vector. The structure of a counter-propagation network may appear to be similar to that of a multilayer perceptron. However, we should treat them as two different kinds of networks because, unlike a multilayer perceptron, a counter-propagation network involves only one hidden layer, and the training rules are quite different.

The term *counter-propagation* indicates the purpose for which this type of network was originally designed. It uses an auto-associative memory (i.e., using an input pattern to recall a pre-efined desired output pattern). In Figure 3.10, the input and output layers each been divided into two parts in order to deal with the input vector pair (A, B) and output pair (A', B'). The activation flow is as follows: the

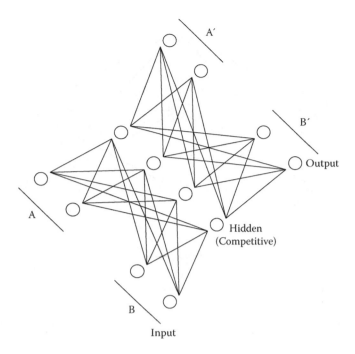

FIGURE 3.10   Counter-propagation network acts as an associated memory. See text for details.

input vector A from the left of the input layer will recall output vector B' on the right of the output layer, and input vector B will be associated with output pattern A' on the left of the output layer. The activation flow is diagonal; hence this type of network is called a *counter-propagation network*.

3.3.1   COUNTER-PROPAGATION NETWORK TRAINING

The counter-propagation network has a particular property: Only those weights connecting to the winning neurone are updated. Other weight values remain unchanged. This is a quite different strategy from the one used in the multilayer perceptron. An example is shown in Figure 3.11, which illustrates a winning neurone (shaded) located in the hidden layer. Only those weights (shown in bold) connecting to this neurone from both the input and output layers are adjusted.

A counter-propagation network is thus made up of two kinds of network: a SOM network and a Grossberg layer. The algorithm used for updating the network weights during the training stage is described in the following paragraphs.

The algorithm for identifying the winning neurone in the hidden layer requires a normalization process in which both the input vector and the weights connecting the input and hidden layers are normalized to a length of 1. More specifically, let $\mathbf{b} = (b_1, b_2, \ldots, b_m)$ denote an input vector, and $\mathbf{w}_j = (w_{j1}, w_{j2}, \ldots, w_{jm})$ denote the weight set from the input layer to hidden neurone j. This algorithm requires

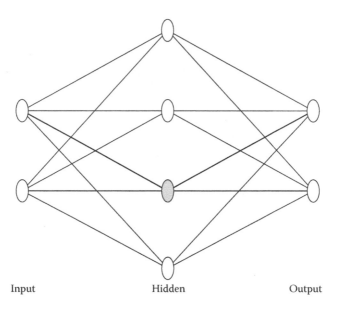

Input   Hidden   Output

FIGURE 3.11   The weight adjustment rule for counter-propagation. Only weights connecting to the winning neurone (shaded) are adjusted.

$$\| \mathbf{b} \| = \| \mathbf{w}_j \| = 1 \tag{3.18}$$

where

$$\| \mathbf{b} \| = \sqrt{\sum_i b_i^2}, \text{ and } \| \mathbf{w}_j \| = \sqrt{\sum_k w_{jk}^2} \tag{3.19}$$

When an input vector is fed into the network, the following weighted sum is calculated for each hidden neurone j:

$$S_j = \sum_i b_i w_{ji} \tag{3.20}$$

The winning neurone v in the hidden layer is the neurone that has the maximum weighted sum S_v. The weights connecting the input layer neurones to the winning neurone v are then updated in terms of following rule:

$$w_{vi}^{n+1} = w_{vi}^n + \alpha \left(b_i - w_{vi} \right) \tag{3.21}$$

where α, $0 < \alpha \le 1$, is the learning rate, and w_{vi}^n denotes the weight state at time n. The other weight sets, $\mathbf{w}_z$, for $z \ne v$, retain their original values. The weight set $\mathbf{w}_v$ is then again subjected to the following normalization process to maintain its normalization state:

$$\frac{\mathbf{w}_v}{\|\mathbf{w}_v\|} \tag{3.22}$$

For the weight set $\mathbf{h}_v$, $\mathbf{h}_v = (h_{1v}, h_{2v}, \ldots, h_{nv})$, connecting the winning neurone v in the hidden layer to the output layer, $\mathbf{h}_v$ is updated in terms of

$$h_{jv}^{n+1} = h_{jv}^{n} + \beta\left(a_j - o_j\right) \tag{3.23}$$

where β is another learning constant, o_j is the desired output for output neurone j, and a_j is the network output derived from

$$a_j = h_{jv} \tag{3.24}$$

Thus, it is clear that the counter-propagation method takes the weights connecting from the winning neurone as output activities. Other weights connecting to losers remain unchanged. Note that the values of learning constants, α and β, should not to be too large. Values of 0.1 or 0.2 are generally used.

3.3.2 TRAINING ISSUES

Since the counter-propagation training algorithm does not contain a feedback flow from the output layer to the weights between the input and hidden layer, the links between the weights of the first two layers and the desired outputs will have a weaker relationship than those trained by back-propagation algorithm. As a result, it will involve a higher probability that the same winning neurone in the hidden layer is triggered by similar input patterns, even though the patterns actually belong to different classes. An improved version of the training algorithm to overcome this problem may include feedback to force the network to trigger different winner neurones in response to differences in input patterns (Reilly et al., 1982).

3.4 HOPFIELD NETWORKS

The Hopfield network (Hopfield, 1982, 1984) has mainly been exploited for auto-associative memory or solving optimization problems, such as the well-known traveling salesman problem, and temporal feature tracking. Hopfield networks belong to the class of recurrent networks. In such networks, neuronal outputs are employed as inputs to feed back to the network itself. This property is quite different from the feed-forward networks described in previous sections, and may be thought to represent a kind of memory in the network.

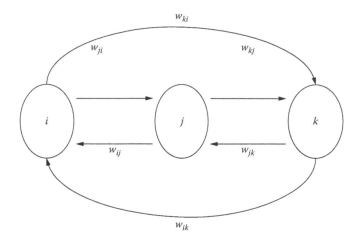

FIGURE 3.12 A simple Hopfield network containing three processing units. Each pair of units is connected by weighted links in both directions.

3.4.1 HOPFIELD NETWORK STRUCTURE

A Hopfield network is constructed by interconnecting a large number of simple processing units. Each pair of units (i, j) is connected in both directions, i.e., from neurone i to j and from neurone j to i. A simple Hopfield network structure with three processing units is illustrated in Figure 3.12.

In general, the ith processing unit is described by two variables. These variables are the input (or current state) from other neurones, denoted by u_i, and the output denoted by v_i. The output value v_i is generated in terms of some predefined output function $f(u_i)$ in order to limit the value of v_i to the range [0, 1] or [−1, 1].

A total of $n(n − 1)$ interconnections is required for a Hopfield network containing n processing units. Let w_{ji} denote the weighted link from neurone i to j. The value of w_{ji} can be considered as a connection that assigns the weight to the data flow from i to j (activity in a neurone i is weighted by the amount w_{ji} and then passed to neurone j). Hopfield (1984) shows that a set of symmetric weights (i.e., $w_{ji} = w_{ij}$) is a sufficient condition to guarantee network convergence to a stable state.

3.4.2 HOPFIELD NETWORK DYNAMICS

There are two kinds of models, termed *discrete* and *continuous*, for characterizing the Hopfield network's dynamics, i.e., the change of network state. In the discrete version, the network output is either 1 or 0. The input to the ith neurone at time n, denoted by u_i^n, is

$$u_i^n = \sum_{j \neq i} w_{ij} v_j + I_i \qquad (3.25)$$

where I_i is an external input to neurone i. The v_i, which is the output of the neurone i, are usually defined in terms of a step function:

$$v_i = 1, \quad u_i > s_i, \text{ and}$$
$$v_i = 0, \quad u_i < s_i. \tag{3.26}$$

in which s_i is a predefined threshold for neurone i. In general, s_i is set to 0.

In the continuous version, the network state for neurone i is described by the differential equation

$$\frac{du_i}{dn} = -\frac{u_i}{\lambda \tau} + \sum_{j \neq i} w_{ij} v_j + I_i \tag{3.27}$$

in which n denotes the iteration number; τ is a constant for neurone I; v_i is generated in terms of a monotonically increasing nonlinear mapping function $f(u_i)$, typically a sigmoid function (see Equation [3.2] and Figure 3.2 for details); and λ is a parameter that scales the input to function $f(u)$ as: $f(u) \rightarrow f(\lambda u)$. The right-hand side of Equation (3.27) specifies the amount of change in u_i following the change in time n.

Each state of the Hopfield network has an associated energy function, which is used to indicate the degree of error. The higher the energy the larger the error, and the aim of the network operations is to minimize the energy. The energy minimization concept is used again in Chapter 8. At each iteration, the change of network state is directed toward a search for a lower energy state until convergence is reached. Hopfield (1984) describes two energy functions for the discrete and continuous cases. In the discrete version, energy E is expressed as

$$E = -\frac{1}{2} \sum_i \sum_{\substack{j \\ j \neq i}} w_{ij} v_i v_j - \sum_i I_i v_i \tag{3.28}$$

while in the continuous version, E is defined as

$$E = -\frac{1}{2} \sum_j \sum_{\substack{i \\ j \neq i}} w_{ij} v_i v_j - \sum_i I_i v_i + \sum_i \frac{1}{\lambda \tau} \int_0^{v_i} f^{-1}(v) dv \tag{3.29}$$

where $f^{-1}(v)$ is the inverse of function $f(u)$. For very high values of λ, which are typical of Hopfield networks, the last term in Equation (3.29) is near zero and can be neglected. Equation (3.29) then becomes equivalent to Equation (3.28).

3.4.3 NETWORK CONVERGENCE

The successive updating of the network state is directed toward the search for a lower energy state. Eventually the network will converge at a point at which energy is at a minimum. This may be a local or global minimum. The following paragraphs illustrate how the network is seeking a lower energy value during the successive changes of network state.

In the discrete version of the Hopfield model, each neurone outputs either 0 or 1. If there is a change of state for neurone i from time n to $n + 1$, the difference in energy, denoted by ΔE, due to the ith neurone can be written as (refer to Equation [3.28])

$$\Delta E = E_i^{n+1} - E_i^n = -\frac{1}{2}\left[\sum_{j\neq i} w_{ij}v_j + I_i\right](v_i^{n+1} - v_i^n) \tag{3.30}$$

According to Equation (3.26), one can ensure that the output described by Equation (3.30) will always be $\Delta E \leq 0$. For example, a change in the state of v_i from 0 to 1 will only occur at time $n + 1$ if

$$\sum_{j\neq i} w_{ij}v_j + I_i > 0 \tag{3.31}$$

ΔE will be negative and thus decrease the energy. The case for the state of v_i changing from 1 to 0 can be inferred in a similar way. One may therefore calculate how the change in neurone state can affect the energy level. In the continuous case, let $\lambda \gg 1$ in order to simplify the problem. First, rewrite Equation (3.29) as

$$E_i = -\frac{1}{2}v_i\left(\sum_j w_{ij}v_j + I_i\right) \tag{3.32}$$

Equation (3.32) denotes the energy change due to neurone i, and its time derivative becomes

$$\frac{dE_i}{dn} = \frac{dE_i}{dv_i}\frac{dv_i}{dn}$$

$$= -\frac{1}{2}\left(\sum_{i\neq j} w_{ij}v_j + I_i\right)\left(\frac{dv_i}{dn}\right) \tag{3.33}$$

$$= -\frac{1}{2}\left(\frac{du_i}{dn}\right)\left(\frac{dv_i}{dn}\right)$$

Since

$$u_i = f^{-1}(v_i) \rightarrow \frac{du_i}{dn} = f^{-1'}(v_i)\frac{dv_i}{dn} \tag{3.34}$$

Equation (3.33) becomes

$$\frac{dE_i}{dn} = -\frac{1}{2} f^{-1'}(v_i) \left(\frac{dv_i}{dn} \right)^2 \tag{3.35}$$

Since $f^{-1}(v_i)$ is a monotonically increasing function indicating $f^{-1'}(v_i) > 0$, it follows that Equation (3.35) will always be negative or equal to zero. Convergence is therefore guaranteed.

$$\frac{dE}{dn} = 0 \rightarrow \frac{dv_i}{dn} = 0 \text{ for all } i. \tag{3.36}$$

3.4.4 ISSUES RELATING TO HOPFIELD NETWORKS

In order to use a Hopfield network for problem solving, one has to determine the number of processing units in the network, and define the energy function and weight matrix. The adequacy of the solution depends upon these choices. The energy function represents the problem to be solved. Careful construction of the energy function can reduce network search time, and help the network converge to an acceptable solution. Definition of a complete energy function generally involves the specification of two elements, called the *objective function* and the *constraint function*. The objective function gives a general measure of cost, such as the similarity between the features or the distance of travel. The constraint function describes limitations to the solution of the problem. For instance, in the traveling salesman problem, the objective function will be a measure of distance traveled by the salesman, while the constraint function addresses restrictions such as the fact that each city may be visited only once, or that only one city can be visited per time period. More specifically, the constraint function acts as a penalty term that associates a high energy or cost with violations of the constraints. An example is presented in Section 3.4.5.

Two steps are required to determine the values of the elements of the weight matrix $\mathbf{W}$. The first step is to define an energy function. The second step uses the specified energy function to obtain the weights w_{ij} in terms of Equation (3.32), which can be reduced to

$$w_{ij} = -2 \frac{\partial^2 E_{ij}}{\partial v_i \partial v_j} \tag{3.37}$$

The relationship between w_{ij} and energy E_{ij} is fully specified. Aiyer et al. (1990) give a theoretical analysis of the effect of the weight matrix W on the performance of Hopfield networks.

As in all optimization problems, one has to consider the question of local and global minima. Since the energy function may contain many local minima, the network usually converges to a local minimum, as illustrated in Figure 3.13. Here, point c is the global minimum of the energy function, while points a and b are local minima. Thus, if the initial network state is located at point d, the network is very likely to

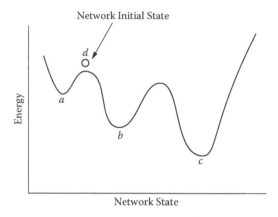

FIGURE 3.13 The energy surface and local and global minima. See text for further discussion.

converge at point a or b, but not c. If this situation arises, the optimization procedure is said to have become trapped in a local minimum. To overcome this problem, a modified Hopfield network can be used, as described by Lee and Sheu (1991).

Another approach for attaining a global minimum is based on the technique of annealing, which can be explained by introducing a temperature parameter to the system. The system convergence process will start at a high temperature then gradually cool down. This technique is described more fully in Chapter 8.

3.4.5 ENERGY AND WEIGHT CODING: AN EXAMPLE

Consider the traveling salesman problem. At the start of his journey, the salesman has to plan a route such that the distance traveled is the smallest possible. A solution based on the Hopfield network was proposed by Hopfield and Tank (1985). A two-dimensional array of units is employed. Figure 3.14 shows an example for a 4 × 4 Hopfield network in which the columns denote the visit sequence and the rows represent the city to be visited. Units with an output value of 1 are shaded, while units with an output value of 0 are unshaded. The solution shown by Figure 3.14 indicates that the city visit sequence is Z2–Z4–Z1–Z3.

In order to obtain a meaningful result from a Hopfield network, the energy function has to be properly modeled (Section 3.4.4). The energy function consists of two parts: *constraint* and *objective*. In the traveling salesman problem, one can establish the following three constraints:

1. Each city can only be visited once; that requires the sum of each row must be exactly 1.
2. Only one city can be visited at a given time, which indicates that each column also must sum to 1.
3. All the cities must be visited, which means the total number of active units m must be equal to the total number of cities (in our case, $m = 4$).

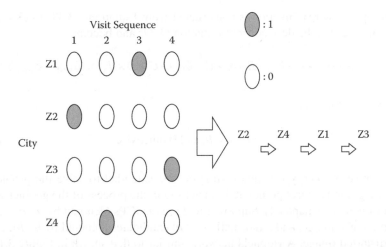

FIGURE 3.14 A Hopfield network constructed of a set of 4 × 4 neurones to solve the traveling salesman problem. See text for details.

According to constraint condition (1), one can code the constraint element of energy E_1 as

$$E_1 = \frac{A}{2} \sum_X \sum_i \sum_{i \neq j} v_{Xi} \times v_{Xj} \qquad (3.38)$$

where v_{Xi} denotes output neurones at location (X, i), i.e., city X and visit sequence i. Constraint (2) can be coded in a similar way as

$$E_2 = \frac{B}{2} \sum_X \sum_i \sum_{X \neq Y} v_{Xi} \times v_{Yi} \qquad (3.39)$$

The third constraint can be expressed as

$$E_3 = \frac{C}{2} \left(\left(\sum_X \sum_i v_{Xi} \right) - m \right)^2 \qquad (3.40)$$

The objective part of the energy function, denoted by E_4, can be coded as

$$E_4 = \frac{D}{2} \sum_X \sum_{X \neq Y} \sum_i d_{XY} \times v_{Xi} \times (v_{Y(i+1)} + v_{Y(i-1)}) \qquad (3.41)$$

where d_{XY} denotes the distance between two cities X and Y, and A, B, C, and D are the weighting factors determined by the user.

A complete energy function is constructed from Equations (3.38) to (3.41). The weights w_{ij} can then be derived from Equation (3.37) and become

$$w_{Xi,Yj} = -A \times \delta_{XY} \times (1 - \delta_{ij}) - B \times \delta_{ij} \times (1 - \delta_{XY}) - C - D \times d_{XY} \times (\delta_{j(i+1)} + \delta_{j(i-1)}) \quad (3.42)$$

where δ_{ab} is defined as

$$\delta_{ab} = 1 \text{ if } a = b, \text{ and } 0 \text{ otherwise.} \quad (3.43)$$

This example of energy function coding can be extended to solve the problem of automatic ground control point (GCP) matching in the process of the geometric correction of images (Chapter 1). Suppose there are m GCPs (denoted by $a_1, a_2, ..., a_m$) in the reference image, and n (normally $m < n$) candidates (denoted by $b_1, b_2, ..., b_n$) in the distorted image. A Hopfield network similar to that shown in Figure 3.14 can be constructed, noting that the rows now represent the GCPs a_i ($1 \le i \le m$) and the columns represent candidates b_j ($1 \le j \le n$). Accordingly, three constraints similar to those used in the traveling salesman problem can be set up:

1. Each GCP a_i can only be matched once; that requires the sum of each rows must be exactly 1,
2. Only one GCP a_i can be matched at a given candidate b_j, which indicates that the sum of each column has to be no more than 1.
3. All the GCP a_i must be matched, which means the total number of active units must be equal to the total number of GCPs (i.e., m).

Equations (3.38) to (3.40) can be adapted to meet these requirements. Regarding the objective function, one simple way for matching GCPs may be according to the absolute difference between the GCPs a_i and candidate b_j. Once the objective function has been defined, the energy function is fully specified. Equation (3.37) can then be used to initialize the Hopfield network weights and to search for the solution.

3.5 ADAPTIVE RESONANCE THEORY (ART)

Adaptive resonance theory (ART), which is a theory of human cognitive information processing (Grossberg, 1976), has led to an evolving series of neural network models for unsupervised (ART1, ART2, fuzzy ART) and supervised (ARTMAP and fuzzy ARTMAP) pattern recognition.

ART1 (Carpenter and Grossberg, 1987a) was the first ART-based model to be used for pattern recognition. It is designed to categorize binary input patterns. ART2 (Carpenter and Grossberg, 1987b) and fuzzy ART (Carpenter et al., 1991a) are extensions of the ART1 model. Both ART2 and Fuzzy ART can categorize either analogue (real number) or binary input signals. ARTMAP (Carpenter et al., 1991b) and fuzzy ARTMAP (Carpenter et al., 1992) are supervised learning architectures. ARTMAP is based on two networks of the binary ART1 model, which learns binary patterns.

Carpenter et al. (1992) use fuzzy ART to replace ART1 in the ARTMAP system; the resulting fuzzy ARTMAP can therefore learn analogue or binary input signals. In the following, we introduce the ART1 model to illustrate the basic concept of the ART network system. The fuzzy ARTMAP model, which is useful in classifying remotely sensed imagery, is described later.

3.5.1 FUNDAMENTALS OF THE ART MODEL

An ART1 network is a self-regulating system containing two layers: the input layer (F1) and the category layer (F2), as shown in Figure 3.15a. The F1 and F2 layers are fully connected by top–down (g_{ji}) and bottom–up (h_{ij}) weighted links, which are also called long-term memory (LTM). The control unit in Figure 3.15a is used to manipulate the system mode, as described in the following paragraphs, and this distinguishes this system from a simple Hopfield network. Each neurone in the F2 layer is self-feeding in that an output linkage connects back to the neurone itself, as shown in Figure 3.15b. A positive weight (generally a value of 1) is used, while negative connections (called the competition parameter, ε) exist to every other neurone in F2, as illustrated in Figure 3.15b. The value ε is subject to the following constraint:

$$\varepsilon < \frac{1}{m_2} \tag{3.44}$$

where m_2 is the total number of neurones in the F2 layer.

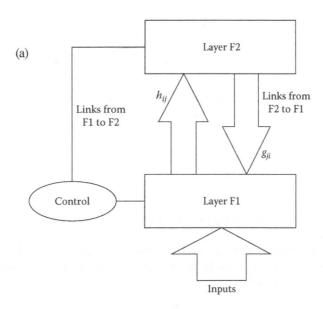

FIGURE 3.15   (a) An overview of an ART1 network.

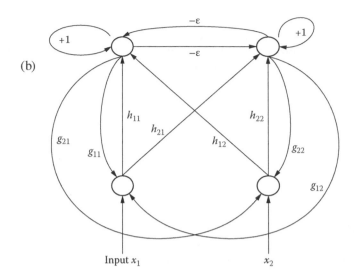

FIGURE 3.15 (continued) (b) Expanded view of ART1 structure in the case of two neurones on each layer. See text for details.

The learning strategy adopted by the ART model is that of competitive learning, as described in Section 3.2. In the initial state, all of the weights g_{ij} are set to 1.0, as described by Carpenter and Grossberg (1987a), while the weights h_{ij} have to obey the restriction that h_{ij} must be greater than 0 and no bigger than $(\alpha + m_1)^{-1}$, where α is a small positive constant and m_1 is the total number of neurones in layer F1. In general, for simplicity, the weights h_{ij} can all be set to

$$h_{ij} = \frac{1}{\alpha + m_1}, \forall i, j \tag{3.45}$$

However, an alternative is to let the h_{ij} decrease following their index order. Therefore, for example, the weights from the first neurone in F1 to all the neurones in F2, denoted by h_{i1}, $\forall\, i$, will have the following relationship:

$$0 < h_{m1} < h_{(m-1)1} < ... < h_{11} \leq \frac{1}{\alpha + m_1} \tag{3.46}$$

The effect of weight setting using either Equation (3.45) or Equation (3.46) is considered in Section 3.5.2.

During the learning process, a pattern vector X, $X = \{x_1, x_2, ..., x_n\}$ is presented, and is converted to activity in the layer F1 and combined with linking weights h_{ij} to feed into the F2 layer in terms of

$$s_i = \sum_j h_{ij} x_j \tag{3.47}$$

where s_i is the input for neurone i in layer F2. An iterative process is then carried out to update the activity of each neurone in the F2 layer (so that neurones start to compete with each other) until finally only one winning neurone u_{win} is identified:

$$u_{win} = \max\{u_i, i = 1 \text{ to } m_2\} \tag{3.48}$$

where u_i is the activity for the neurone i in F2, and is computed by

$$u_i^{n+1} = f\left(u_i^n - \varepsilon \sum_{i \neq k} u_k^n\right) \tag{3.49}$$

where n denotes the nth iteration, ε is the competition parameter introduced earlier (Figure 3.15b), and $f(y)$ is generally a nonlinear, monotonically nondecreasing function, for example the sigmoid function, defined as

$$0 \leq f(y) \leq 1,$$

and

$$f(y) = 0 \quad \text{for} \quad y \leq 0 \tag{3.50}$$

The activities present in the F1 and F2 layers are also called short-term memory (STM) because they only exist when a signal (e.g., a training pattern) is passing through the layer.

After the identity of the winning neurone in layer F2 is determined, a process then starts to test system "vigilance" according to

$$\frac{\sum_{j=1}^{n} g_{j,win} x_j}{\sum_{j=1}^{n} |x_j|} > \rho \tag{3.51}$$

where ρ $(\in [0, 1])$ is a user-defined vigilance parameter.

If the test in Equation (3.51) fails, the current winning neurone is considered invalid and ruled out. The competitive process then starts again in F2 to search for an alternative winning neurone, which is again subject to the vigilance test. If the test in Equation (3.51) succeeds, only the weights, $h_{win,j}$ and $g_{j,win}$, linking to the winning neurone j and layer F1 are adjusted by the following rules:

$$h_{win,j}^{n+1} = \frac{g_{j,win}^n x_j}{\alpha + \sum_{j} g_{j,win}^n x_j}$$

Step:
- (1) Input the pattern vector X and determine the most closest cluster (i.e., winning neurone in layer F2) to X.
  This is done by Eq. (3.47) to (3.49).
- (2) Test if the pattern X and the cluster center (i.e., weights $g_{win,j}$) are close enough (by measuring distance between both). This is done by matching a user defined criterion (i.e., vigilance ρ) as shown in (3.51). If the test failed, choose another cluster (i.e., another winning neurone) and again subject the result to criterion match. If the test succeeds, go to step (3).
- (3) Modify the cluster center so as to make it closer to the pattern X. This is done by (3.52).

FIGURE 3.16   Summary of ART1 clustering dynamics.

$$g_{j,win}^{n+1} = \beta(g_{j,win}^n x_j) + (1-\beta)g_{j,win}^n \tag{3.52}$$

where β ($\in [0, 1]$) is the learning rate parameter. The concept of the ART1 clustering algorithm can be summarized in terms of three steps as shown in Figure 3.16.

3.5.2 CHOICE OF PARAMETERS

The arrangement of weights h_{ij} in Equation (3.46), the choice of values for the vigilance parameter ρ in Equation (3.51), parameter α, and learning rate β in Equation (3.52) are issues that require more elaboration. As the weights h_{ij} are decreasing following their spatial index (as shown in Equation [3.46]), it follows that neurones with a low ordinal number (1, 2, ...) are more likely to be selected than neurones with a high ordinal number. Thus, the set of winning neurones chosen by the system will follow a sequential order.

The vigilance parameter ρ in the ART model controls the tightness of a cluster. In the case of small values of the vigilance parameter ρ, Equation (3.51) allows more patterns to be associated with the same neurone in F2, and the result is a loose cluster. A high value of the vigilance ρ will cause the network system to perform only exemplar learning (i.e., one pattern, one cluster) rather than category learning.

Parameter α must be sufficiently large to affect the weights h and subsequently s in Equation (3.47). Normally, α is set to be greater than 0.001. The learning rate β controls how the system learns or adapts (Carpenter and Grossberg, 1987). Learning rates that are too fast, in the extreme case, will make the system learn every input pattern as a new class, while learning rates that are too slow will make the system insensitive to new input patterns.

ART1 (Binary)	Fuzzy ART (Analog)

Choose Cluster (i.e., $\max\{s_j, \text{ for } \forall j\}$)

$$s_j = \sum_j h_{ij} \cap x_j = \sum_j h_{ij} x_j \qquad s_j = \sum_j h_{ij} \wedge x_j = \sum_j \min\left(h_{ij}, x_j\right)$$

Vigilance Text ($>\rho$)

$$\frac{\sum_{j=1}^{n} g_{j,win} \cap x_j}{\sum_{j=1}^{n} |x_j|} = \frac{\sum_{j=1}^{n} g_{j,win} x_j}{\sum_{j=1}^{n} |x_j|} \qquad \frac{\sum_{j=1}^{n} g_{j,win} \wedge x_j}{\sum_{j=1}^{n} |x_j|} = \frac{\sum_{j=1}^{n} \min\left(g_{j,win}, x_j\right)}{\sum_{j=1}^{n} |x_j|}$$

Move Cluster Center (Fast Learning)

$$g_{j,win}^{n+1} = g_{j,win}^{n} \cap x_j = g_{j,win}^{n} x_j \qquad g_{j,win}^{n+1} = g_{j,win}^{n} \wedge x_j = \min\left(g_{j,win}^{n}, x_j\right)$$

FIGURE 3.17 Comparison between ART1 and fuzzy ART (modified from Carpenter, G. A., S. Grossberg, N. Markuzon, J. H. Reynolds, and D. B. Rosen. 1992. "Fuzzy ARTMAP: A Neural Network Architecture for Incremental Supervised Learning of Analogue Multidimensional Maps." *IEEE Transactions on Neural Networks* 3 (1992):698–713.

3.5.3 Fuzzy ARTMAP

Fuzzy ARTMAP is a supervised neural network that differs from ARTMAP in that fuzzy ARTMAP uses two fuzzy ART neural networks instead of two ART1 networks as its basic structure, and can therefore deal with either binary or analogue values. Before going on to consider the fuzzy ARTMAP network, it is necessary to introduce the basic properties of the fuzzy ART network.

Fuzzy ART substitutes the concept of the fuzzy set (described in detail in Chapter 5) for the crisp set as used in ART1 system. Both Equations (3.47) and (3.51) can be regarded as the crisp set intersection operation by the intersection operator ($\cap$). In fuzzy ART, the operator ($\cap$) is replaced by the fuzzy *AND* operator ($\wedge$), where $p \cap q = p \times q$ for the logical *AND* operation, and $p \wedge q = \min(p, q)$ for the fuzzy *AND* operation. A comparison between ART1 and fuzzy ART is given in Figure 3.17. The third step (i.e., moving the cluster center) shown in Figure 3.17 needs some comment. According to Equation (3.52), the update of the cluster center vector for fuzzy ART can be written as

$$g_{j,win}^{n+1} = \beta \times (g_{j,win}^{n} \wedge x_j) + (1-\beta) \times g_{j,win}^{n} \qquad (3.53)$$

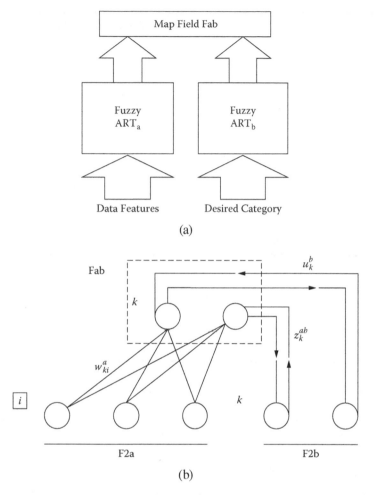

FIGURE 3.18 (a) An overview of a fuzzy ARTMAP architecture containing two fuzzy ART systems. (b) The map field is connected to $F2_b$ with one-to-one, nonadaptive pathways in both directions. Each $F2_a$ neurone is connected to all map field F_{ab} nodes via adaptive weights. See text for further explanation.

where the operator ($\cap$) is replaced by ($\wedge$) in the case of fuzzy logic. Fast learning corresponds to the value of the learning rate parameter β being set equal to 1.

The simplified architecture of a fuzzy ARTMAP is shown in Figure 3.18. Figure 3.18a illustrates a fuzzy ARTMAP network formed by two fuzzy ART systems, ART_a and ART_b, which are linked by an inter-ART module called the *map field*, denoted by F_{ab}. ART_a is used to input the values of the data features (e.g., pixel values), while ART_b is used to receive the desired category for the current data feature presented to ART_a. The map field F_{ab} is used to form predictive associations between categories and match tracking (defined below). The number of neurones in F_{ab} is equal to the number of desired classes. Let variables in ART_a and ART_b be designated by the subscripts or superscripts a and b. Figure 3.18b illustrates the linkages between F_{ab} and both $F2_a$ (in the case of three neurones) and $F2_b$ (in the case of two desired categories). The structure of the $F2_a$ and $F2_b$ layers is the same as shown in Figure 3.15b. However, for simplicity, both self-linkage and interneurone linkages are ignored. Each neurone in $F2_a$ is linked to all F_{ab} neurones by way of $w_{ki}{}^a$, while neurones in $F2_b$ are connected to F_{ab} in terms of one-to-one pathways in both directions ($F2_b \rightarrow F_{ab}$ and $F2_b \rightarrow F_{ab}$). The variables u_k^b and z_k^{ab} each denote the output of $F2_b$ and F_{ab}, respectively. The inputs for both ART_a and ART_b are normalized in terms of complement coding. An input vector X, $X = \{x_1, x_2, ..., x_n\}$ becomes $X = \{x_1, x_2, ..., x_{2n}\}$, where $x_{n+i} = (1 - x_i)$, for $i = 1$ to n, while the rest of the elements remain the same. Complement coding provides a useful means of solving a potential fuzzy ART category proliferation problem (Carpenter et al., 1991a).

During training, the initialization process for weights within ART_a and ART_b is the same as described in Section 3.5.1. The weights $w_{ki}{}^a$ ($F2_a \rightarrow F_{ab}$) are initialized to the value of unity. Once a training pair (i.e., data features and desired category) is presented to the fuzzy ARTMAP, ART_a and ART_b are triggered to determine their own winning neurones on $F2_a$ and $F2_b$ using the procedures described in Figure 3.16, steps 1 and 2. Let I denote the winning neurone in $F2_a$. The variable z_k^{ab} in F_{ab} is then computed by

$$z_k^{ab} = u_k^b \wedge w_{kI}, \text{ for all } k \tag{3.54}$$

and subjected to the vigilance test:

$$\sum_k z_k^{ab} \geq \rho^{ab} \tag{3.55}$$

where ρ^{ab} is the vigilance parameter for F_{ab}. The purpose of Equation (3.55) is in fact to test if ART_a favors the same category as shown in ART_b. If the test in Equation (3.55) does not succeed, the vigilance in ART_a, denoted by ρ^a, is increased by an amount thar is sufficiently large so as to make the current ART_a winner neurone invalid, that is

$$\frac{\sum_{j=1}^{n} g_{j,win} \wedge x_j}{\sum_{j=1}^{n} |x_j|} < \rho^a \tag{3.56}$$

The variables in Equation (3.56) are the same as those defined in Section 3.5.1. ART_a then starts again to search for another winning neurone, which is again tested using Equation (3.55). Such a process is called *match tracking*. If the winning neurone in ART_a satisfies Equation (3.55), the weights in ART_a are adjusted as shown by Equation (3.52) except that $g_{j,win}^n x_j$ is replaced by $g_{j,win}^n \wedge x_j$, and the weights w_{ki}^a between $F2_a$ and F_{ab} are changed to $w_{ki}^a \wedge u_k^b$. The vigilance ρ^a is then set to its initial value before the next training pair is presented. Note that in both ART1 and fuzzy ART systems, the vigilance ρ is a fixed parameter. However, in ARTMAP and fuzzy ARTMAP, the vigilance parameter ρ^a is dynamically changed in order to perform match tracking.

3.6 NEURAL NETWORKS IN REMOTE SENSING IMAGE CLASSIFICATION

Over the past decade, the use of neural networks for classifying remotely sensed imagery has developed rapidly, mainly because neural network classifiers are believed to outperform standard statistical classifiers such as maximum likelihood. The lack of assumptions concerning data distribution is another factor that makes neural networks more attractive than statistical classifiers, especially when the size of training data is limited so that adequate estimates of statistical parameters are difficult to obtain.

3.6.1 An Overview

The most popular neural network classifier in remote sensing is the multilayer perceptron. Many classification experiments using the multilayer perceptron are found in the literature. Good surveys are provided by Paola and Schowengerdt (1995), Kanellopoulos and Wilkinson (1997), and Atkinson and Tatnall (1997). Kanellopoulos et al. (1992) conducted a 20-class classification experiment on SPOT high-resolution visible (HRV) imagery, and the result is satisfactory. Cloud identification is reported by Lee et al. (1990) and Welch et al. (1992); classification using synthetic aperture radar (SAR) is described by Decatur (1989), and a comparison between statistical classifier and multilayer perceptron is provided by Benediktsson et al. (1990). Note that all the previously mentioned classification experiments used the multilayer perceptron network. The same classification can also be carried out by other main types of neural networks such as SOM and fuzzy ARTMAP system, though the results might reveal some variation (a comparative study is performed later). In addition to the examples cited previously, multilayer perceptrons have also been used for other kinds of classification purposes. For example, Foody et al. (1997) use a multilayer perceptron to explore the subpixel mixture problem, and Jin and Liu (1997) estimate biomass from microwave imagery. A method for forest stand parameter estimation is demonstrated by Wang and Dong (1997). Aires et al. (2004) use Bayesian technique to evaluate uncertainties in ANN parameters in order to monitor the robustness of the ANN model. The theoretical developments are further applied to the issue of retrieving surface skin temperature, microwave surface emission, and integrated water vapor content over land.

The multilayer perceptron can also be used to detect line features on SPOT panchromatic images. Each training pattern is formed by the nine pixel values falling within a 3×3 window. We selected 31 training patterns and used a 9|30|16|2 multilayer perceptron for training. The results are shown in Figure 3.19. Although

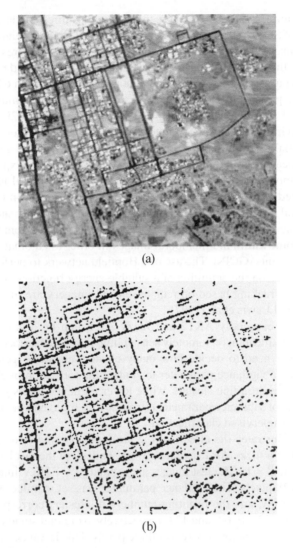

(a)

(b)

FIGURE 3.19 (a) Test SPOT image. (b) Lines detected by multilayer perceptron.

some line fragments are present on the image, the overall performance seems to be moderately successful.

In comparison with the multilayer perceptron, the Kohonen SOM is not used as frequently for classifying remotely sensed imagery. Besides its use in image classification, SOM can be used for other purposes, such as input data dimension reduction. For example, the number of input features could be very large in a remote sensing classification exercise. The number of input features could be reduced using the divergence index or B distance (Chapter 2), which aim to select the "best" feature subset. An alternative is to use a SOM to perform a mapping for different multidimensional feature groups, and then use the resultant features for classification. Note that by means of SOM, reduction in the number of input channels and gray levels can be done simultaneously. For instance, if the number of input features is 4, each with 256 gray levels, the user may define 4 neurones in the input layer and 8×8 (= 64) neurones in the output layer. The SOM will output a single image with 64 gray levels.

As described previously, Hopfield networks converge to a minimum of an energy function defined as a goal. Using such a concept, Tatem et al. (2002) propose an approach involving designing the Hopfield energy function to produce the prediction of the spatial distribution of class components in each pixel so as to perform mixture classification. Although Hopfield networks are mainly exploited for auto-associative memory, a new direction of investigation for temporal feature tracking has also emerged (Bersini et al., 1994; Côté and Tatnall, 1997). In remote sensing, there is a special need for matching features that vary over time, such as the automatic monitoring of the direction of cloud rotation or iceberg movement, or identifying the same point within multitemporal images in order to automatically construct ground control points (GCPs). The use of a Hopfield network to perform such kinds of operations relies on the definition of a suitable energy function. Studies of multi-temporal feature tracking based on remote sensing imagery are given by Côté and Tatnall (1997) and Lewis et al. (1997).

ART systems, especially ART2 and fuzzy ART, can be useful in performing unsupervised classification on remotely sensed imagery. As described in Section 3.5, the user does not need to decide how many clusters are to be generated; rather, one has to define a vigilance parameter to control the formation of clusters. Higher vigilance levels result in fine clusters, while lower vigilance levels generate coarser clusters. A range of vigilance parameters (from low to high) can be used to perform a hierarchical unsupervised classification.

Figure 3.20 illustrates the effects of the vigilance parameter on clustering performance. We use a fuzzy ART neural network, and adopt the fast learning strategy (i.e., learning rate $\beta = 1$). The test images are 3-band SPOT HRV images, each containing 128×128 pixels. The vigilance parameter was set to values of 0.6, 0.7, 0.8, and 0.9, respectively, and the resulting numbers of clusters corresponding to each vigilance level are 19, 40, 64, and 124, respectively. Note that some clusters formed by the ART network contain only two or three pixels. This is a specific characteristic of ART in comparison with other traditional clustering algorithms, in which clusters are unlikely to be made up of such a small number of pixels.

An example of the use of fuzzy ARTMAP for remote sensing image classification is presented by Carpenter et al. (1997) in which the results output by a fuzzy ARTMAP classifier are compared to those produced by the maximum likelihood, nearest neighbor, and multilayer perceptron methods. Carpenter et al. (1997) conclude that the fuzzy ARTMAP is faster and more stable. The same observation is also made by Mannan et al. (1998) who found that the fuzzy ARTMAP outperforms both the multilayer perceptron and maximum likelihood methods in terms of classifier accuracy. Liu et al. (2004) present an ARTMAP-based model called ART mixture MAP (ART-MMAP) for estimating land cover fractions within a pixel. However, one should note that in order to obtain good results, one might have to test

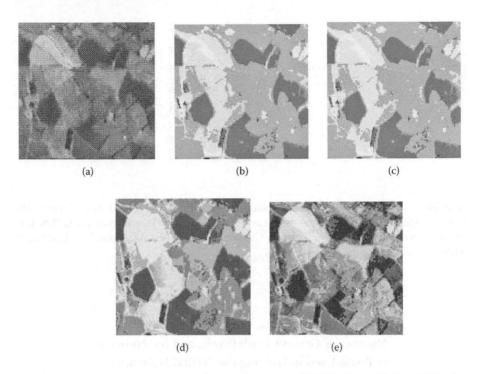

(a)   (b)   (c)

(d)   (e)

FIGURE 3.20   (a) Combined 3-band SPOT HRV image (©CNES [Centre National d'Etudes Spatiales] 1994—SPOT Image Distribution). (b) Image formed by fuzzy ART with 0.6 vigilance. (c) Image with 0.7 vigilance. (d) Image with 0.8 vigilance. (e) Image with 0.9 vigilance.

a range of ART model-associated parameters. Other studies using ARTMAP are Borak and Strahler (1999), Muchoney et al. (2000), Karen et al. (2003), and Alilat et al. (2006).

3.6.2   A COMPARATIVE STUDY

In this section, the classification performance of four types of neural networks (multilayer perceptron, Kohonen SOM, counter-propagation, and fuzzy ARTMAP) is compared. The study area is located near Feltwell, Norfolk, in eastern England. The input data consist of seven ERS-1 three-look radar images obtained for the 1993 crop-growing season (April 20, May 9, May 25, June 16, June 29, July 18, and August 3). Figure 3.21a shows a composite image made from the first three multi-temporal images. It is clear that the images are seriously affected by speckle. A Lee filter (Lee, 1981) with a window size of 5 × 5 was applied to each image in order to reduce this speckle effect. The resulting filtered color-composite image is shown in Figure 3.21b. The classification experiment is based on seven speckle-filtered images, and seven information classes (crop categories) were chosen for identification. The legend and numbers of test pixels are shown in Table 3.1. The ground reference image is displayed in Figure 3.21c. A total of 2297 training samples were selected,

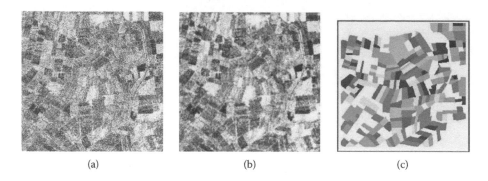

(a)   (b)   (c)

FIGURE 3.21   (a) Composite of three multitemporal ERS-1 three-look radar images. (b) Color-composite of filtered images using Lee speckle filter. (c) Ground truth image. The legend is shown in Table 3.1. Original data Copyright ESA (1993). Distributed by Eurimage/NRSC.

TABLE 3.1

Classification Legends, the Corresponding Number of Ground Truth Pixels, and the Number of Pixels Used in Training the Neural Networks

No.	Information Classes	Number of Ground Truth Pixels	Number of Training Pixels
1	Grass	11868	395
2	Winter wheat	34038	309
3	Spring wheat	9774	325
4	Potato	15378	341
5	Sugar beet	30142	287
6	Carrot	10341	344
7	Spring barley	1776	296
Total		**113317**	**2297**

---

using a stratified sampling strategy. The reader is referred to Table 3.1 for details of the number of training pixels for each information class.

A multilayer perceptron with the structure 7|28|48|7 was arbitrarily chosen and trained with the learning rate set to 0.2. The network converged after around 5000 iterations, which took nearly 30 CPU minutes. Three Kohonen SOM networks, with (a) 7×7, (b) 12×12, and (c) 25×25 output neurones, respectively, forming the mapping cortex, were investigated. The reason for investigating the effect of using different number of output neurones is that the SOM classification mechanism is a concept similar to clustering and labeling. Each output neurone denotes a cluster and we wish to examine the effect of increasing the number of output neurones (i.e., clusters) on classification accuracy. Similar experiments were performed using counter-propagation and fuzzy ARTMAP networks. The initial learning rate was set to 0.1, and 8000 iterations were arbitrarily chosen for the first stage unsupervised training (note that the number of iterations must be sufficiently large to allow the SOM to converge). The unsupervised training time for SOM candidates (a), (b), and (c) are around 50, 200, and 900 CPU minutes, respectively. A supervised training process using the learning vector quantization (LVQ) algorithm was then applied. For SOM candidates (a) and (b), the process converged after 100 iterations, which took around 10 CPU minutes, while in the case of SOM candidate (c), it converged after 200 iterations, which took around 40 CPU minutes to complete.

Three counter-propagation networks with 49, 144, and 625 hidden neurons (the same number as used in the three SOM networks), respectively, were tested. The learning rate parameters α and β in Equations (3.21) and (3.23) were both set to 0.2. All of the counter-propagation networks converged quickly, requiring around 500 iterations, and the corresponding running times were around 6 CPU minutes for the network containing 49 hidden neurones, and 15 and 130 CPU minutes for the networks containing 144 and 625 hidden neurones, respectively.

Three fuzzy ARTMAP networks with different vigilance parameters were investigated. For all three of these networks, the parameter α in Equation (3.45) was chosen as 0.01, the map field vigilance parameter ρ^{ab} in Equation (3.55) was set to 1.0 to guarantee that the input patterns and desired output category show the tightest linkage. In the first fuzzy ARTMAP network, the learning rates β^a and β^b for the fuzzy ART systems $F2_a$ and $F2_b$ (Figure 3.18) were set to 1.0 (*i.e.* fast learning). The vigilance values ρ^a and ρ^b were set to 0.8, and a total of 487 neurones (clusters) in $F2_a$ was triggered. In the second fuzzy ARTMAP network, the learning rates β^a and β^b were again set to 1.0 while the vigilance values ρ^a and ρ^b were set to 0.97, and a total of 987 neurones in $F2_a$ were generated. In the third fuzzy ARTMAP network, the learning rates β^a and β^b were set to 0.2. The vigilance values ρ^a and ρ^b were set to 0.99, and a total of 2039 neurones in $F2_a$ were triggered. The training stages for all of the tested fuzzy ARTMAP networks were very fast. The networks converged within 3 CPU minutes for the first and the second fuzzy ARTMAP candidates. The third fuzzy ARTMAP took around 20 CPU minutes.

The resulting classification accuracies are shown in Table 3.2. The best classification images (in terms of overall classification accuracy) derived from each type of neural network are shown in Figure 3.22, and the corresponding confusion matrices are shown in Table 3.3. The multilayer perceptron and the SOM networks (candidates b and c) achieved the highest classification accuracy of around 60% (with Kappa values of around 0.52), a value that is roughly 3% higher than the best-performing fuzzy ARTMAP, and around 10% better than the counter-propagation network.

Different networks show different reactions to changes in the number of clusters (i.e., neurones in the mapping cortex in SOM, hidden neurones in counter-propagation, and vigilance levels in fuzzy ARTMAP). In the case of the SOM, a network with 7×7 output neurones generates the accuracy of 45.83% (Kappa = 0.376). If a 12×12 layer of output neurones is used, the accuracy increases considerably to 60.91% (Kappa = 0.531). However, a network with a layer of 25×25 output neurones showed no significant improvement in classification accuracy, although training time rose considerably, to around 950 CPU minutes for both unsupervised and supervised training. The same behavior is also exhibited by the fuzzy ARTMAP.

TABLE 3.2

Comparison of Accuracy of Classification Using Different Types of Neural Networks

Network Type	(%) Total Accuracy	Kappa	Parameter Values and Network Structure
Multilayer perceptron	60.28	0.521	Learning rate: 0.2 Structure: 7\28\48\7
SOM	45.83	0.376	Initial learning rate: 0.1 Structure: 7\(7 × 7)
	60.91	0.531	Structure: 7\(12 × 12)
	61.17	0.531	Structure: 7\(25 × 25)
Counter-propagation	48.33	0.382	Learning rate: 0.2 Structure: 7\49\7
	46.19	0.369	Structure: 7\144\7
	49.17	0.397	Structure: 7\625\7
Fuzzy ARTMAP	49.39	0.394	Vigilance: $\rho^a = \rho^b = 0.8$ Learn. Rate: $\beta^a = \beta^b = 1.0$
	57.31	0.488	Vigilance: $\rho^a = \rho^b = 0.97$ Learn. Rate: $\beta^a = \beta^b = 1.0$
	57.01	0.487	Vigilance: $\rho^a = \rho^b = 0.99$ Learn. Rate: $\beta^a = \beta^b = 0.2$

When the vigilance level was set to 0.8, a classification accuracy of 49.39% (Kappa = 0.394) was obtained. When the vigilance value was set to 0.97, accuracy rose to 57.31% (Kappa = 0.488). A further increase in the vigilance value to 0.99 and a decrease in the learning rate to 0.2 showed no improvement. In contrast, the classification accuracies associated with the three counter-propagation networks showed little variation, with output values in the range 46% ~ 49%.

The results of these experiments indicate that the multilayer perceptron is a suitable tool for image classification in terms both of classification accuracy and training time. Although the performance of the SOM networks is comparable to that of the multilayer perceptron, its considerably longer training time may decrease its attractiveness. The performance of the counter-propagation networks was the least satisfactory in our study. The training stage for fuzzy ARTMAP network is significantly faster than the other network types tested. However, in contrast to the observations made by Carpenter et al. (1997) and Mannan et al. (1998), the performance of the fuzzy ARTMAP networks in our experiments did not show outstanding results. Further comparative studies are required to completely investigate the sensitivity of the performance of fuzzy ARTMAP to factors such as variations in the model-associated parameters and training patterns.

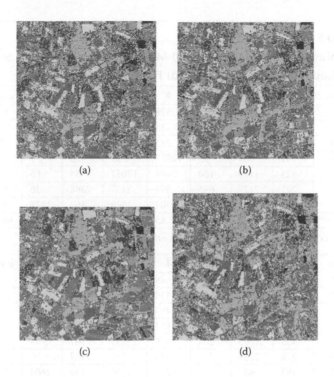

(a)

(b)

(c)

(d)

FIGURE 3.22 Classification results performed by (a) multilayer perceptron, (b) Kohonen's SOM, (c) counter-propagation network, and (d) Fuzzy ARTMAP.

TABLE 3.3
Confusion Matrix Output by (a) Multilayer Perceptron, (b) SOM, (c) Counter-propagation, and (d) Fuzzy ARTMAP Network

No.	1	2	3	4	5	6	7	u.a%.
1	**9580**	5171	852	578	2205	871	61	49.59
2	790	**16644**	2229	439	1021	569	26	76.64
3	236	7302	**5554**	283	914	541	22	37.40
4	147	607	93	**10587**	5285	305	7	62.16
5	525	1149	180	2489	**17051**	585	66	77.35
6	297	2125	695	599	2137	**7308**	16	55.46
7	293	1040	171	403	1529	162	**1578**	30.49
p.a.%	80.72	48.90	56.82	68.85	56.57	70.67	88.85	**68302**

Note: (a) Total Accuracy: 60.28%. Kappa coefficient: 0.521

No.	1	2	3	4	5	6	7	u.a%.
1	**9340**	4255	621	648	2664	720	51	51.04
2	752	**16750**	1984	376	1464	597	31	76.30
3	468	7916	**5947**	483	1252	599	11	35.66
4	313	1129	216	**11114**	4298	344	10	63.79
5	435	1065	204	1956	**17397**	767	48	79.54
6	378	2253	731	514	1762	**7165**	20	55.88
7	182	670	71	287	1305	149	**1605**	37.60
p.a.%	78.70	49.21	60.85	72.27	57.72	69.29	90.37	**69318**

Note: (b) Total Accuracy: 61.17%. Kappa coefficient: 0.531

No.	1	2	3	4	5	6	7	u.a%.
1	**8139**	4799	741	981	4568	796	55	40.53
2	810	**12084**	1971	408	872	441	25	72.75
3	345	9111	**5138**	481	1429	713	45	29.76
4	444	1695	249	**9626**	5645	646	12	52.55
5	1206	1456	354	2124	**12611**	862	74	67.49
6	557	2995	915	1517	3951	**6621**	64	39.84
7	367	1898	406	241	1066	262	**1501**	26.15
p.a.%	68.58	35.50	52.57	62.60	41.84	64.03	84.52	**55720**

Note: (c) Total Accuracy: 49.17%. Kappa coefficient: 0.397

No.	1	2	3	4	5	6	7	u.a%.
1	**8056**	3120	392	316	1737	360	36	57.48
2	824	**14837**	1678	425	1506	525	16	74.89
3	591	8584	**5718**	405	898	535	10	34.16
4	442	1139	265	**10480**	4227	247	13	62.33
5	706	1593	336	2513	**16420**	638	35	73.83
6	892	3737	1149	904	3475	**7809**	39	43.37
7	357	1028	237	335	1879	227	**1627**	28.59
p.a.%	67.88	43.59	58.50	68.15	54.48	75.51	91.61	**64947**

Note: (d) Total Accuracy: 57.31%. Kappa coefficient: 0.488

4 Support Vector Machines

Support vector machines (SVMs) represent a group of theoretically superior machine learning algorithms. The development of SVM was initially triggered by the exploration and formalization of learning machine capacity control and overfitting issues (Vapnik, 1979, 1995, 1998). Although SVMs can be said to have emerged in the late 1970s (Vapnik, 1979), they have not received significant attention until recent years. The attractiveness of SVMs is their ability to minimize the so-called structural risk, or classification errors, when solving the classification problem. Unlike conventional statistical maximum likelihood methods that minimize misclassification error in an empirical sense, which is directly determined by the distribution of training samples, the structural risk minimization concept adopted by SVMs is to minimize the probability of misclassifying a previously unseen data point drawn randomly from a fixed but unknown probability distribution (Vapnik, 1995, 1998). Such a property is also different from the decision boundary–forming logic of ANN, as addressed in Chapter 3. Specifically, SVM training always finds a global minimum. Their simple geometric interpretation, as shown later in this chapter, provides fertile ground for further investigation (Burges, 1998).

SVMs were originally linear binary classifiers, which allocate the labels +1 and −1. The core operation of SVMs is to construct a separating hyperplane (i.e., a decision boundary) on the basis of the properties of the training samples, specifically their distribution in feature space. Such a separating hyperplane is subject to the condition that the margin of separation between class +1 and class −1 samples is maximized (Vapnik, 1979). Figure 4.1 illustrates this idea in which hyperplane b separates the two classes with maximal margin, while hyperplanes a, c, and d do not fulfill the requirements of minimizing the structural risk. As the reader will see in the later mathematical derivations, not necessarily do all the training samples contribute to the building of the hyperplane, but normally only a subset of training samples are chosen as support vectors. This attribute is unique to SVMs. As shown in Figure 4.1, only the shaded points are support vectors that define the hyperplane b separating the two classes with maximal distance.

Recent applications and extensions of SVMs have been introduced into a variety of research domains, such as pattern recognition (Burges, 1998), handwriting identification (Burges and Schölkopf, 1997), object tracing (Blanz et al., 1996), document categorization (Joachims, 1999), clustering (Hur et al., 2001; Kriegel et al., 2004), and regression (Hong and Hwang, 2003). In the field of remote sensing, the implementation of SVMs is also gradually spreading and results show that the SVM generally presents an improved classification result compared with traditional classifiers such as maximum likelihood (Huang et al., 2002; Foody and Mathur, 2004, 2006; Pal and Mather, 2005; and Pal, 2006).

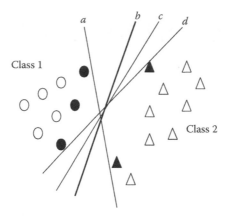

FIGURE 4.1 Hyperplane *b* separates the two classes with the maximal margin.

This chapter provides the level of knowledge of SVMs needed to allow readers to be able to intelligently use SVMs to perform remotely sensed imagery classification. We start from the introductory linear case, which is then followed by solving the nonlinear classification issue in terms of the well-known kernel function methods. The ways of dealing with multiclass case and parameters assignment are also taken into account. A comparative study using SVMs with different types of kernels for classifying remotely sensed data is presented in the last section of this chapter.

4.1 LINEAR CLASSIFICATION

4.1.1 THE SEPARABLE CASE

Assume there are two linearly separable classes. The training data set is represented by pairs $\{\mathbf{x}_i, y_i\}$, $i = 1, \ldots, n$, $y_i \in \{1, -1\}$, $\mathbf{x}_i \in \mathbf{R}^d$, where $\mathbf{x}_i$ are the observed multispectral features and y_i is the label of the information class for training case i. The label is either +1 or −1, representing class 1 and class 2. The aim of the support vector classifier is to build an optimal *hyperplane* that separates the two classes in such a way that the distance from the hyperplane to the closest training data points in each of the classes is as large as possible. This distance is called the *margin*. The hyperplane can be represented as the following decision function:

$$\mathbf{w}^T \mathbf{x} + b = 0 \tag{4.1}$$

where $\mathbf{x}$ is a point lying on the hyperplane, $\mathbf{w}$ is normal to the hyperplane, T denotes matrix transposition, and the parameter b denotes the *bias*. The perpendicular distance from the hyperplane to the origin is $|b|/\|\mathbf{w}\|$, where $\|\mathbf{w}\|$ is the Euclidean norm of $\mathbf{w}$. Figure 4.2 illustrates these concepts.

Suppose that all the training data satisfy the following constraints:

$$\mathbf{w}^T \mathbf{x}_i + b \geq +1, \text{ for } y_i = +1 \tag{4.2}$$

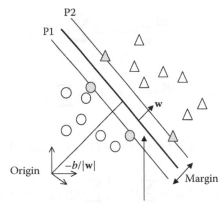

Optimal separating hyperplane: $\mathbf{w}^T \mathbf{x} + b = 0$

FIGURE 4.2 Linearly separable case.

$$\mathbf{w}^T \mathbf{x}_i + b \leq -1, \text{ for } y_i = -1 \tag{4.3}$$

These two equations can be further combined to give

$$y_i(\mathbf{w}^T \times \mathbf{x}_i + b) - 1 \geq 0. \tag{4.4}$$

One can then implicitly define a scale for $(\mathbf{w} \times b)$ to generate two canonical hyperplanes (i.e., $P1$ and $P2$ in Figure 4.2), namely, $\mathbf{w}^T \times \mathbf{x}_i + b = 1$ for the closest training points lying on one side of the hyperplane with normal $\mathbf{w}$ and perpendicular distance from the origin $|1-b|/\|\mathbf{w}\|$, and second, $\mathbf{w}^T \times \mathbf{x}_i + b = -1$ for the closest training points on the other side with perpendicular distance from the origin $|-1-b|/\|\mathbf{w}\|$. These training points are referred to as support vectors. In other words, those vectors are central to the establishment of the optimal separating hyperplane. Accordingly, the margin between these two hyperplanes is $2/\|\mathbf{w}\|$. The maximization of this margin in turn leads to

$$\min \left\{ \frac{\|\mathbf{w}\|^2}{2} \right\} \tag{4.5}$$

subject to the inequality constraints shown in Equation (4.4).

Equation (4.4) becomes easier to handle if a primal Lagrangian formulation is used:

$$L_{primal} = \frac{1}{2} \|\mathbf{w}\|^2 - \sum_{i=1}^{n} \alpha_i \left(y_i \left(\mathbf{w} \times \mathbf{x}_i + b \right) - 1 \right) + \sum_{i=1}^{n} \alpha_i \tag{4.6}$$

where α_i are positive Lagrangian multipliers. One should then minimize L_{primal} with respect to $\mathbf{w}$ and b. From Wolfe's theorem (Fletcher, 1987), one can take the derivative of L_{primal} with respect to b and $\mathbf{w}$ to obtain

$$\frac{\partial L_{primal}}{\partial b} = 0 \Rightarrow \sum_{i=1}^{n} \alpha_i y_i = 0 \tag{4.7}$$

$$\frac{\partial L_{primal}}{\partial \mathbf{w}} = 0 \Rightarrow \mathbf{w} = \sum_{i=1}^{n} \alpha_i y_i \mathbf{x}_i \tag{4.8}$$

and substitute to the primal formulation (4.6) to give the Wolfe dual Lagrangian:

$$L_{dual} = \sum_{i=1}^{n} \alpha_i - \frac{1}{2} \sum_{i=1}^{n} \alpha_i \alpha_j y_i y_j \mathbf{x}_i \times \mathbf{x}_j . \tag{4.9}$$

The training of the SVM now involves the maximization of Equation (4.9) with respect to $\alpha_i \geq 0$, subject to the constraint shown in Equation (4.7) with the solution given by (4.8). Notice that there is a Lagrange multiplier α_i for every training point. Those points with $\alpha_i > 0$ are support vectors, and lie on one of the parallel hyperplanes $P1$ or $P2$ (Figure 4.2). All other training points have $\alpha_i = 0$ and lie either on (or that side of) $P1$ or $P2$. Specifically, the solutions for $\mathbf{w}$ and b are formulated as

$$\mathbf{w} = \sum_{i=1}^{nsv} \alpha_i y_i \mathbf{x}_i \tag{4.10}$$

$$b = -\frac{1}{2} \mathbf{w} \times \left(\mathbf{x}_r + \mathbf{x}_s \right) \tag{4.11}$$

where nsv denotes the number of support vectors, $\mathbf{x}_r$ is the support vector belonging to class $y_r = 1$, and $y_s = -1$ for $\mathbf{x}_s$. Note that bias b may be computed using all the support vectors on the margin for stability concern. Accordingly, in the case of a two-class classification problem, the decision rule shown in Equation (4.1), which separates the two information classes, can be derived as

$$f(\mathbf{x}) = sign\left(\sum_{i=1}^{nsv} \alpha_i y_i \left(\mathbf{x} \times \mathbf{x}_i \right) + b \right) . \tag{4.12}$$

This formulation of the SVM optimization problem is called the *hard margin* formulation, since no training errors are allowed. All the training samples satisfy the inequality $y_i \times f(\mathbf{x}_i) \geq 1$. In certain cases, one may require the separating hyperplane to pass through the origin by choosing a fixed $b = 0$. This variant is called the hard margin SVM *without threshold*. In this case, the optimization problem remains the same as that in Equation 4.12 except that the constraint $\alpha_i \times y_i = 0$ in Equation (4.7) no longer exists.

4.1.2 THE NONSEPARABLE CASE

Information classes derived from remotely sensed data are not usually totally separated by linear boundaries. Thus, the constraints of Equation (4.4) cannot be satisfied in practice, and so the slack variables ξ_i, $i = 1, ..., n$, which are proportional to some measure of cost, are introduced to relax the constraints (Cortes and Vapnik, 1995). This is sometimes called the *soft margin* method. The reader is referred to Figure 4.3 for further illustration of this idea. When slack variables are included, Equation (4.4) becomes

$$y_i(\mathbf{w}^T \times \mathbf{x}_i + b) \geq 1 - \xi_i; \ \xi_i \geq 0, \forall i. \tag{4.13}$$

The optimal problem then resolves into

$$\min\left\{ \frac{\|\mathbf{w}\|^2}{2} + C\sum_{i=1}^{n}\xi_i \right\} \tag{4.14}$$

where C is a penalty parameter to be determined by the user. A larger C means "assign a higher penalty to errors." The first part of Equation (4.14) aims to maximize the margin, while the second part seeks to penalize the training samples located on the "wrong" side of the decision boundary. In other words, a large value of C corresponds to overfitting to the training data and thus reduces generalization capability. Belousov et al. (2002), however, show that the SVMs can show a high degree of robustness to variations in parameter values.

The primal Lagrangian formulation in Equation (4.6) then becomes

$$L_{primal} = \frac{1}{2}\|\mathbf{w}\|^2 - \sum_{i=1}^{n}\alpha_i\left(y_i\left(\mathbf{w}^T \times \mathbf{x}_i + b\right) - 1 + \xi_i\right) + C\sum_{i=1}^{n}\xi_i - \sum_{i=1}^{n}\mu_i\xi_i \tag{4.15}$$

where the μ_i are the Lagrange multipliers introduced to enforce positivity of the slack variable ξ_i. The solution to Equation (4.15) is determined by the saddle points of the

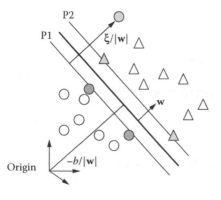

FIGURE 4.3 Linear hyperplanes for the partially separable case.

Lagrangian, by minimizing with respect to $\mathbf{w}$, ξ and b, and maximizing with respect to $C \geq \alpha_i \geq 0$ and $\mu_i \geq 0$. The Lagrangian multipliers α_i now have an upper bound of C. The upper bound $\alpha_i \leq C$ corresponds to an upper bound on the force that any given support vector is allowed to exert on the hyperplanes $P1$ or $P2$. Note that, in the nonseparable case, the Wolfe dual Lagrangian formulation remains the same as shown in Equation (4.9).

4.2 NONLINEAR CLASSIFICATION AND KERNEL FUNCTIONS

In some cases where a linear hyperplane is unable to separate the classes appropriately, SVMs then adapt a rather old strategy (Aizerman et al., 1964), which maps the raw input data into a higher dimensional space so as to improve the separability between classes. The SVMs then perform manipulation within this newly mapped feature space. This method is referred to as that of nonlinear SVMs.

4.2.1 NONLINEAR SVMS

For nonlinear decision surfaces, Boser et al. (1992) propose that a feature vector, $\mathbf{x}_i \in \mathbf{R}^d$, is mapped into a higher dimensional Euclidean space (or a generalization of Euclidean space called Hilbert space (Halmos, 1967; Kolmogorov and Formin, 1970), in order to spread the distribution of the training samples in a way that facilitates the fitting of a linear hyperplane. Specifically, the training samples are projected into a higher dimensional space $\mathcal{H}$, via a nonlinear vector mapping function called $\Phi: \mathbf{R}^d \to \mathcal{H}$. Recall Equation (4.12), which can then be further derived as

$$f(\mathbf{x}) = sign\left(\sum_{i=1}^{nsv} \alpha_i y_i \Phi(\mathbf{x}) \cdot \Phi(\mathbf{x}_i) + b\right). \tag{4.16}$$

The computational burden of $(\Phi(\mathbf{x}) \times \Phi(\mathbf{x}_i))$ can be quite expensive in a higher dimensional space. Vapnik (1995) proposes an alternative to reduce the computational loading via a positive definite kernel function denoted as $K(\mathbf{x},\mathbf{y})$, such that $K(\mathbf{x},\mathbf{y}) = \Phi(\mathbf{x}) \times \Phi(\mathbf{y})$. To illustrate such a relationship further, let $K(\mathbf{x},\mathbf{y}) = (\mathbf{x} \times \mathbf{y})^2$, where $\mathbf{x} = (x_1, x_2)$, and $\mathbf{y} = (y_1, y_2)$. One can then expand the kernel function as

$$\begin{aligned}
K(\mathbf{x},\mathbf{y}) &= \left(x_1 y_1 + x_2 y_2\right)^2 \\
&= x_1^2 y_1^2 + 2 x_1 y_1 x_2 y_2 + x_2^2 y_2^2 \\
&= \left(x_1^2, x_2^2, \sqrt{2} x_1 x_2\right) \times \left(y_1^2, y_2^2, \sqrt{2} y_1 y_2\right) \\
&= \Phi(\mathbf{x}) \times \Phi(\mathbf{y})
\end{aligned} \tag{4.17}$$

Clearly, according to Equation (4.17), by using the kernel function $K(\mathbf{x},\mathbf{y}) = (\mathbf{x} \times \mathbf{y})^2$ instead of

$$\Phi(x)\times\Phi(y)=\left(x_1^2,x_2^2,\sqrt{2}x_1x_2\right)\times\left(y_1^2,y_2^2,\sqrt{2}y_1y_2\right),$$

the representation of the data is greatly simplified and the computational burden is reduced. One would only need to use function K in the training algorithm, and there is probably no need to explicitly know what Φ actually is. More examples of kernel functions are discussed in Section 4.2.2. The optimization problem of Equation (4.9) can then be rewritten as

$$L_{dual}=\sum_{i=1}^{n}\alpha_i-\frac{1}{2}\sum_{i=1}^{n}\sum_{j=1}^{n}\alpha_i\alpha_jy_iy_jK\left(\mathbf{x}_i\times\mathbf{x}_j\right)\qquad(4.18)$$

and the decision rule expressed in Equation (4.16) now generalizes into

$$f\left(\mathbf{x}\right)=sign\left(\sum_{i=1}^{nsv}\alpha_iy_iK\left(\mathbf{x},\mathbf{x}_i\right)+b\right).\qquad(4.19)$$

In effect, the role of a kernel function consists of mapping the training samples into a higher dimensional space where the training samples may be spread farther apart. Figure 4.4 illustrates this idea in which data originally are linearly nonseparable in a two-dimensional space, but once mapped into higher dimension feature space $\Phi(\ )$, data may be more easily separated in terms of a hyperplane.

Note that kernel functions being used to fit SVMs should satisfy Mercer's condition (i.e., a kernel function $K(x,y)$ should meet the constraint

$$\iint K(x,y)g(x)g(y)d(x)d(y)\geq 0,$$

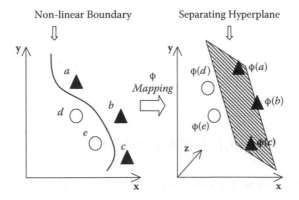

FIGURE 4.4 In the linearly nonseparable case, SVMs map the raw data into a higher dimension to increase the separability between classes.

where $g(x)$ is a square integrable function) so as to guarantee that a mapping pair $\{\Phi, \mathcal{H}\}$ exists, and consequently the training process converges (Vapnik, 1995). It is also noted that for the nonseparable case, the soft margin SVMs with quadratic penalization of errors can be regarded as a special case of the hard margin version with the modified kernel $K = K + \mathbf{I}/C$, where $\mathbf{I}$ denotes the identity matrix (Cortes and Vapnik, 1995).

4.2.2 KERNEL FUNCTIONS

The discussion in the previous paragraph demonstrates that the purpose of a kernel function is to enable operations to be performed in the current domain (i.e., $K(x,y)$ in Equation [4.19]) rather than the potentially high dimensional feature space (i.e., $\Phi(x) \times \Phi(y)$ in Equation [4.16]). This provides a smart means of resolving the computation issue caused by high dimensionality. In the following discussion, five common choices of kernel functions are introduced:

- Polynomial (homogeneous)

$$K\left(\mathbf{x}_i,\mathbf{x}_j\right) = \gamma\left(\mathbf{x}_i \times \mathbf{x}_j\right)^d \tag{4.20}$$

- Polynomial (inhomogeneous):

$$K\left(\mathbf{x}_i,\mathbf{x}_j\right) = \left(\gamma\left(\mathbf{x}_i \times \mathbf{x}_j\right) + \delta\right)^d, \ \gamma > 0, \ \delta > 0 \tag{4.21}$$

- Radial basis function:

$$K\left(\mathbf{x}_i,\mathbf{x}_j\right) = \exp\left(-\gamma\left\|\mathbf{x}_i - \mathbf{x}_j\right\|^2\right), \ \gamma > 0 \tag{4.22}$$

- Gaussian radial basis function:

$$K\left(\mathbf{x}_i,\mathbf{x}_j\right) = \exp\left(-\left\|\mathbf{x}_i - \mathbf{x}_j\right\|^2 / 2\sigma^2\right) \tag{4.23}$$

- Sigmoid:

$$K\left(\mathbf{x}_i,\mathbf{x}_j\right) = \tanh\left(\gamma\left(\mathbf{x}_i \times \mathbf{x}_j\right) - \delta\right), \ \gamma > 0, \ \delta > 0 \tag{4.24}$$

Huang et al. (2002) show that both polynomial Equations (4.20) and (4.21) are popular methods for nonlinear mapping with polynomial order $d > 1$, and there is an obvious trend of improved accuracy as d increases. It is also suggested that the value of d has different impacts on kernel performance when different numbers of input variables are used. When dealing with radial basis kernel functions, Schölkopf et al. (1997) show that a Gaussian radial basis function (Equation [4.23]) gives excellent results compared to classical radial basis functions (Equation [4.22]). The sigmoid kernel function, as shown in Equation (4.24), can be regarded as a particular kind of sigmoid neural network. The architecture (number of weights) of the neural network is determined by the number of support vectors used in SVM training. Here, the support vectors correspond to the first layer and the Lagrange multipliers to the weights. Note that the sigmoid kernel function may not satisfy Mercer's condition for certain values of γ and δ (Vapnik, 1995).

From the above descriptions it is clear that the performance of SVMs is closely related to the choice of the kernel function. Although certain studies have worked on choosing suitable kernels based on prior knowledge (Burges, 1998), the best choice of kernel for a given problem still requires further investigation. Once the type of kernel is determined, the user has also to carefully choose the kernel-relating parameters (such as d and γ in Equations [4.21] and [4.22]) and the error penalty C to ensure the most satisfactory classification results. Figure 4.5 addresses the effects of parameter selection. Figure 4.5a and 4.5b show the effect of polynomial degree on the shape of the decision boundary between two classes, namely white nodes and black nodes (note that the white line denotes the hyperplane). Figure 4.5c and 4.5d present the decision boundary formed by using linear kernel and Gaussian radial basis functions. It is clear that the position of the decision boundary varies with the degree of the polynomial and also with different kernels (Gunn, 1997).

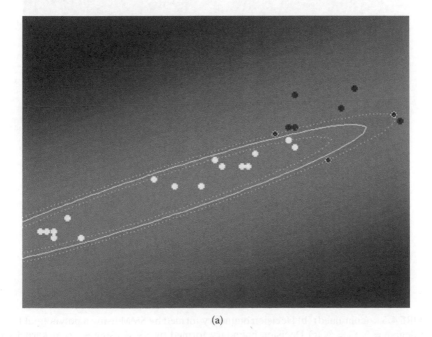

(a)

FIGURE 4.5 (a) Decision boundary formed by SVM using a polynomial kernel (Equation [4.21]) with degree = 2, $C = \infty$.

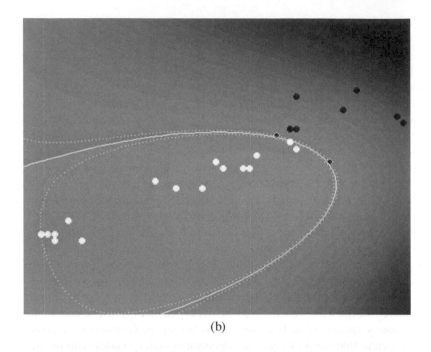

(b)

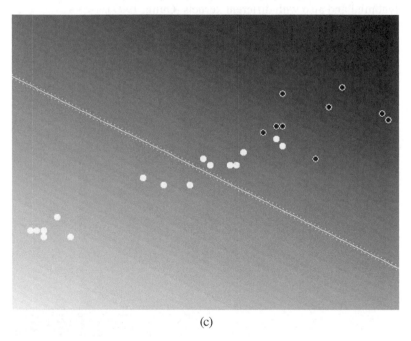

(c)

FIGURE 4.5   (continued) (b) Decision boundary formed by SVM using a polynomial kernel with degree = 5, C = ∞. (c) Decision boundary formed by SVM using a linear kernel with C = ∞.

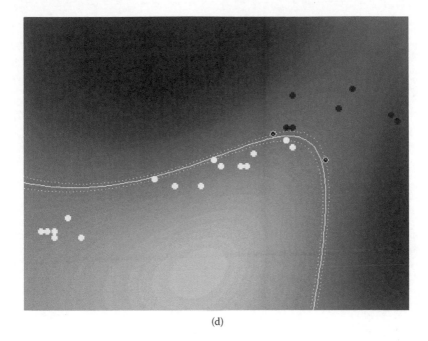

(d)

FIGURE 4.5 (continued) (d) Decision boundary formed by SVM using a Gaussian radial basis function kernel with $\sigma = 2$, $C = \infty$.

Further aspects of the relationship between the location and nature of the decision boundary and penalty value C are shown in Figure 4.6, which presents certain results in the case of polynomial kernel functions. It is shown that when the value C is small ($C = 0.1$, Figure 4.6a), the decision boundary is less clear-cut in contrast to the case when a large C (i.e., a heavier penalty) is used (Figure 4.6b and 4.6c). However, based on Figure 4.5 and 4.6, it is hard to determine which decision boundary pattern is optimal. Clearly, the answer is case-dependent. However, there is clear evidence that the choice of parameters does affect classification results. Ways of resolving the parameter determination issue are thus worth our further attention.

4.3 PARAMETER DETERMINATION

It is clear from the preceding discussion that the main limitation on the use of SVMs is the need to select suitable parameter values so as to improve the capability of the well-trained SVMs to accurately predict unknown data. To achieve the aim of selecting appropriate parameter values, one has first to ensure that the estimate of validation errors is unbiased. Then, using this unbiased error estimate, one can perform parameter adjustments that enhance the generalization capability of the classifier. This section illustrates ways of determining unbiased estimates of the errors, and this is followed by a consideration of two kinds of parameter adjustment methodology, known as the *grid search* and the *gradient descent technique*, respectively.

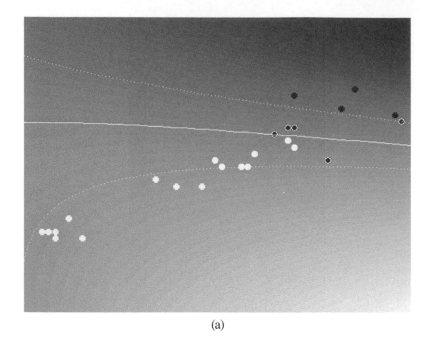

(a)

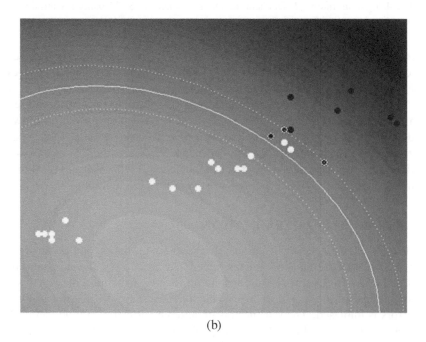

(b)

FIGURE 4.6 (a) Decision boundary formed by SVM using a polynomial kernel with degree = 2, C = 0.1. (b) Decision boundary formed by SVM using a polynomial kernel with degree = 2, C = 2. Note that dotted lines parallel to the decision boundary are the lines respectively formed by the support vectors of two classes.

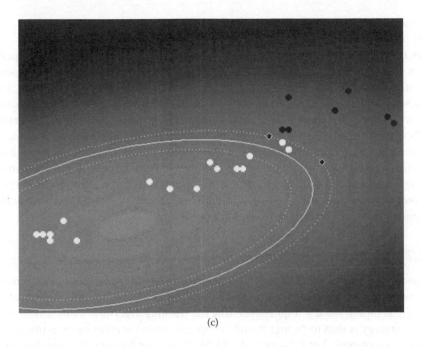

(c)

FIGURE 4.6   (continued) (c) Decision boundary formed by SVM using a polynomial kernel with degree = 2, C = 20.

4.3.1   *t*-FOLD CROSS-VALIDATIONS

As illustrated in previous chapters, the available ground samples are randomly allocated to one of two sets, one set being used for training purposes and the other treated as unknown test data for validation purposes. If the available samples are sufficient, the estimate of the validation error will be less biased. In the case of SVMs, the estimate of the error can be written as

$$P_{error}^{t} = \frac{1}{t}\sum_{i=1}^{t}\Gamma\left(-y_{i}f\left(\mathbf{x}_{i}\right)\right)$$

(4.25)

where t is the number of samples for validation, $f(\mathbf{x}_{i})$ is the decision function as addressed in Equation (4.19), and Γ is the step function, $\Gamma(\beta) = 1$ when $\beta > 0$ and $\Gamma(\beta) = 0$ otherwise. The parameters that achieve the highest validation accuracy would be the ideal candidates for specifying the final classifier.

The *overfitting* problem is considered in Section 3.1.2. The same problem can be encountered more seriously when the error estimation procedure outlined previously in this section is applied. Since the SVMs' decision boundaries are defined by the support vectors, such *one-trial* training and validation may not realistically reflect the actual distribution of the unclassified samples, especially in cases where the number of samples is quite limited. In other words, the resulting SVMs that achieve the highest training accuracy might not successfully generalize to the unknown samples. A technique called *t*-fold cross-validation is generally implemented to overcome this issue.

In the *t*-fold cross-validation approach, the training samples are separated equally into a user-defined number of subsets, *t*. During the training phase, the user chooses one of the *t* subsets for validation and the remaining $(t-1)$ subsets for training. This process is sequentially repeated until all the data sets have been tested. The classification errors computed from validation sets are then averaged over *t* trials. This method is also called *leave-one-out* cross-validation. The *t*-fold cross-validation accuracy is the percentage of validation samples that are correctly classified, and this measure provides a more objective and accurate estimate of classification accuracy than the *one-trial* method. According to the study performed by Luntz and Brailovsky (1969), the *leave-one-out* procedure can give an almost unbiased estimate of the expected generalization error regarding validation samples.

4.3.2 BOUND ON LEAVE-ONE-OUT ERROR

Although the use of *t*-fold *leave-one-out* cross-validation to estimate the generalization error is generally recommended, it may nevertheless become very costly to actually compute since it requires running the training algorithm *t* times. The alternative strategy is thus to "upper bound" or approximate the error by an estimate that is easier to compute. Let f^{all} denote the SVM trained for forming decision boundary using all the training samples, and f^i the decision boundary obtained when sample *i* is removed, one can derive the expected generalization error to the SVM as

$$E\left[P_{error}^{t-1}\right]=\frac{1}{t}\sum_{i=1}^{t}\Gamma\left(-y_i f^i\left(\mathbf{x}_i\right)\right) \tag{4.26}$$

where P_{error}^{t-1} denotes the actual error for an SVM trained on $t-1$ samples, $E\left[P_{error}^{t-1}\right]$ is the expectation of the actual error over all choices of training set of size $t-1$, and the expectation E is taken over a random choice of training samples. The right-hand side of Equation (4.26) can be further derived as

$$\frac{1}{t}\sum_{i=1}^{t}\Gamma\left(-y_i f^i\left(\mathbf{x}_i\right)\right)=\frac{1}{t}\sum_{i=1}^{t}\Gamma\left(-y_i f^{all}\left(\mathbf{x}_i\right)+y_i\left(f^{all}\left(\mathbf{x}_i\right)-f^i\left(\mathbf{x}_i\right)\right)\right). \tag{4.27}$$

Assume the term $y_i(f^{all}(\mathbf{x}_i)-f^i(\mathbf{x}_i))$ is bounded by U_i. One can then express the following upper bound on the *leave-one-out* error as

$$E\left[P_{error}^{t-1}\right]\le\frac{1}{t}\sum_{i=1}^{t}\Gamma\left(U_i-1\right). \tag{4.28}$$

Several studies develop estimates of the error bound in different ways as addressed in the following portion of this section.

Based on the theory of SVMs, it is known that removing a nonsupport vector from the training set does not change the solution calculated by the SVM (i.e., $U_i = 0$, for nonsupport vector $\mathbf{x}_i$), Vapnik (1995) accordingly derives the bound on the *leave-one-out* error of SVMs as

$$E\left[P_{error}^{t-1}\right] \leq \frac{E\left[\text{number of support vectors}\right]}{t}.$$

(4.29)

For SVMs without threshold b (cf. Equation [4.12]), Jaakkola and Haussler (1999) propose another bound on the *leave-one-out* error:

$$E\left[P_{error}^{t-1}\right] \leq \frac{1}{t}\sum_{i=1}^{t}\Gamma\left(\alpha_i K\left(\mathbf{x}_i,\mathbf{x}_j\right)-1\right)$$

(4.30)

where α_i is the Lagrangian multiplier for support vector i and $K(\mathbf{x}_i,\mathbf{x}_j)$ is the kernel function. Opper and Winter (2000) adopt the concept of linear response theory to propose the following generalization error bound under the assumption that the set of support vectors does not change when removing sample i:

$$E\left[P_{error}^{t-1}\right] \leq \frac{1}{t}\sum_{i=1}^{t}\Gamma\left(\frac{\alpha_i}{\left(\mathbf{K}_{SV}^{-1}\right)}-1\right)$$

(4.31)

where $\mathbf{K}_{SV}$ is the matrix of dot products between support vectors.

Another expression defining the bound on the leave-one-out error is based on the concept of sphere and radius. Let $\Phi(\mathbf{x}_1)$, $\Phi(\mathbf{x}_2)$, ..., $\Phi(\mathbf{x}_t)$ be the mappings of the training samples in feature space (note that there is a total of t sets, and where $\mathbf{x}_i$ denotes the training set with the ith set excluded), and lying within a sphere of radius R. Here, m is the margin between two hyperplanes. Vapnik (1998) suggests the following bound for the generalization error of the *leave-one-out* procedure:

$$E\left[P_{error}^{t-1}\right] \leq \frac{1}{t}\frac{R^2}{m^2}$$

(4.32)

where the radius R^2 can be derived in terms of maximizing the following Wolfe dual (Burges, 1998):

$$R^2 = \sum_{i=1}^{t}\beta_i K\left(\mathbf{x}_i \times \mathbf{x}_i\right)-\sum_{i=1}^{t}\sum_{j=1}^{t}\beta_i\beta_j K\left(\mathbf{x}_i \times \mathbf{x}_j\right)$$

(4.33)

subject to

$$\sum_{i=1}^{t} \beta_i = 1,$$

$\beta_i \geq 0$ for all i. Note that kernel function $K(\mathbf{x}_i, \mathbf{x}_j)$ is again used to replace $\Phi(\mathbf{x}_i) \times \Phi(\mathbf{x}_j)$.

Under the assumption that the set of support vectors remains the same during the *leave-one-out* procedure, Vapnik and Chapelle (2000) derive the following error estimate based on the concept of the *span* of support vectors:

$$E\left[P_{error}^{t-1}\right] \leq \frac{1}{t} \sum_{i=1}^{t} \Gamma\left(\alpha_i S_i^2 - 1\right) \tag{4.34}$$

where S_i is the Euclidean distance between the sample $\Phi(\mathbf{x}_j)$ and the set Ω_i, where Ω_i is defined as

$$\Omega_i = \left\{ \sum_{i \neq j, \alpha_j > 0} \lambda_j \Phi\left(\mathbf{x}_j\right) \right\} \tag{4.35}$$

subject to

$$\sum_{i \neq j} \lambda_j = 1 \text{ and } \forall i - j, \, \alpha_j + y_i y_j \alpha_i \lambda_j \geq 0. \tag{4.36}$$

Note that values of λ_i less than 0 are allowed. The user can then try to minimize the bound on the errors by using one of the error-bounded formulae, that is, Equations (4.29), (4.30), (4.31), (4.32), and (4.34), introduced previously (Pal, 2006).

4.3.3 GRID SEARCH

The question now remaining is how to seek the optimal way of choosing suitable parameter values. The parameters to be determined are the kernel function–related parameters, as shown in Equations (4.20) to (4.24), and the penalty term C in Equations (4.14) and (4.15), respectively. A method called *grid search* can be implemented to estimate parameter values (Chang and Lin, 2007). The concept of grid search is similar to that of the exhaustive search within a solution space. The user starts by randomly choosing the parameter values that are input to the classifier to evaluate the t-fold cross-validation performance. Then the parameter values are increased or decreased, and the performance is reevaluated until all the chosen parameters have been evaluated. If the radial basis kernel function (Equation [4.22]) is used to fit SVMs, one has to determine the possible parameter pairs of (γ, C) values and test the corresponding performances. The parameter pair (γ, C) that produces the best cross-validation accuracy is then selected.

Traditionally, a grid search over a bounded space (x, y) begins at one corner of the grid and is evaluated at every grid point separated by a value δ until the opposite corner of the grid is reached. The parameter values that generate the highest classification accuracy are then selected. To conduct a complete grid search using such a step-by-step method may not be practical as the computational demands are considerable. One alternative is initially to evaluate the parameter values over a coarse grid (i.e., bigger δ), and then focus the search on a finer grid (smaller δ) consisting of a small part of the coarse grid. For instance, according to the recommendations made by Chang and Lin (2007), one may choose exponentially growing sequences of pairs (γ, C), $\gamma = 2^{-20}, 2^{-18}, ..., 2^6$, and $C = 2^{-10}, 2^{-8}, ..., 2^{10}$ (i.e., $\delta = 2^2$), in order to roughly identify the best parameter pair. If the pair $(2^{-2}, 2^4)$ is identified to generate the better cross-validation outcome, a finer search with (γ, C), $\gamma = 2^{-0.8}, 2^{-1.0} ... 2^{3.2}$, and $C = 2^{2.8}, 2^{3.0}, ..., 2^{5.2}$, is then used to trace out the suitable values. Figure 4.7 demonstrates the result of grid search using a radial basis function kernel (Equation 4.22) accompanied by a cross-validation technique. The algorithm is part of the SVM package LIBSVM (Chang and Lin, 2007) and the dataset being tested contains 6 classes and 300 randomly selected subsamples of Landsat MSS four-band imagery generated by the publicly accessible StatLog project (Michie et al., 1994) and preprocessed by Hsu and Lin (2002). A more detailed description of the data set is given later in Section 4.6. The grid search algorithm converges quickly (within a minute), and the best classification accuracy is found to be around 89% with both parameters C (penalty value) and γ locating within the log domain of [1, −1] (Figure 4.7). The performances shown vary by around 3% in accuracy. However, as the later experiments show, the parameters can considerably affect the SVM's classification performance.

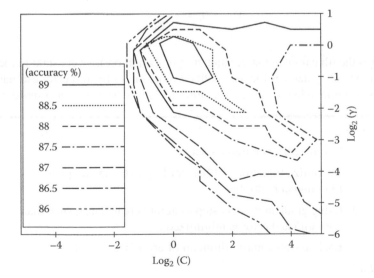

FIGURE 4.7 Results of a parameter determination for both C and γ using the grid search technique.

4.3.4 Gradient Descent Method

If the number of parameters exceeds two, the grid search method becomes unfeasible. It is recognized that as the number of parameters grows sequentially, so the solution space will expand exponentially. It is because of the difficulty in efficiently estimating the SVM parameters that there are only limited types of kernels currently available. In the following paragraphs, a technique based on the concept of gradient descent for dealing with large numbers of parameters is introduced (Chapelle et al., 2002). Once the parameters selection issue is resolved, it is possible that new and possibly more complex kernels may be introduced for classification purposes.

The gradient descent approach searches parameter space by starting at a random point defined by random values of the parameters and proceeding by changing the parameter values in accordance with the slope (derivative) and the direction of slope of the error surface. Position on the error surface is defined by parameter values, and the value of the error surface at a given position is the classification error obtained with those parameters. For illustrative purposes, let θ denote the set involving both the SVM kernel parameters and the penalty term C. The elements of θ can be best estimated by the iterative procedure shown in Figure 4.8 (Chapelle et al., 2002). Note that if one determines to use the bound on the *leave-one-out* error $E[P_{error}^{t-1}]$, shown in Equations (4.30), (4.31) or (4.34), as the error estimator in Figure 4.8 step 3, the step function Γ may cause a problem because Γ is not differentiable. To overcome this problem, Chapelle et al. (2002) recommend the use of a sigmoid function instead. The reader is referred to Section 3.1.1 in Chapter 3 for the illustration of the characteristic of sigmoid function.

To perform the gradient descent procedure in Figure 4.8 step 3, parameter θ_i can be updated in terms of

$$\theta_i^{n+1} = \theta_i^n - \eta \frac{\partial P_{error}}{\partial \theta_i^n} \tag{4.37}$$

where θ_i^n is the ith parameter at iteration n, η is a small positive constant in the range $0 \le \eta \le 1$, and P_{error} denotes the user-defined function for the test error estimate. The reader is recommended to review Equation (3.5) in Chapter 3 for a more detailed

---

Begin

1. Initialize SVM parametr set θ;

2. Maximize Equation (4.19) according to θ and compute Lagrangian multiplier α;

3. Using gradient descent step to iteratively update q such that the estimated error is minimized.

4. Back to step 2 until minimum of error is achieved.

End

FIGURE 4.8   Gradient descent algorithm for SVM parameter estimates.

description of the gradient approach. In what follows, according to Chapelle et al. (2002), the different parameter derivative methods regarding various error estimate function P_{error} are addressed.

If Equation (4.25) is selected as the error estimate function P_{error}, one has to compute the derivative of function $f(\mathbf{x}_i)$ (Equation [4.19]), which in turn determines the derivative of α and b with respect to parameter θ_i. Under the assumption that the samples of nonsupport vectors are removed from the training set, all the remaining samples must lie on the margin and this situation can be expressed as

$$\underbrace{\begin{pmatrix} K^Y & Y \\ Y^T & 0 \end{pmatrix}}_{H}\begin{pmatrix} \alpha \\ b \end{pmatrix} = \begin{pmatrix} 1 \\ 0 \end{pmatrix} \Rightarrow (\alpha, b)^T = H^{-1}(1, 1, ..., 1, 0)^T \qquad (4.38)$$

where $K_{ij}^Y = y_i y_j K(\mathbf{x}_i, \mathbf{x}_j)$. The derivatives of those parameters with respect to θ_i can be written as

$$\frac{\partial(\alpha, b)}{\partial \theta_i} = -H^{-1}\frac{\partial H}{\partial \theta_i}H^{-1}(1, 1, ..., 0)^T = -H^{-1}\frac{\partial H}{\partial \theta_i}(\alpha, b)^T. \qquad (4.39)$$

Equation 4.39 can also be used for resolving the derivative of the *leave-one-out* error bound proposed by Jaakkola and Haussler (1999) (Equation [4.30]) and Opper and Winter (2000) (Equation [4.31]) with $b = 0$.

If one uses the bound R^2/m^2 (Vapnik, 1998) in Equation (4.32) for the error estimate, one has to compute the derivative of square radius R^2 and square margin m^2. According to Equation (4.33), the derivative of the radius with respect to θ_i can be written as

$$\frac{\partial R^2}{\partial \theta_i} = \sum_{i=1}^{n}\beta_i^0\frac{\partial K(\mathbf{x}_i \times \mathbf{x}_i)}{\partial \theta_i} - \sum_{i=1}^{n}\sum_{j=1}^{n}\beta_i\beta_j\frac{K(\mathbf{x}_i \times \mathbf{x}_j)}{\partial \theta_i} \qquad (4.40)$$

where β_i^0 maximizes the Wolfe dual (Equation [4.33]). Margin m is defined by

$$m = \frac{1}{\|\mathbf{w}\|} \Rightarrow \frac{R^2}{m^2} = R^2\|\mathbf{w}\|^2. \qquad (4.41)$$

The term $\mathbf{w}$ is the surface normal defined in Equation (4.10). Vapnik (1998) shows that $\frac{1}{2}\|\mathbf{w}\|^2$ is equal to the optimization of Equation (4.19) and so the following formulation holds

$$\frac{1}{2}\|\mathbf{w}\|^2 = \sum_{i=1}^{n}\alpha_i - \frac{1}{2}\sum_{i=1}^{n}\sum_{j=1}^{n}\alpha_i\alpha_j y_i y_j K(\mathbf{x}_i \times \mathbf{x}_j). \qquad (4.42)$$

According to Equation (4.41), the derivative of the square margin m^2 with respect to θ_i then leads to

$$\frac{\partial \left\| \mathbf{w} \right\|^2}{\partial \theta_i} = -\sum_{i=1}^{n} \sum_{j=1}^{n} \alpha_i \alpha_j y_i y_j \frac{\partial K\left(\mathbf{x}_i \times \mathbf{x}_j\right)}{\partial \theta_i}. \tag{4.43}$$

Finally, if the error estimate is based on the concept of the *span* of support vectors (Equation [4.34]), the derivative of the square span S_i^2 can be expressed as (Chapelle et al., 2002)

$$\frac{\partial S_i^2}{\partial \theta_i} = S_i^4 \left(\tilde{\mathbf{K}}_{SV}^{-1} \frac{\partial \tilde{\mathbf{K}}_{SV}}{\theta_i} \tilde{\mathbf{K}}_{SV}^{-1} \right) \tag{4.44}$$

where $\tilde{\mathbf{K}}_{SV}$ is the extended matrix of the dot products between the support vectors expressed as

$$\tilde{\mathbf{K}}_{SV} = \begin{pmatrix} \mathbf{K}_{SV} & 1 \\ 1^T & 0 \end{pmatrix} \tag{4.45}$$

where T denotes the transpose of a matrix.

The derivative of each error bound is illustrated above. The reader can then choose the error function of interest and use Equation (4.37) and the algorithm shown in Figure 4.8 to sequentially tune the SVM's parameters so as to achieve improved classification accuracy.

4.4   MULTICLASS CLASSIFICATION

SVMs were originally designed for binary classification, i.e., one SVM can only separate two classes, yet most remote sensing applications require dealing with multiple classes. Several approaches, including one-against-one, one-against-others, directed acyclic graph (DAG) strategies, and multiclass SVM have been proposed for applying SVMs to multiclass classifications. These methods are described in the following pages.

4.4.1   ONE-AGAINST-ONE, ONE-AGAINST-OTHERS, AND DAG

The one-against-one classification strategy was proposed by Knerr et al. (1990) who used pair-wise comparisons between all k classes. Thus, each classifier is trained exclusively on two of the k classes. All possible one-against-one class combinations are evaluated from the training set of k classes. A total of $k(k-1)/2$ classifiers will be needed to perform the task. The decision rule for labeling a pixel is in terms of majority voting. The pixel will be given the label of the class having the most votes. In case of a tie, a tie-breaking strategy may be adopted. One may select at random one of the classes that are tied, or refer to the nearest neighbor pixel's class for breaking the tie.

The one-against-others strategy is to build a total of k SVMs for the k classes. Each of the SVMs is trained to classify one class against all the other classes. Pixel **x** will then be given the label of the class with the highest value of $\mathbf{w}_i \Phi(\mathbf{x}) + b_i$, $I = 1$ to k. (Vapnik, 1995).

Platt et al. (2000) propose the use of directed acyclic graph (DAG) SVMs for multiclass classification. The training phase of a DAG-SVM is the same as the one-against-one method, i.e., by building $k(k-1)/2$ classifiers, where k is the number of classes. These $k(k-1)/2$ classifiers then form the internal nodes of a DAG with each node being a binary SVM using the ith and jth classes. Starting from the root of the DAG, a classified test sample is compared at the node and then, depending on the output value of the node, directed to either the left or right node at the next layer, until a leaf is achieved. Note that the choice of the class order to form a DAG-SVM can be arbitrary without significantly affecting classification accuracy. One advantage of using DAG-SVMs is the substantial improvements in both training and evaluation time in comparison with the one-against-one and one-against-others algorithms (Platt et al., 2000). Figure 4.9 illustrates a four-class DAG-SVM for classifying roads, trees, bare soil, and vegetation; in total, six binary classifiers are built. It can be seen that at each level, the DAG-SVM expels one of the classes and then heads to the node containing the *survival* class pair, until a final decision is achieved.

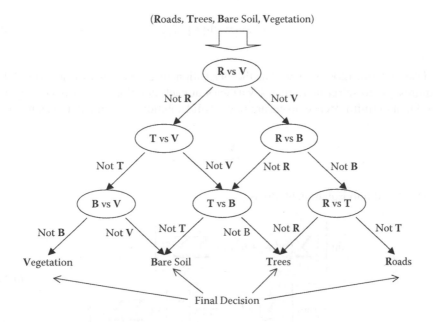

FIGURE 4.9 A DAG-SVM for classifying roads, trees, bare soil, and vegetation.

4.4.2 MULTICLASS SVMs

4.4.2.1 Vapnik's Approach

Vapnik (1998) developed an approach to SVM multiclass classification problems. Such a multiclass SVM is similar to the one-against-others approach by using a single optimization approach. Suppose that we have a total of k classes. The multiclass SVM constructs k two-class rules in which the ith hyperplane $\mathbf{w}_i^T \Phi(\mathbf{x}) + b$ separates the training samples of class i from other classes. Therefore, although there are k decision functions or hyperplanes, the solution may be obtained by solving one problem. The formulation of a multiclass SVM is expressed as

$$\min\left\{\frac{1}{2}\sum_{i=1}^{k}\mathbf{w}_i^T\mathbf{w}_i + \sum_{j=1}^{n}\sum_{i\neq y_j}\xi_j^i\right\} \tag{4.46}$$

under the constraints that

$$\mathbf{w}_{y_j}^T\Phi(\mathbf{x}_j) + b_{y_j} \geq \mathbf{w}_i^T\Phi(\mathbf{x}_j) + b_i + 2 - \xi_j^i$$

$$\xi_j^i \geq 0, j = 1, \ldots, n;\ i \in \{1, \ldots, k\}/y_j \tag{4.47}$$

where n is the number of training samples. The decision function is then

$$\arg\max_{i=1\,\text{to}\,k}\left(\mathbf{w}_i^T\Phi(\mathbf{x}) + b_i\right). \tag{4.48}$$

Hsu and Lin (2002) show that the derivation of a solution to Equation (4.48) requires one to select more than k variables at each iteration, which is not an easy condition to fulfill. Hence, to ensure that a valid solution is obtained, the element

$$\sum_{i=1}^{k} b_i^2$$

is introduced into Equation (4.46) as

$$\min\left\{\frac{1}{2}\sum_{i=1}^{k}\left(\mathbf{w}_i^T\mathbf{w}_i + b_i^2\right) + C\sum_{j=1}^{n}\sum_{i\neq y_j}\xi_j^i\right\}$$

$$= \min\left\{\frac{1}{2}\sum_{i=1}^{k}\begin{bmatrix}\mathbf{w}_i^T & b_i\end{bmatrix}\times\begin{bmatrix}\mathbf{w}_i \\ b_i\end{bmatrix} + C\sum_{j=1}^{n}\sum_{i\neq y_j}\xi_j^i\right\} \tag{4.49}$$

under the constraints

$$\left[\mathbf{w}_{y_j}^T \quad b_{y_j} \right] \times \left[\begin{array}{c} \Phi(\mathbf{x}_j) \\ 1 \end{array} \right] \geq \left[\mathbf{w}_i^T \quad b_i \right] \times \left[\begin{array}{c} \Phi(\mathbf{x}_j) \\ 1 \end{array} \right] + 2 - \xi_j^i$$

$$\xi_j^i \geq 0, j = 1, \ldots, n; \; i \in \{1, \ldots, k\}/y_j. \tag{4.50}$$

Equation (4.49) can be transferred into the following formulation (Hsu and Lin, 2002):

$$\min \left\{ \sum_{r=1}^{n} \sum_{s=1}^{n} \left(\frac{1}{2} c_s^{y_r} A_r A_s - \sum_{i=1}^{k} \alpha_r^i \alpha_s^{y_r} + \frac{1}{2} \sum_{i=1}^{k} \alpha_r^i \alpha_s^i \right) (K_{r,s}+1) - 2 \sum_{r=1}^{n} \sum_{i=1}^{k} \alpha_r^i \right\} \tag{4.51}$$

with $0 \leq \alpha_r^i \leq C, \alpha_r^{y_r} = 0$, where

$$A_r = \sum_{i=1}^{k} \alpha_r^i, \quad c_s^{y_r} = \begin{cases} 1 & \text{if } y_r = y_s \\ 0 & \text{if } y_r \neq y_s \end{cases}. \tag{4.52}$$

The decision function then becomes

$$f(\mathbf{x}) = \arg\max_{i=1,\ldots,k} \left(\sum_{r=1}^{n} \left(c_r^i A_r - \alpha_r^i \right) \times \left(K(\mathbf{x}_r, \mathbf{x}) + 1 \right) \right). \tag{4.53}$$

4.4.2.2 Methodology of Crammer and Singer

Crammer and Singer (2002) propose an approach to the solution of SVMs multiclass classification issues based on the following optimization function:

$$\min \left\{ \frac{1}{2} \sum_{i=1}^{k} \mathbf{x}_i^T \mathbf{x}_i + C \sum_{j=1}^{n} \xi_j \right\} \tag{4.54}$$

under the constraints that

$$\mathbf{w}_{y_j}^T \Phi(\mathbf{x}_j) - \mathbf{w}_i^T \Phi(\mathbf{x}_j) \geq e_j^i - \xi_j \quad j = 1, \ldots, n \tag{4.55}$$

where $e_j^i = 1 - \delta_{y_j,i}$, $\xi_j \geq 0$, and $\delta_{y_j,i}$ is defined as

$$\delta_{y_j,i} = \begin{cases} 1 & \text{if } y_j = i \\ 0 & \text{if } y_j \neq i \end{cases}. \tag{4.56}$$

The decision function is

$$\underset{i=1,\,\dots,\,k}{\arg\max} = \mathbf{w}_i^T \Phi(\mathbf{x}) \tag{4.57}$$

The dual formulation of Equation (4.54) is

$$\min f(\alpha) = \frac{1}{2} \sum_{r=1}^{n} \sum_{s=1}^{n} K_{r,s} \bar{\alpha}_r^T \bar{\alpha}_s + \sum_{r=1}^{n} \bar{\alpha}_r^T \bar{e}_i \tag{4.58}$$

under the constraints that

$$\sum_{i=1}^{k} \alpha_r^i = 0, r = 1, \dots, n, \ \alpha_r^i \leq C, \text{ if } y_i \neq m, \text{ and } \alpha_r^i \leq 0, \text{ otherwise} \tag{4.59}$$

where

$$\bar{\alpha}_r = \left[\alpha_r^1, \dots, \alpha_r^k \right]^T, \text{ and } \bar{e}_r = \left[e_r^1, \dots, e_r^k \right]^T. \tag{4.60}$$

Then one obtains

$$\mathbf{w}_i = \sum_{r=1}^{n} \alpha_r^i \Phi(\mathbf{x}_r) \tag{4.61}$$

and the decision function in Equation (4.57) can be written as

$$\underset{i=1,\,\dots,\,k}{\arg\max} \sum_{r=1}^{n} \alpha_r^i K(\mathbf{x}_r, \mathbf{x}). \tag{4.62}$$

The multiclass solution proposed by Crammer and Singer (2002) is much simpler than that of Vapnik (1998) (Equation 4.53). Even so, according to the comparative study presented by Hsu and Lin (2002), both multiclass SVMs (Crammer and Singer, 2002; Vapnik, 1998) suffer the drawback of slow convergence, and classification accuracy is less ideal than that generated by the one-against-one and DAG-SVM approaches.

4.5 FEATURE SELECTION

Hyperspectral imagery is becoming more widely used in remote sensing classification (Mather, 2004). As noted in Chapter 1, the main difficulty in dealing with hyperspectral data is the large number of spectral bands. There is an obvious link between the number of input features and computational burden, yet the classification accuracy may not show any significant improvement and, indeed, may reduce as the number of input bands increases (Guyon et al., 2002). It therefore becomes attractive to select the *optimal* subset of features from all those available in order to achieve higher accuracy and reduce computational cost. SVMs can be used to solve the feature selection problem, as described below (Chapelle et al., 2002).

One way to use SVMs for feature selection is to adopt the concept of weight elimination as in the case of artificial neural networks (ANN). According to the ANN learning theory introduced in Chapter 3, within an ANN, the weights close to zero can be eliminated without significantly affecting the classification performance (Kavzoglu and Mather, 2002). These small weights, even though multiplied by the activities at the preceding layer, may contribute nearly nothing to the node in the next layer. A similar idea can be applied using the scaling parameters of the SVM's kernel functions (Equation [4.20] to [4.24]). If one of the input features does not significantly contribute to the classification problem, the corresponding scaling parameters will be close to zero. Thus, the classification performance is not likely to be affected by removing such insignificant features. Based on such an idea, Chapelle et al. (2002) propose an approach in combination with the gradient descent algorithm (Section 4.3.4) for feature selection by eliminating those features whose scaling parameters are the smallest. The algorithm is shown in Figure 4.10. Note that both the error estimate functions and the corresponding gradient descent procedures are discussed in Section 4.3. The algorithm terminates when the error estimate cannot be further reduced by the elimination of further spectral bands or features. Rather than base feature selection on the raw features, the algorithm can also be applied to the features transformed by principal component analysis (Section 2.1.2). In this way the feature space can be condensed further.

Guyon et al. (2002) propose another idea for feature selection by recursively minimizing the cost function $\frac{1}{2}\|\mathbf{w}\|^2$ and eventually finding a subset of features that generates the largest margin between classes. The algorithm is shown in Figure 4.11, and is an instance of backward feature elimination (Kohavi and John, 1997). Note that the features that are eliminated last are not necessarily the ones that are individually most closely controlling the classification accuracy. However, they are likely to be the best performing subset of features. Also, in order to reduce the computational burden, the algorithm may simultaneously eliminate several features at each iteration. However, this may be at the expense of classification performance degradation.

The feature selection algorithms described above are for solving two-class classification problems. For the multiclass classification problem, the algorithms can be naturally extended in combination with one-against-one or one-against-others strategies to perform the selection for suitable features.

Begin

 1. Choose an error etimate function P_{error};

 2. Randomly initialise the SVM parameters;

 3. Optimise SVM formulation (Equation (4.19)) to obtain α;

 4. Minimize the error estimate P_{error} with respect to kernel scaling parameters in terms of gradient descent method;

 5. If a local minimum of P_{error} is not reached, return to step 3;

 6. Eliminate spectral bands of small scaling parameters and return to step 3.

End

FIGURE 4.10 Algorithm for feature selection by Chapelle, O., V. Vapnik, O. Bousquet, and S. Mukherjee. 2002. "Choosing Multiple Parameters for Support Vector Machines." *Machine Learning* 46 (2002):131–159.

Begin

 1. Define subset of surviving spectral bands as $s = \{1, \ldots, n\}$;

 2. Optimise SVM formulation (Equation (4.19)) to obtain α;

 3. Calculate $d\mathbf{w}_r = \left| \|\mathbf{w}_s\|^2 - \|\mathbf{w}_{s-r}\|^2 \right|$, for all r, according to Equation (4.42), where $\mathbf{w}_{s-r}$ denotes the spectral band r being removed from s;

 4. Find band t, where $t = \arg \min\{d\mathbf{w}_r, \forall r\}$, then put band t to a ranking list, and eliminate t from s, return to step 1 until $s = \{\}$;

 5. Select a subset from the ranking list to perform classification.

End

FIGURE 4.11 Algorithm for feature selection by Guyon, I., J. Weston, S. Barnhill, and V. Vapnik. "Gene Selection for Cancer Classification Using Support Vector Machines." *Machine Learning* 46 (2002):389–422. See text for discussion.

4.6 SVM CLASSIFICATION OF REMOTELY SENSED DATA

Recent studies have shown that using SVMs to deal with classification issues may result in higher accuracy than other classifiers and also require fewer training samples. For instance, Huang et al. (2002) analyze four kinds of classifiers, including SVMs, maximum likelihood (ML), ANN, and the decision tree classifier, and conclude that the SVMs generate more stable overall accuracies. Pal and Mather (2005) compare SVMs with ML and ANN methods. The results show that the SVMs achieve a higher level of classification accuracy than either the ML or the ANN classifier, and that the SVM can be used with small training data sets and high-dimensional data. Foody and Mathur (2004, 2006) investigate both the characteristics and the size of training samples in SVMs, and illustrate the training data acquisition strategies (such as targeting on decision boundary or spectral mixed pixels) to allow efficient and accurate image classification by small training samples. The studies made by Keuchel et al. (2003) and Liu et al. (2006) also show that SVMs are useful in classifying remotely sensed imagery.

For demonstration purposes, a set of four-band Landsat MSS data (Michie et al., 1994), which is public accessible and frequently used for SVM validations, is implemented for testing the performances of different kernel functions and the effects of relating parameters. The data set contains 6435 samples, of which 4435 are used as training samples and 2000 samples are used for testing purposes. Each input sample to the SVM consists of 36 attributes indicating a pixel centered within a 3×3 window to be classified (where 36 attributes are the four Landsat MSS bands measured at each of the nine pixels within a 3×3 neighborhood centered on the input sample position). The class names and the corresponding number of training and test samples are shown in Table 4.1. Evaluation of SVM performance is based on the measurement of overall accuracy.

The SVM package used here is LIBSVM, developed by Chang and Lin (2007) (http://www.csie.ntu.edu.tw/~cjlin/libsvm/index.html). LIBSVM is an easy-to-use freeware module for the MATLAB environment. In preparing the inputs for LIBSVM, one can use the *read-sparse* command to turn raw data into a LIBSVM-recognized format. For practical usage, LIBSVM mainly provides two simple commands: *Svm-train* is used to train an SVM, and *Svm-predict* is used for class predictions. Both commands require two files as input. One contains pixel attributes (an $m \times n$ matrix with m samples and n attributes), and the other is the pixel class (an $m \times 1$ matrix), in such a way that both training error and testing error can be automatically computed and displayed by LIBSVM. The user may also use the relevant parameter selection facilities provided by LIBSVM to test various SVM kernels and corresponding parameters. For further description see Chang and Lin (2007).

Figure 4.12 shows the results of classifications using polynomial (Equation [4.21], $\delta = 1$, $d = 3$), radial basis function (RBF; Equation [4.22]), and sigmoid (Equation

TABLE 4.1
Landsat MSS Samples for SVM Training and Testing

Class Name	No. of Training Samples	No. of Test Samples
Red soil	1072	461
Cotton crop	479	224
Gray soil	961	397
Damp gray soil	415	211
Soil with vegetation stubble	470	237
Very damp gray soil	1038	470

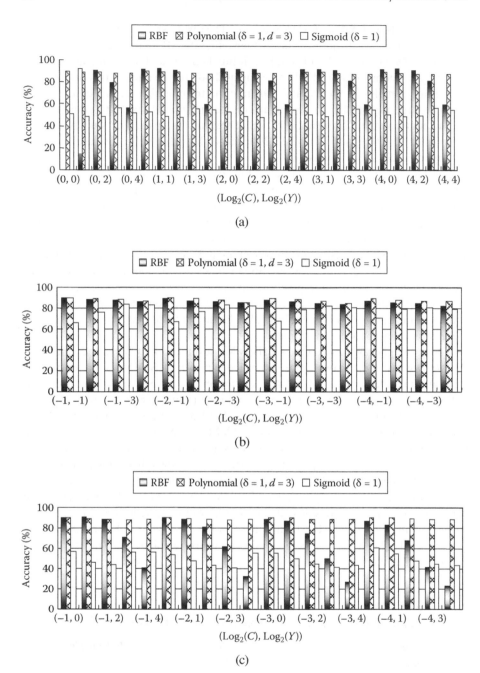

FIGURE 4.12 (a) Classification results based on RBF, polynomial, and sigmoid kernels, respectively, with Log(C) and Log(γ) range from 0 to 4. (b) Classification results based on RBF, polynomial, and sigmoid kernels, respectively, with Log(C) and Log(γ) ranging from −1 to −4. (c) Classification results based on RBF, polynomial, and sigmoid kernels, respectively, with Log2(C) ranging from −1 to −4, and Log2 (γ) from 0 to 4.

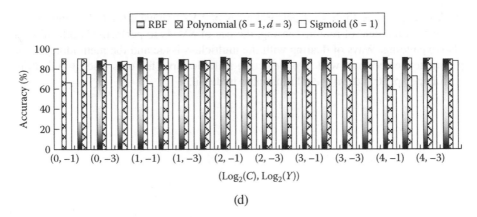

FIGURE 4.12   (continued) (d) Classification results based on RBF, polynomial, and sigmoid kernels, respectively, with Log2(C) ranging from 0 to 4, and Log2 (γ) from -1 to -4.

[4.24], $\delta = 1$) kernels, respectively. The penalty C and γ parameters being tested in the log 2 domain are arbitrarily chosen within the range of [-4 to 4] stepped by 1, and therefore a total of 81 trials are made for each kernel as shown in the figure. It is found that the performances obtained by SVMs with RBF kernel and polynomial kernel are generally comparable. The sigmoid kernel within the denoted parameter range, however, is found not to be ideal in the classification experiments. According to the experiments shown, the highest classification accuracy achieved by sigmoid function is 87.85% ($C = 16$, $\gamma = 0.0625$). For the polynomial kernel, the highest accuracy is 90.2% ($C = 4$, $\gamma = 0.25$), while by using a radial basis function kernel, the highest accuracy of 91.95%($C = 16$, $\gamma = 2$) was achieved. The worst classification performance using these kernel functions can be as low as 23.05% for the radial basis function kernel ($C = 0.0625$, $\gamma = 16$), and 40.7% for the sigmoid kernel ($C = 0.25$, $\gamma = 8$), while for the polynomial kernel, 84.75% accuracy ($C = 0.125$, $\gamma = 0.0625$) is the worst performance in this experimental case. Note that in this experiment, the degree parameter d in polynomial function and offset parameter δ for both polynomial and sigmoid functions are fixed. Adjustment of those parameters might lead to different results. However, from Figure 4.12 it is obvious that the accuracy varies considerably as kernel-related parameters change, and there is not a clear trend of classification accuracy increasing or decreasing as the kernel-relating parameters vary. This case once again demonstrates the importance of parameter assignment when using the SVM classifier.

4.7 CONCLUDING REMARKS

This chapter contains the knowledge necessary for readers to implement SVMs as classifiers in a robust way. It is clear that the major disadvantage of the SVM is that like the ANN classifiers, the user-defined parameters have a strong effect on its performance. This may restrict the capabilities of SVM to resolve more complex classification problems. However, based on the experiments and studies mentioned in this chapter, the high-level performances of SVMs have attracted attention and interest. In the field of remote sensing, the use of SVMs to perform classification is also expanding. Ways of dealing with the multiclass issue and the methodologies of parameter selection as well as of selecting suitable input features (useful in dealing with hyperspectral data) are all illustrated in detail.

5 Methods Based on Fuzzy Set Theory

Everyday language and decision making are not generally deterministic but are usually characterized by some level of fuzziness or uncertainty. Concepts such as *hot*, *cold*, *good*, or *difficult* contain elements of subjectivity, which is another way of saying that they cannot be completely (or deterministically) specified. One person's *hot* may well overlap with another person's *warm*, for example.

The same problem can also occur in the classification of remotely sensed imagery. A considerable number of identification errors are due to pixels that show an affinity with several information classes. This type of pixel is often described as *mixed*, and it may be more realistic to consider an approach that acknowledges this problem, although it is not clear whether the term *classification* applies to these methods (Mather, 2004). Traditional classification methods (e.g., the k-means clustering algorithm or the parallelepiped classifier, described in Chapter 2) do not provide a good mechanism for coping with uncertainty. For example, in the case of the k-means clustering algorithm, the formation of each cluster is in terms of competitive logic, so that once a pixel has been assigned to one cluster, its effect on other clusters is nil. For those pixels located in an interclass overlapping area of feature space, there will be a high probability that one cluster may incorrectly include some pixels properly belonging to some other cluster. These outliers will shift the mean of the cluster and result in clustering bias.

Fuzzy set theory (Zadeh, 1965), which was triggered by these considerations, provides a conceptual framework for solving knowledge representation and classification problems in an ambiguous environment. The fuzzy concept has been adopted in different fields such as fuzzy logic control (Wang et al., 2005; Lam and Leung, 2007), fuzzy neural networks (Chang et al., 2005; Mali and Mitra, 2005), and fuzzy rule base (Bardossy and Samaniego, 2002; Pal et al., 2005). The fuzzy concept is also a valuable tool for dealing with classification problems. In remote sensing classification, fuzzy-based classifiers are becoming increasingly popular. This chapter describes the main fuzzy-based classifiers. The introduction to fuzzy methodology provides the basis for the development of more robust approaches to the remote sensing classification problem. An important field of remote sensing image analysis close to fuzzy methodology is *spectral unmixing* (Schowengerdt, 1996; Asner and Heidebrecht, 2002; Li et al., 2005; Lee and Lathrop, 2005; Quintano et al., 2006).

5.1 INTRODUCTION TO FUZZY SET THEORY

The difference between crisp and fuzzy sets can be characterized by means of a membership function. The membership function in a crisp set can only output two choices, {yes, no} or {0, 1}. In other words, an element of a crisp set can be a member of only one group, for which it has a membership grade of 1. The concept of the fuzzy set softens this constraint and allows the concept of partial membership, such that one data element may simultaneously hold several nonzero membership grades for different groups or clusters. Clearly, in comparison with the crisp approach, the fuzzy concept allows greater flexibility.

5.1.1 FUZZY SETS: DEFINITION

Let S represent a universe of discourse composed of generic elements denoted by s. A fuzzy subset G of S is determined by a membership function μ_G, which assigns a membership grade within the interval [0, 1] to each element s. The membership grade can be expressed by

$$\mu_G : s \rightarrow [0,1] \tag{5.1}$$

Such a membership mapping mechanism is quite different from the traditional crisp set approach in which the membership grade must be either 0 or 1. The fuzzy set approach is much more flexible for handling problems of indistinct boundaries, which are common in the natural world.

Figure 5.1a shows the traditional crisp set concept in which membership grade is normally a step function that outputs values of either 0 or 1, meaning that no intersection is allowed between the clusters, $a1$ and $a2$. However, in fuzzy set terms, as shown in Figure 5.1b, different clusters, e.g., $a1$ and $a2$, may share some units, and the membership grade of a unit will generally decrease as the distance between the unit and the cluster center increases.

Let $\mu_G(s)$ denote the grade of membership of s in a fuzzy subset G that can be expressed as

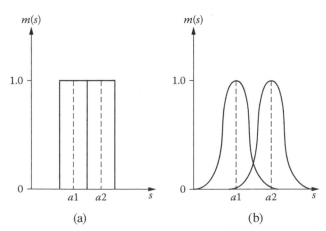

(a)   (b)

FIGURE 5.1   (a) In the traditional crisp set concept, the membership grade of cluster $a1$ or $a2$ is either 0 or 1. (b) In fuzzy set theory, overlap between the two clusters is allowed. See text for details.

$$G = \sum_i \mu_G(s_i)/s_i.$$ (5.2)

In the continuous case, G becomes

$$G = \int \mu_G(s)/s \, ds.$$ (5.3)

Note that in Equation (5.3), the symbol "/" does not refer to the division operator. It is used to represent the link between the value of s and its corresponding membership grade $\mu_G(s)$ in the fuzzy subset G. As the value of $\mu_G(s)$ approaches unity, the greater is the chance that s belongs to G. For example, suppose that a very simple universe of discourse S contains only three pixels and the membership function μ_G for the fuzzy subset G is defined by

$$\mu_G(s) = \frac{10}{0.5s^2 + 10}, \; s \in \Lambda$$ (5.4)

where $\Lambda = \{0, 5, 27\}$ denotes the gray value of each pixel. The fuzzy subset G can be represented as

$$G = (1/0) + (0.44/5) + (0.027/27).$$ (5.5)

The term $0.44/5$ is interpreted as "a pixel with grey value 5 has a membership grade of 0.44 in G."

The height of a fuzzy subset G denoted by height(G) is defined as the highest membership value contained in that fuzzy subset. For example, Equation (5.5) shows that the height of fuzzy subset G is height(G) = 1.

The α-cut of a fuzzy subset generates a crisp set in which the universe of discourse has the membership grades equal to or greater than α. Thus, the α-cut of a fuzzy subset G, denoted by G_α, can be expressed by

$$G_\alpha = \left\{ s \in S \middle| \mu_\alpha(s) \geq \alpha \right\}.$$ (5.6)

For example, in the case of Equation (5.5), if one chooses $\alpha = 0.1$, then the α-cut of G will be $G_{0.1} = \{0, 5\}$.

5.1.2 Fuzzy Set Operations

Once a fuzzy subset has been generated, various operations may be carried out to scale the membership grade of each discourse. In general, two operations known as *dilution* and *concentration* are used.

Dilution is an operation, which when applied to a fuzzy subset, increases the magnitudes of the membership grades in such a way that lower membership grades increase more than higher membership grades. Therefore, the difference between all the membership grades will decrease. A typical dilution operation for a fuzzy subset G is

$$Dilution(\mu_G(s)) = \sqrt{\mu_G(s)}. \tag{5.7}$$

For instance, suppose a fuzzy subset involves only two membership values, namely, 0.9 and 0.4. After the dilution operation using Equation (5.7), these values become 0.948 and 0.632, respectively. Clearly, the difference between the two membership values is reduced.

The *concentration* operation is the converse of dilution. All the membership grades are decreased when the concentration operation is applied. However, in such cases, higher membership grades decrease much less than do the lower membership grades. Therefore, the concentration operation makes the membership grades in a fuzzy subset more concentrated. A common concentration function is the square operator:

$$Concentration(\mu_G(s)) = \mu_G(s)^2. \tag{5.8}$$

Two membership values, 0.9 and 0.4, are used as an example. After the concentration operation using Equation (5.8), these membership grades become 0.81 and 0.16, respectively. The difference between the membership values is increased.

The complement of a fuzzy subset G is denoted by $\overline{G}$, and is defined as

$$\mu_{\overline{G}}(s) = 1 - \mu_G(s). \tag{5.9}$$

For instance, the complement of the fuzzy subset shown in Equation (5.5) is $\overline{G} = (0/0) + (0.56/5) + (0.973/27)$.

In the case of two fuzzy subsets denoted by G and H, the *union* ($\vee$) of G and H denoted by $\mu_{G \vee H}$, is defined as

$$\mu_{G \vee H}(s) = \max\{\mu_G(s), \mu_H(s)\}, \text{ for all } s. \tag{5.10}$$

That is, the union operation compares two membership grades of two fuzzy subsets and chooses the larger one. For instance, if there are two fuzzy subsets denoted by A and B, each containing two membership grades as $A = \{0.2, 0.9\}$, and $B = \{0.5, 0.4\}$, then the union of A and B will become $A \vee B = \{0.5, 0.9\}$.

Intersection ($\wedge$) of G and H is denoted by $\mu_{G \wedge H}$ and is expressed by

$$\mu_{G \wedge H}(s) = \min\{\mu_G(s), \mu_H(s)\}, \forall s. \tag{5.11}$$

This is the converse of the union operation. Using the same example as used previously, the intersection operation $A \wedge B$ gives the result $\{0.2, 0.4\}$. Both the union and interaction operations have their particular usage in manipulating a fuzzy system, as described in Section 5.4.

5.2 FUZZY *C-MEANS* CLUSTERING ALGORITHM

The k-means clustering algorithm (Chapter 2) is a crisp clustering procedure because it outputs membership grades of either 0 or 1 for each pixel in respect of each class. Methods such as k-means assume that clusters can be adequately represented as spherical patterns in feature space, and that the mean of each cluster can be estimated reasonably. Figure 5.2 illustrates an example of a possible invalid clustering. A more robust clustering method is clearly needed in this and similar situations. The *fuzzy c-means algorithm* (Bezdek, 1981; Bezdek et al., 1984) provides a means of overcoming the false clustering problem. This algorithm separates data clusters with *fuzzy* means and *fuzzy* boundaries, and is less dependent on the initial state of clustering. The algorithm is described as follows.

Let $X = \{x_1, x_2, ..., x_n\}$, be a finite subset of R^d, the d dimensional real number vector space (e.g., where three features are used for clustering, $d = 3$). Let the integer $c, n \geq c > 2$, denote the number of fuzzy subsets. Thus, a fuzzy c partition of X can be represented by a $(c \times n)$ matrix U in which each entry of U, denoted by u_{ik}, satisfies the following two constraints:

$$u_{ik} \in \left[0, 1\right]$$

and

$$\sum_{i=1}^{c} u_{ik} = 1, \quad \text{for all } k \tag{5.12}$$

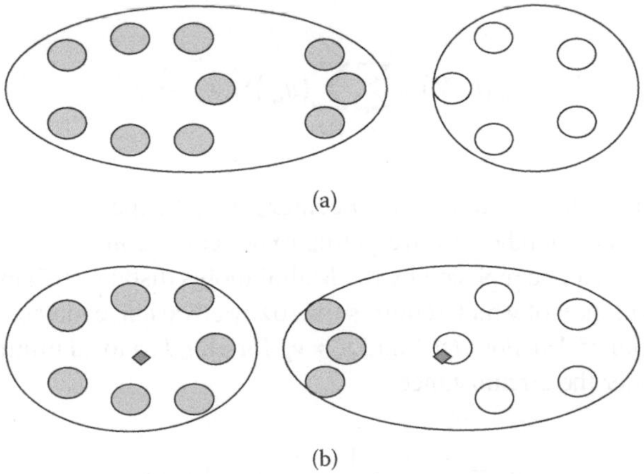

(a)

(b)

FIGURE 5.2 (a) Two clusters shown by blank and shaded balls. (b) Possible invalid clustering result generated by ISODATA. The lozenges (in grey) denote two cluster centers.

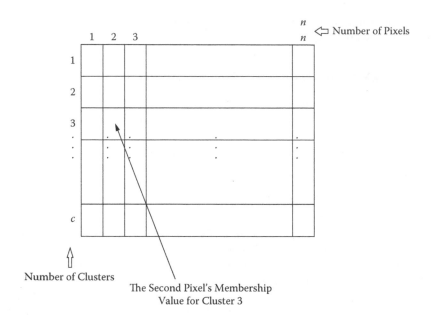

FIGURE 5.3   The fuzzy c-means clustering membership matrix.

In the case of image classification, n will be the number of pixels, and c is the number of clusters. The resulting matrix is shown in Figure 5.3. The value u_{ik} corresponding to the entry at the location (i, k) stores the kth pixel's membership value for class i (see Figure 5.3). Note that all the entries in a given column must sum to 1 as specified in Equation (5.12).

The clustering criterion used in the *fuzzy c-means* algorithm is based on minimizing the generalized within-groups sum of the square error function J_m:

$$J_m(U,V) = \sum_{k=1}^{n}\sum_{i=1}^{c}(u_{ik}) \times (x_k - v_i)^2 . \tag{5.13}$$

$V = (v_1, v_2, \ldots, v_c)$ is the vector of cluster centers (i.e., the means of the clusters), with $v_i \in R^d$, and m is the membership weighting exponent, $1 \leq m < \infty$. The term $(x_k - v_i)^2$ in Equation (5.13) can be replaced by the Mahalanobis distance (Chapter 2, Equation [2.10]), the calculation of which requires the fuzzy covariance matrix, which is introduced in Equation (5.18). For $m > 1$ and $x_k \neq v_i$, for all i, k, a local minimum of J_m may be achieved under the circumstance

$$u_{ik} = \frac{1}{\sum\limits_{j=1}^{c}\left(\dfrac{|x_k - v_i|}{|x_k - v_j|}\right)^{2/(m-1)}} \quad \text{for all } k \tag{5.14}$$

and the ith cluster mean is calculated from

$$v_i = \frac{\sum_k (u_{ik})^m \times x_k}{\sum_k (u_{ik})^m} \quad \text{for all } i .$$ (5.15)

The fuzzy c-means clustering is thus performed by iteratively applying Equations (5.14) and (5.15). A complete algorithm is presented in Figure 5.4.

The effect of the exponent is worthy of further consideration. As $m \to 1$, the solution of the fuzzy c-means procedure converges to hardness or crispness (i.e., a pixel's membership grades become closer to 1 or 0, respectively), while the greater the value of m (e.g., two or more), the fuzzier the membership assignments will be (i.e., the membership grades for all clusters are close to each other). Cannon et al. (1986) suggested that, for any value of m between 1.3 and 1.8, a high performance–oriented segmentation could be obtained. Nikhil and Bezdek (1995) suggested that the choice of the m value in the interval [1.5, 2.5] gives an acceptable clustering result, whereas Chen and Lee (2001) conclude that an m value of 2.5 gives the best results. Once the clustering process is finished, each pixel can be assigned to the cluster for which the pixel's membership grade is largest, if one wants to perform a crisp interpretation. Alternatively, one can scale membership grades to the interval [0, 255] and produce membership grade images for each class.

The choice of the number of clusters is not always easy, especially when the user does not have any knowledge about the number of information classes. An alternative procedure to solve this issue is to design specific indices to measure clustering performance with a range of cluster numbers. The solution generating the optimal performance indices will usually be adopted. For instance, Pal and Bezdek (1995, 1997) measure clustering validity for different cluster numbers in terms of a within-clusters compactness index and an among-clusters separation index. Kim et al. (2004) propose an approach for determining a suitable number of clusters by minimizing an overlapping index and maximizing a between-clusters separation index. Another interesting algorithm called *fuzzy partition—optimum number of clusters* (Gath and

---

Begin
    Choose the number of clusters and mean exponent m,
      and the false tolerance ε,
    Randomly initialise clustering membership matrix U;
    Do
      Compute cluster mean v_i, for all i, using Eq. (5.15);
      Update matrix U using Eq. (5.14);
    Until ($\|U^{new} - U^{old}\| < ε$)
End

FIGURE 5.4 The fuzzy c-means clustering algorithm.

Geva, 1989) uses a combination of the fuzzy c-means and fuzzy maximum likelihood algorithms (Section 5.3) and is illustrated here as an example. The algorithm is similar to the fuzzy c-means algorithm described, except that the distance measure used to define similarity between pixel x_k and a cluster center v_i is defined as

$$d^2(x_k, v_i) = \frac{\sqrt{|F_i|}}{P_i} \exp\left[\frac{(x_k - v_i)^T F_i^{-1}(x_k - v_i)}{2} \right] \tag{5.16}$$

where P_i is the *a priori* probability of the ith cluster, defined as

$$P_i = \frac{\sum_k u_{ik}}{n} \tag{5.17}$$

n is the number of pixels, F_i is the fuzzy covariance, expressed by

$$F_i = \sum_k \frac{u_{ik}^m (x_k - v_i)(x_k - v_i)^T}{\sum_k u_{ik}^m} \tag{5.18}$$

and $|F_i|$ denotes the determinant of the ith cluster covariance matrix.

Note that, during the clustering process, the pixel membership grades, the cluster means, and the covariance matrix F_i must be modified at each iteration. Once the algorithm has converged, the process is repeated using a different number of clusters. Finally, the optimal clustering scheme is chosen in terms of the minimization of the overall determinants of the fuzzy covariance matrices (i.e., $|F_i|$) for all i.

The standard fuzzy c-means method assumes that every information class has been specified, so that each pixel can be completely described in terms of its membership in the given set of categories. This is not always the case, as Foody (2000b) points out. He suggests that the standard fuzzy c-means method produces membership grades that are analogous to relative proportions of the specified classes (with the possibility of error if some classes are omitted). The possibilistic c-means algorithm (Krishnapuram and Keller, 1993, 1996; Zhang and Leung, 2004) is proposed as an alternative, as it provides a means of estimating the absolute membership grade of each class independently of all other classes.

5.3   FUZZY MAXIMUM LIKELIHOOD CLASSIFICATION

The crisp maximum likelihood classification algorithm is described in Chapter 2. It uses the hard mean and hard covariance matrix of each cluster. In what follows, we show that fuzzy set theory can be extended to the maximum likelihood algorithm to measure the membership grade of the pixels (Wang, 1990a; Maselli et al., 1995; Foody, 1996; Frizzelle and Moody, 2001).

Based on probability theory, if an event A is a precisely (hard) defined set of elements in the universe of discourse Ψ, the probability density function of A denoted by $P(A)$ can be expressed by

$$P(A) = \int_{\Psi} H_A(s) \qquad (5.19)$$

where s denotes the element in Ψ, and H_A is a hard membership function, i.e., $H_A(s) = 0$ or 1. In the case of image classification, event A is the cluster or class, and s is a vector of feature values associated with a specific pixel. The membership function H_A thus describes s either as belonging to A (i.e., membership grade is 1) or not (i.e., membership grade is 0).

If A is regarded as a fuzzy event, which means that set A is a fuzzy subset in Ψ, a probability measure of A becomes

$$P(A) = \int_{\Psi} \mu_A(s). \qquad (5.20)$$

The term μ_A is the fuzzy membership function as defined in Equation (5.1). Equation (5.20) is an extension and generalization of Equation (5.19). Even a partial membership value of observation s in A can provide a contribution to the total probability $P(A)$.

The mean and variance of fuzzy set A relative to a probability measure can similarly be quantified as

$$v_A = \frac{1}{P(A)} \int_{\Psi} s\mu_A(s) \qquad (5.21)$$

and

$$\sigma_A^{\,2} = \int_{\Psi} (s - v_A)^2 \mu_A(s). \qquad (5.22)$$

Equations (5.21) and (5.22) determine the fuzzy mean and fuzzy variance, respectively, both of which are derived for the continuous case. In practice, the discrete fuzzy mean v_i and fuzzy covariance matrix F_i for class i can be expressed as

$$v_i = \frac{\sum_j \mu_i(x_j)x_j}{\sum_j \mu_i(x_j)} \qquad (5.23)$$

and

$$F_i = \frac{\sum_j \mu_i(x_j) \times (x_j - v_i) \times (x_j - v_i)^T}{\sum_j \mu_i(x_j)} \qquad (5.24)$$

where x_j denotes the feature vector for pixel j. If the exponent m in Equation (5.18) is set to 1, Equation (5.24) becomes equivalent to Equation (5.18), which is used in the optimum clustering algorithm. It can be inferred that when the value of $\mu_i(x_j)$ becomes either 0 or 1 (i.e., crisp membership function), Equations (5.23) and (5.24) become the conventional crisp mean and covariance matrix formulations.

A fuzzy set is characterized by its membership function. Wang (1990a) defines the membership grade for each land-cover class based on the maximum likelihood classification algorithm with a fuzzy mean and fuzzy covariance matrix as shown in Equations (5.23) and (5.24) as follows:

$$\mu_k(x_j) = \frac{P_k(x_j)}{\sum_i P_i(x_j)} \tag{5.25}$$

where k is the land-cover class and probability $P_i(x_j)$ denotes the class-conditional probability for class i given the observation x_j, which is described in Chapter 2, Equation (2.16), except that the crisp mean and covariance matrix in Equation (2.16) are replaced by the fuzzy mean and fuzzy covariance matrix. This is the core of the fuzzy maximum likelihood algorithm. Calculating membership grades in terms of Equation (5.25) is equivalent to normalizing the probabilities of the pixel to all of the information classes. Although such a method is quite straightforward, its validity requires further investigation.

5.4 FUZZY RULE BASE

Rule-based classification methodology is an attempt to gather the human operator's knowledge that is relevant to a classification task. Traditional rule-based methods can be regarded as crisp versions of fuzzy methods. That is, each rule has equal weight (or effectiveness). This kind of crisp rule-based classifier may face some problems. For instance, in the case of a hierarchical decision rule base, the classification is implemented by traversing the decision tree until an end node is reached. This traversing process is equivalent to gradually defining crisp partitions of the solution space. If the boundaries between classes are well defined, a hierarchical decision rule base should perform well. In remotely sensed imagery, information classes often overlap with each other in feature space. For those pixels lying in the overlap area of feature space, the use of the crisp decision tree methodology can result in classification error. If the crisp rule base is inferred simultaneously rather than hierarchically, the problem of rule collision (i.e., rules being triggered against each other) is also likely to happen. One solution for solving these problems is to use a fuzzy rule base, and to apply a fuzzy inference mechanism. The idea is illustrated below.

The difference between the fuzzy rule and crisp rule is that each rule in a fuzzy rule base contains a strength (also known as a weighting or a certainty) parameter. In a fuzzy rule base, the simultaneous triggering of several rules is allowed, even though these triggered rules may act against each other (for instance, rule a may classify the pixel as class 1, while rule b may classify the pixel as class 2). However,

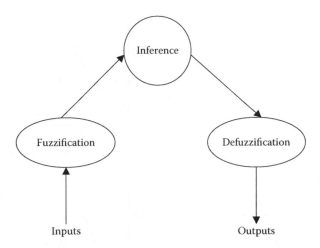

FIGURE 5.5 Basic architecture of the fuzzy system includes fuzzification, inference, and defuzzification. See text for explanation.

since each rule is triggered by the level of strength (or certainty), a decision can be made in favor of the rule that contains the greatest strength.

A fuzzy rule base (or fuzzy system) used for classification generally comprise three principal steps, as shown in Figure 5.5. The first step, *fuzzification*, involves the division of the input feature space into fuzzy subspaces, each specified by a fuzzy membership function (defined in Section 5.4.1). Fuzzy rules are then generated from each fuzzy subspace. The second step, *inference*, requires the calculation of the strength of each rule being triggered. The final step, *defuzzification*, combines all triggered rules and generates a nonfuzzy (i.e., crisp) outcome.

5.4.1 FUZZIFICATION

The purpose of fuzzification is to partition the feature space into fuzzy subspaces and generate rules for each fuzzy subspace. Note that all fuzzy subspaces normally overlap each other to some degree (see below). To carry out the process of fuzzification, one must first define membership functions in order to calculate the membership grade for the input pixels. Equation (5.4) is one kind of membership function that maps the input s to a grade of fuzzy membership. Although the fuzzy membership function can take any form (as long as the function can map the inputs onto the range [0, 1]), four kinds of fuzzy membership functions, known as *monotonic, triangular, trapezoidal*, and *bell shaped*, are the most frequently used in fuzzy rule base experiments. Figure 5.6 illustrates the geometric shapes of these four types of membership functions, and the corresponding mathematical expressions are given in Equations (5.26) to (5.29), respectively.

Monotonic:

$$\mu(s) = 1 - \frac{a-s}{\lambda}, \text{ for } 0 \le a - s \le \lambda; \mu(s) = 0 \text{ otherwise.} \tag{5.26}$$

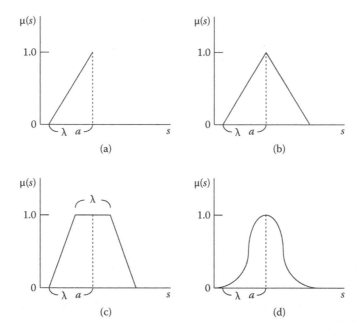

FIGURE 5.6 Four types of commonly used fuzzy membership functions: (a) Monotonic. (b) Triangular. (c) Trapezoidal. (d) Bell shaped.

Triangular:

$$\mu(s) = 1 - \frac{|s-a|}{\lambda}, \text{ for } 0 \le |s-a| \le \lambda; \mu(s) = 0 \text{ otherwise.} \qquad (5.27)$$

Trapezoidal:

$$\mu(s) = \min\left\{2 - \frac{2|s-a|}{\lambda}, 1\right\}, \text{ for } c\text{-}\lambda \le |s-a| \le a+\lambda; \qquad (5.28)$$

$$\mu(s) = 0 \text{ otherwise.}$$

Bell-shaped:

$$\mu(s) = 2\left(1 - \frac{|s-a|}{\lambda}\right)^2, \text{ for } \frac{\lambda}{2} \le |s-a| \le \lambda;$$

$$\mu(s) = 1 - 2\left(1 - \frac{|s-a|}{\lambda}\right)^2, \text{ for } 0 \le |s-a| \le \frac{\lambda}{2}. \qquad (5.29)$$

The parameter s denotes a measurement (e.g., a pixel value), $\mu(s)$ is the membership grade, a denotes the center of the fuzzy membership function, and λ controls the width of the membership function.

Each membership function shown above generates different membership grades. For instance, a trapezoidal membership function outputs a membership grade of 1 for a range of inputs s (i.e., $0 \leq |s - a| \leq \lambda$), while other membership functions allow only unique values of s (i.e., $s = a$) to hold the full membership value. As the distance between element s and the membership function center c increases, the decreasing level of membership grade varies with different membership functions. Also, the greater the value of parameter λ, the wider the range of s that is given a nonzero membership grade. Examples are shown in Figure 5.7. Here, two kinds of membership functions, triangular (Figures 5.7a and 5.7b based on Equation [5.27]) and trapezoidal (Figures 5.7c and 5.7d based on Equation [5.28]) are presented. Membership functions shown in Figures 5.7a and 5.7c use the value $\lambda = 10$, and can output nonzero membership grades for $0 < s < 20$. Membership functions shown in Figures 5.7b and 5.7d use the value $\lambda = 20$, and both can thus output nonzero membership grades for $0 < s < 40$. The areas of the trapezoidal membership functions that generate membership grades of 1 are shown as the shaded area of Figures 5.7c and 5.7d.

The resulting fuzzy partitions on one-dimensional space using the membership functions shown in Equations (5.26) to (5.29) alone are illustrated in Figure 5.8a. Note that the shape of the membership function and the size of the overlapped area may vary. Besides using the same membership function to partition the space, one may also perform fuzzy partitions using combinations of different membership

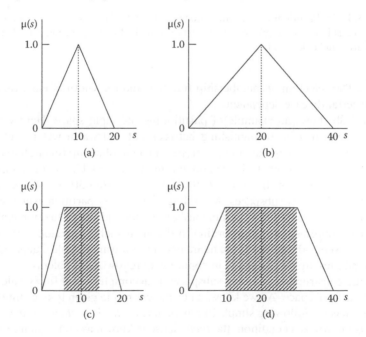

FIGURE 5.7   (a) and (b): Triangular membership functions. (c) and (d): Trapezoidal membership functions. See text for discussion.

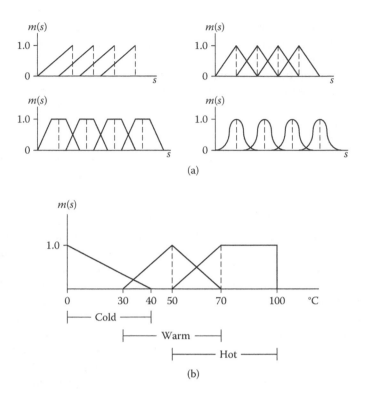

(a)

(b)

FIGURE 5.8 (a) Partitioning one-dimensional space into fuzzy subspaces using a single fuzzy membership function. (b) The fuzzy partitions to deal with the problem of distinguishing hot, warm, and cold water.

functions. The selection of membership function and the width of each fuzzy subspace are certainly case dependent.

Figure 5.8b shows an example of partitioning the input space. Here we take a simple problem, that of distinguishing between hot and warm water, and we use one monotonic, one triangular, and one trapezoidal membership function to perform fuzzy partitioning. The input s is the temperature measure (°C) and the temperature range is divided into hot (from 50° to 100°C), warm (from 30° to 70°C), and cold (from 0° to 40°C) fuzzy subspaces. According to this fuzzy partition, water temperatures between 50° and 70°C simultaneously hold hot and warm fuzzy membership (the membership grade varies according to the membership functions). Water temperatures between 30° and 40°C are members of the warm and cold categories. Such fuzzy boundaries are more flexible in coping with real-world problems.

Once the construction of fuzzy subspaces is accomplished, a fuzzy rule is then set up for each subspace. As we know, a normal crisp rule given to one-dimensional input space has the following simple linguistic form—*If s is A, then Z*—in which the *if* clause is known as a condition, the *then* clause is known as consequent, s denotes a measurement, A describes the decision boundaries, and Z denotes the result or action. For example, in case of image classification, the rule could become

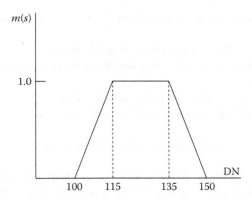

FIGURE 5.9   A trapezoidal membership function to model the concept middle. See text for description.

> *IF the pixel value is between 100 and 150,*
> *THEN the pixel is assigned to the class "Potato."*

Once a crisp rule is triggered, it results in a unique decision. The fuzzy rule concept softens the crisp rule description, so that the previous crisp rule becomes

> *IF pixel value is middle,*
> *THEN the pixel is assigned to the class "Potato" with strength* **w**.

The term *middle* is modeled by a fuzzy membership function, and the strength **w** is determined by how well the pixel value belongs to *middle*. For instance, if we use a trapezoidal membership function to model *middle* as shown in Figure 5.9, then pixel values between 115 and 135 will trigger the rule at full strength giving a membership grade of 1. Pixel values in the ranges [100, 115] and [135, 150] trigger the rule at lower strength, and pixel values outside the range [100, 150] have a membership grade of 0.

The example given above is an extremely simple case. In practical situations, a multidimensional input space and multiple classes are involved. In addition, the rules being triggered can be numerous because the fuzzy membership functions normally overlap. Hence, a pixel value falling within the overlap area will simultaneously trigger several rules. A final solution requires the use of inference and defuzzification.

5.4.2   INFERENCE

The inference stage computes the strength contributed by the triggered rules and aggregates those triggered rules. The process can be illustrated by the following example showing an automatic air conditioner controller with two-dimensional inputs: one is a measure of the current room temperature, denoted by $s1$; the other is a measure of how quickly the room temperature increases, denoted by $s2$. Assume that the automatic air conditioner controller is handled by the following three rules:

Rule 1: *IF room temperature is high and the increase in room temperature is high,*
         THEN the control should be turned to high.
Rule 2: *IF room temperature is low and the increase in room temperature is middle,*
         THEN the control should be turned to low.
Rule 3: *IF room temperature is middle and the increase in room temperature is low,*
         THEN the control should be turned to medium.

where *if* clauses *high, middle, low,* and *then* clauses *high, low, medium* are all modeled by a triangular membership function as shown in Figure 5.10. For each rule, the inference engine first generates the membership grades for the measurement *s*1 and *s*2, respectively. Both membership grades are then combined through an interaction (∧) operator, which is defined in Section 5.1.2 to be equal to the minimum operator.

If both membership grades are equal to one (i.e., the rule condition is fully satisfied) the *then* clause in the rule should be fully adopted (i.e., with full strength). On

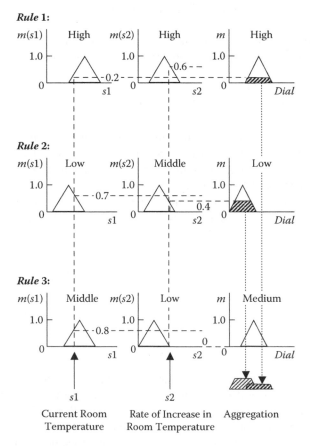

FIGURE 5.10   An aspect of fuzzy inference in a three-rule case.

the other hand, if the rule condition is only partially satisfied, the *then* clause should be partially weighted. Two weighting approaches, known as *multiplication* and *minimization*, are commonly used. In mathematical terms, the multiplication weighting method for Rule 1 can be expressed as

$$strength(\text{Rule 1}) = \beta1 \times large \qquad (5.30)$$

where *large* is the fuzzy membership function described in the Rule 1 *then* clause, and $\beta1$ is a weighting factor calculated by

$$\beta1 = \mu(s1) \wedge \mu(s2) \qquad (5.31)$$

where $\mu(s1)$ and $\mu(s2)$ are membership grades for measures $s1$ and $s2$, respectively. The use of minimization as a weighting method is given by

$$strength(\text{Rule 1}) = \beta1 \wedge large. \qquad (5.32)$$

The calculation of the strength of Rules 2 and 3 is carried out in a similar manner.

In Figure 5.10, the condition given by Rule 1 is matched (given membership grade μ) to the degree of 0.2 (for input $s1$) and 0.6 (for input $s2$), respectively. After Equation (5.31) is applied (i.e., $\min\{0.2, 0.6\} = 0.2$), the degree of match to the condition in Rule 1 is eventually determined as 0.2 (i.e., $w1 = 0.2$). Here, a minimization method (shown in Equation [5.32]) is adopted for calculating the rule strength. Thus, the strength of Rule 1 should be truncated (minimized) to 0.2 as shown in the shaded area. Similarly, the condition of Rule 2 is partially satisfied to the degree of 0.4, thus Rule 2 only contributes 0.4 of the strength (also shown in the shaded area of Rule 2 in Figure 5.10). Since the condition defined by Rule 3 is not matched, Rule 3 is not triggered.

Once the strength of each rule is determined, all of the triggered rules are then aggregated in terms of the union ($\vee$) operator as defined in Section 5.1.2. The aggregation of Rule 1 and Rule 2 (Figure 5.10) is expressed as

$$Rule\ Aggregation = strength\ (\text{Rule 1}) \vee strength\ (\text{Rule 2})$$
$$= (\beta1 \wedge large) \vee (\beta2 \wedge large) \qquad (5.33)$$

where $\beta1$ and $\beta2$ are defined in Equation (5.31). Since the result of rule aggregation is a membership function, a defuzzification process has to be implemented in order to obtain a deterministic value.

5.4.3 DEFUZZIFICATION

Several kinds of defuzzification strategies have been suggested in the literature. The most popular methods of defuzzification are the center-of-gravity and the mean-of-maximum methods (Pedrycz, 1989; Kosko, 1992). A membership function is often

represented in terms of discrete data. The center-of -gravity method can be calculated from the following equation:

$$\text{center-of-gravity} = \frac{\sum_{s=1}^{n} s \times \mu(s)}{\sum_{s=1}^{n} \mu(s)} \qquad (5.34)$$

where n is the number of elements of the sampled membership function, and $\mu(s)$ is
the membership grade of measurement s.

The mean-of-maximum method generates a deterministic value by taking the
average of some measurement s whose membership grades $\mu(s)$ are a maximum. In
the discrete case, this is calculated by

$$\text{mean-of-maximum} = \sum_{s=1}^{m} \frac{s}{m} \qquad (5.35)$$

where m denotes the number of s reaching the relatively maximal membership grade
(described below). An example is given in Figure 5.11, which follows the result of
Figure 5.10. Assume that the air conditioner controller contains seven choices; for
example, the values 0.1, 0.2, ... 0.7 (refer to Figure 5.11). Using the center-of-gravity
method, the defuzzification value is calculated as

$$\text{center-of-gravity} = \frac{(0.1+0.2+0.3) \times 0.4 + (0.4+0.5+0.6+0.7) \times 0.2}{0.4+0.4+0.4+0.2+0.2+0.2+0.2} = 0.34 \quad (5.36)$$

Using the mean-of-maximum defuzzification method, only three sample elements
reach the relatively maximal membership grade, i.e., elements 0.1, 0.2, and 0.3 with
the membership grade of 0.4, which results in

$$\text{mean-of-maximum} = \frac{0.1+0.2+0.3}{3} = 0.2 . \qquad (5.37)$$

Using these three steps (fuzzification, inference, and defuzzification), one eventually
reaches a solution. In the case of the air conditioner dial, the setting should be 0.34
if the center-of-gravity method is applied.

In comparison with traditional crisp set classification, the valuable property of
the three-step fuzzy classification method is that it recognizes the fuzziness of the
real world, and provides a means of accommodating such fuzziness. Sections 5.2
(fuzzy c-means) and 5.3 (fuzzy maximum likelihood) show that the spirit of the
fuzzy methodology is to try to involve any possible contribution (even it is small)

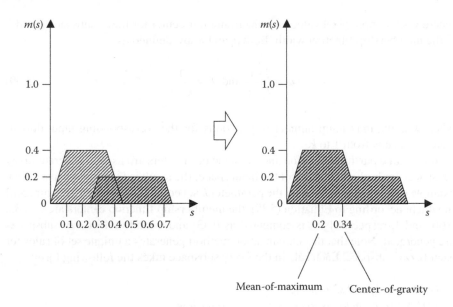

FIGURE 5.11 Defuzzification methods on a discrete domain result in deterministic values, 0.2 (mean-of-maximum) and 0.34 (center-of-gravity).

from the universe of discourse, and then generate the solution. To deal with classification problems in this way is often more realistic than using crisp concepts.

5.5 IMAGE CLASSIFICATION USING FUZZY RULES

This section introduces an image classification methodology using fuzzy rules, as mainly proposed by Ishibuchi et al. (1992, 1999). Further studies of the effect of rule weights and rule selections can be found in Ishibuchi and Nakashima (2001) and Ishibuchi and Yamamoto (2002, 2004). For other rule-generation approaches for classifying remotely sensed data, see Bardossy and Samaniego (2002) and Pal et al. (2005). Note that although the ways of generating fuzzy rules might be different, the core of those fuzzy rule methods is based on the three steps described in Section 5.4, namely fuzzification, inference, and defuzzification.

5.5.1 INTRODUCTORY METHODOLOGY

For simplicity, a two-dimensional input case is examined. The input features are normalized onto the range [0, 1]. Each dimension in the input space is then partitioned into k fuzzy subspaces denoted by $\{A_1, A_2, \ldots A_k\}$, where A_i indicates the ith fuzzy subspace. A symmetric triangular membership function is adopted, and the fuzzy subspace for A_i is then defined by

$$\mu_i(s) = \max\left\{1 - \frac{|s - a_i|}{\lambda}, 0\right\} \tag{5.38}$$

where s is the input pixel value, a_i is the triangular center for fuzzy subspace i, and λ is the membership function width. Both a_i and λ are defined as

$$a_i = \frac{i-1}{k-1} \text{ and } \lambda = \frac{1}{k-1} \tag{5.39}$$

where k is the maximum number of partitions for the corresponding input dimension, and i runs from 1 to k.

If the same partition function and the same parameters are used to generate fuzzy subspaces for both dimensions of the input space, the result is k^2 fuzzy subspaces. An example of fuzzy partitions with the parameter k set equal to 4 is shown in Figure 5.12 in which, according to Equation (5.39), the membership function centers are 0, 0.33, 0.66, and 1, respectively, λ is computed as 0.33, and a total of 16 fuzzy subspaces are generated. Note that this classification method generates a unique set of rules for each fuzzy subspace. Each rule in the fuzzy subspace takes the following form:

IF s1 is in A_i and s2 is in A_j,
THEN pixel s belongs to class c_{ij} with strength w_{ij}

where $s1$ and $s2$ are the input features for pixel s, and A_i and A_j ($1 \leq i, j \leq k$) are fuzzy subspaces on both dimensions, respectively. The class c_{ij} and rule strength

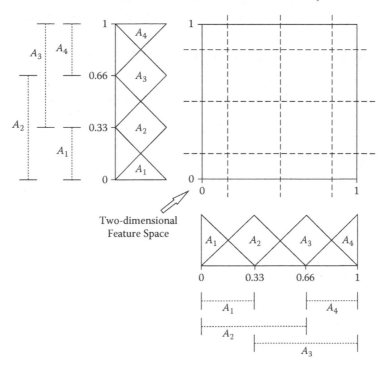

FIGURE 5.12   An example of fuzzy partitions with $k = 4$ results in total 16 fuzzy subspaces.

w_{ij} based upon the fuzzy rule are determined by the training data. The procedure for generating fuzzy rules in the subspace ij is shown in Figure 5.13a. The first step in Figure 5.13a is to calculate the weighting parameter β_x for each class x using the

---

Let l denotes the number of information classes, and nx denotes the number of training patterns for class x currently falling within the fuzzy subspace ij.

(1) Calculate weighting β_x for each class x $(x = 1$ to $l)$ as

$$\beta_x = \sum_{s=1}^{n_x} \mu_i(s1)\,??\,\mu_j(s2) \quad or \quad \beta_x = \sum_{s=1}^{n_x} \mu_i(s1)\,??\,\mu_j(s2)$$

(2) Choose class c_{ij} such that

$$\beta_{c_{ij}} = \max\{\beta_1, \beta_2, ..., \beta_l\}$$

(3) Calculate rule strength w_{ij} as

$$w_{ij} = \frac{(\beta_{c_{ij}} - \beta)}{\sum_{x=1}^{l} \beta_x}, \ where \ \beta = \sum_{x=1,\,x\neq c_{ij}}^{l} \frac{\beta_x}{l-1}$$

(a)

For each pixel.

(1) Calculate α_x for each class x as

$$\alpha_x = \max\left\{\left[\mu_i(s1) - \mu_j(s2)\right] - w_{ij}\middle| c_{ij} = x, +, j\right\} \quad or$$

$$\alpha_x = \max\left\{\left[\mu_i(s1) - \mu_j(s2)\right] - w_{ij}\middle| c_{ij} = x, -i, j\right\}$$

(3) Classify the pixel as class c such that

$$\alpha_c = \max\{\alpha_1, \alpha_2, ..., \alpha_l\}$$

(b)

FIGURE 5.13 (a) Procedure for the construction of fuzzy rule for each fuzzy subspace ij. (b) Procedure for classifying the image.

TABLE 5.1

The Fuzzy Rule Strength w_{ij} Calculated for Each Fuzzy Subspace in Terms of Two Training Pixels from Equations (5.38), (5.39), and Procedures Shown in Figure 5.13a. See text for further details.

Fuzzy Subspace ij	Membership Grade (μ_1, μ_2)	β_2 and β_2 by $\wedge$ Operator	Class c_{ij}	Rule Strength w_{ij}
$i = 1, j = 1$	Class 1 = 0.3, 0.2)	$\beta_1 = 0.2$	$c_{11} = 2$	$w_{11} = (0.5 - 0.2)/(0.5 + 0.2)$
	Class 2 = 0.5, 0.6)	$\beta_2 = 0.5$		$= 0.43$
$i = 1, j = 2$	Class 1 = 0.3, 0.8)	$\beta_1 = 0.3$	$c_{12} = 2$	$w_{12} = (0.4 - 0.3)/(0.4 + 0.3)$
	Class 2 = 0.5, 0.4)	$\beta_2 = 0.4$		$= 0.14$
$i = 2, j = 1$	Class 1 = 0.7, 0.2)	$\beta_1 = 0.2$	$c_{21} = 2$	$w_{21} = (0.5 - 0.2)/(0.5 + 0.2)$
	Class 2 = 0.5, 0.6)	$\beta_2 = 0.5$		$= 0.43$
$i = 2, j = 2$	Class 1 = 0.7, 0.8)	$\beta_1 = 0.7$	$c_{22} = 2$	$w_{22} = (0.7 - 0.4)/(0.7 + 0.4)$
	Class 2 = 0.5, 0.4)	$\beta_2 = 0.4$		$= 0.27$

procedure described in Equation (5.31). We then choose the class c_{ij} according to its maximal weights among all classes. The rule in fuzzy subspace ij is then generated with strength w_{ij} calculated in step 3.

For example, in the two-dimensional case, two classes and two fuzzy partitions are generated for each dimension, using Equations (5.38) and (5.39) (thus $k = 2$, $\lambda = 1$, $a_1 = 0$, $a_2 = 1$). If we use two training pixels, with one having values (0.7, 0.8) for class 1 and the other containing the value (0.5, 0.4) for class 2, the resulting parameters using Equations (5.38), (5.39), and the procedures shown in Figure 5.13a are shown in Table 5.1. The fuzzy rule for subspace $i = 1, j = 1$ can therefore be determined as

IF s1 is in A_1 and s2 is in A_1,
THEN pixel s belongs to class 2 with strength 0.43.

The remaining three rules for fuzzy subspaces $ij = 12$, 21, and 22 can be inferred in a similar manner.

After class labels c_{ij} and strengths w_{ij} in the rule have been determined, a rule for the fuzzy subspace ij is completely specified. The resulting rule base can be applied to classify the image in terms of the procedure illustrated in Figure 5.13b (Ishibuchi et al., 1992, 1999). When an unknown pixel is input to the fuzzy rule base, we first calculate the membership grades in all fuzzy subspaces for the pixel. The resulting membership grades are then combined with the rule strength as shown in step 1 of Figure 5.13b and the membership grade of an unknown pixel with respect to class x is represented by α_x (refer to step 1). After the parameter α_x has been determined for each class x, we then allocate the pixel to the class **c** if α_c is the maximal, as shown in Figure 5.13b, step 2. For instance, if the unknown pixel takes values (0.6, 0.9) on the two input features, then according to Figure 5.13b, step 1 (using the minimum operator $\wedge$), we obtain for class 1: $\alpha_1 = [\mu_2(0.6)\wedge\mu_2(0.9)] \times w_{22} = [0.6\wedge0.9] \times 0.27 = 0.162$

for class 2: $\alpha_2 = \max\{[[\mu_1(0.6)\wedge\mu_1(0.9)] \times w_{11} = [0.4\wedge0.1] \times 0.43 = 0.043$,

$[\mu_1(0.6)\wedge\mu_2(0.9)] \times w_{12} = [0.4\wedge0.9] \times 0.14 = 0.056$,

$[\mu_2(0.6)\wedge\mu_1(0.9)] \times w_{21} = [0.6\wedge0.1] \times 0.43 = 0.043\}$

$= 0.056$ 　　　　　　　　　　　　　　　　　　　　　　　　(5.40)

where μ_i, $i \in \{1, 2\}$ is computed using Equation (5.38). This unknown pixel is placed in class 1 because $\alpha_1 > \alpha_2$ (Figure 5.13b, step 2).

The procedures shown in Figure 5.13 can be extended to deal with higher input dimensions. However, two points should be noted. In Figure 5.13a, step 2, if several classes take the maximum value or the $\beta_x = 0$, for class x, the rule for the corresponding fuzzy subspace cannot be generated (one may adopt different training pixels to resolve such a problem). Similarly, in Figure 5.13b, step 2, if two or more classes take the maximum value or all α_x are zero, the input pattern will be treated as unclassified.

The strength w_{ij} specified in Figure 5.13a is based on the following intuitive property: If all of the training patterns within a fuzzy subspace belong to the same class, the w_{ij} will then reach the maximum permitted value (i.e., $w_{ij} = 1$). However, if the value β_x, for all class x, are close to each other (i.e., $\beta_1 \approx \beta_2 \approx \ldots \approx \beta_l$) then the rule will fire at its lowest strength (i.e., $w_{ij} \approx 0$). The determination of rule strength may be achieved in other ways. One possible choice for specifying the rule strength can be in terms of an entropy measure, and the strength w_{ij} may be defined as

$$w_{ij} = \frac{1}{-\sum_{x=1}^{l}\beta_x \times \ln\beta_x}. \qquad (5.41)$$

High entropy indicates that the values of all β_x are all relatively similar, hence rule strength should be decreased, while low entropy indicates that rule strength should be larger. It is likely that rule strength will vary with different measuring methods, and will therefore affect classification accuracy. One may test different approaches for determining rule strength and compare the classification results.

In determining the methods for generating fuzzy rules from the training data, the effect of the size of fuzzy subspace should be considered. If the fuzzy subspace is too small, there will be a high probability that some fuzzy subspaces (called dummy subspaces) may contain no training data and therefore cannot generate any fuzzy rules. In such a situation, the image pixels falling within these dummy subspaces cannot be classified. On the other hand, if the fuzzy subspace is too large, some fuzzy subspaces may contain a variety of different classes of data, and this will decrease the classification power of the fuzzy rule classifier.

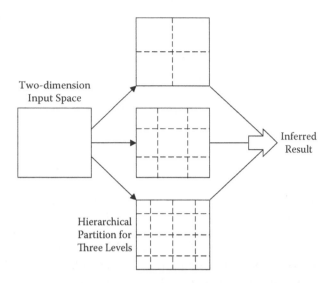

FIGURE 5.14 A hierarchical (three-level) partition from a two-dimension input space.

Ishibuchi et al. (1992, 1999) suggested a hierarchical strategy for solving this space-partitioning problem. Their method is quite straightforward, and involves the partitioning of the input feature space into different sizes of fuzzy subspaces, from large to small, as a hierarchical structure (shown in Figure 5.14). The inference engine then simultaneously employs all rules for determining α_x and α_c in Figure 5.13b. Although this method can improve the classification performance, the resulting computation time will increase substantially, and so will the rule storage requirements. For example, if the maximum number of fuzzy partitions for each input dimension is determined as 20, the total number of rules will be $2^2 + 3^2 + \ldots + 20^2$. Such a large number of rules will generate a considerable computation burden. Ishibuchi and Yamamoto (2004) propose an approach involving the selection of a subset of "significant" fuzzy rules to perform a classification, so that the number of fuzzy rules can be reduced. Another solution to the fuzzy subspace partition problem is based on the density of the training data. Large fuzzy partitions are suitable for areas of feature space that have a low density of training data, while for areas containing a high density of training patterns, the smaller partition of fuzzy subspaces should be applied (Abe and Lan, 1995).

5.5.2 Experimental Results

First, the following nonlinear function is used to generate a two-dimensional test image:

$$f(s) = -0.2 \times \sin(2\pi \times s1) + 0.8 \times s2 - 0.44 \qquad (5.42)$$

where $s1$ and $s2$ denote normalized input features (i.e., within the range [0, 1]). If the input pattern s gives $f(s) \geq 0$, then s is placed in class 1, otherwise, s belongs to class 2. Figure 5.15a shows the resulting test (truth) image of size 100×100 in which

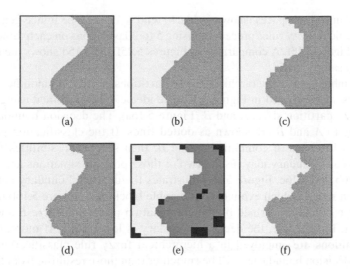

(a)   (b)   (c)

(d)   (e)   (f)

FIGURE 5.15 (a) Test image of size 100 × 100. (b) Classification using 5 fuzzy partitions on each dimension using 400 training patterns. (c) Classification generated by 20 fuzzy partitions on each dimension. (d) Hierarchical (partitions form 5 to 20 on each dimension) fuzzy rule approach. (e) Result generated by 20 fuzzy partitions on each dimension using 50 training samples. (f) Result of classification using hierarchical fuzzy rules with 50 training samples.

class 1 is displayed in grey and class 2 is left unshaded. Standard statistical classifiers find difficulty in determining the location of a decision boundary in a problem such as this. Although the class centers for both classes are clearly located, the decision boundary will be a vertical line located at the center of the horizontal axis.

In order to compare the classification behavior for different sizes of fuzzy partitions, two fuzzy rule classifiers, each with 5 and 20 partitions on each dimension (i.e., $5^2 = 25$ and $20^2 = 400$ fuzzy rules in total), were chosen and 400 training samples (selected at random) were used to train the fuzzy rule classifiers. The resulting images classified by both fuzzy rule classifiers are illustrated in Figure 5.15b and 5.15c, respectively. It is clear that in comparison with Figure 5.15b, Figure 5.15c is closer to the original image (Figure 5.15a), but some blockiness can be seen on the boundary. The classification resulting from the use of hierarchical fuzzy rules (partitions from 5 to 20 on each dimension, i.e., a total of $5^2 + 6^2 + \ldots + 20^2$ rules) is shown in Figure 5.15d. The resulting classification decision boundary is smoother than that shown in Figure 5.15b and 5.15c. This interesting property of the hierarchical fuzzy rule methodology is considered further in the following pararaphs.

The hierarchical fuzzy rule approach is used to overcome the membership size selection problem for fuzzy partitions. As the fuzzy partitions become smaller, and if there are insufficient training patterns, the result is the presence of dummy subspaces (i.e., subspaces for which no fuzzy rules are available). Pixels falling within these dummy subspaces cannot be classified. This difficulty was mentioned earlier. An example is illustrated in Figure 5.15e in which the number of training patterns is 50 (selected at random) and the number of fuzzy partitions in each dimension is set to 20. With such fine fuzzy partitions and with a small number of training patterns, dummy subspaces (shown in black) tend to occur. The image generated by the hierarchical fuzzy rules procedure (using 5 to 20 partitions on each dimension) is shown in Figure 5.15f. A comparison of Figures 5.15f and 5.15d shows the influence of training sample size.

The combination of hierarchical fuzzy partitions results in smoother decision boundaries, as illustrated in Figure 5.16. The idea can be explained using two different fuzzy partition sizes, A and B (Figure 5.16a). The decision boundaries corresponding to A and B are shown as dotted lines. If the classification process is carried out in terms of combining A and B, then several possibilities for forming a decision boundary may result. Two of those possible situations are shown in Figure 5.16b and 5.16c. Figure 5.16b illustrates the decision boundary determined by partition size A, which exhibits considerable blockiness. Figure 5.16b illustrates the decision boundary formed by the combination of partition size B and part of partition A. In this case, the decision boundary is less blocky. If more, different, fuzzy partitions are employed in a hierarchical fuzzy rule classification mechanism, the decision boundaries will be smoother than those resulting from the use of a single fuzzy partition.

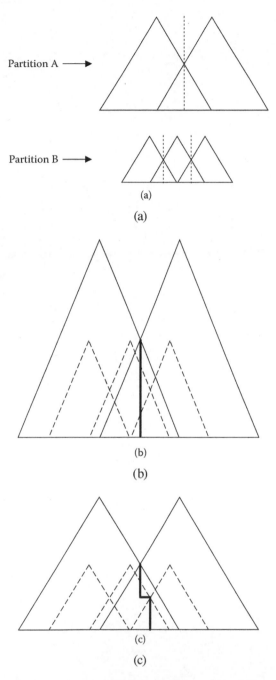

(a)

(b)

(c)

FIGURE 5.16 (a) Two sides of fuzzy partition and their corresponding decision boundaries (b and c). Two possible shapes of decision boundary are shown in bold.

6 Decision Trees

Categorization of data using a hierarchical splitting (or top–down) mechanism has been widely used in the environmental and life sciences. The purpose of using a hierarchical structure for labeling objects is to gain a more comprehensive understanding of relationships between objects at different scales of observation or at different levels of detail. Its simplest representation takes the form of an inverted tree in which the different levels of classification are represented by the separate levels of a hierarchy. When applied to multispectral image data, the design of a decision tree is based on knowledge of the spectral properties of each class and of the relationships among the classes. The main benefit of using a hierarchical tree structure to perform classification decisions is that the tree structure can be viewed as a *white box*, which in comparison with artificial neural networks (ANN) (Chapter 3), is easier to interpret and understand the relation between the inputs and outputs.

A hierarchical decision tree classifier is an algorithm for the labeling of an unknown pattern using a sequence of decisions. A decision tree is composed of a root node, a set of interior nodes, and terminal nodes called *leaf nodes*. The root node and interior nodes, referred to collectively as *nonterminal nodes*, are linked into decision stages. The terminal nodes represent the final classification. The classification process is implemented by a set of rules that determine the path to be followed, starting from the root node and ending at one terminal node, which represents the label for the object being classified. At each nonterminal node, a decision has to be made about the path to the next node. Figure 6.1 illustrates a simple hierarchical decision tree classifier using a pixel's reflectance value as input to identify that pixel as either trees, shrubs, swamp, or water. It is obvious that the nature of the decisions being set and the sequence of attributes occurring within a tree will affect classification results. Thus one knows that the efficiency and performance of this approach is strongly affected by the algorithm for inducting a decision tree.

Two approaches to the design of a decision tree can be considered. One is based on the user's knowledge and relies solely on user interaction. This is called the *manual design approach*. The second approach uses an automatic procedure. In the manual design procedure, statistics for all classes are first computed, and a graph of the spectral range in each band is constructed. From the graph and the statistical parameters, estimates of the decision boundaries are derived and the tree is designed manually to separate classes in a hierarchical fashion. The construction of a tree by manual design methods is time-consuming and may not provide satisfactory results, particularly when the number of classes is large and there is spectral overlap between classes.

Swain and Hauska (1977) proposed a heuristic search technique based on a mathematical evaluation function for solving problems that are more complex. Other strategies for optimal hierarchical tree design are described by Kulkarni and Laveen (1976) and Kurzynski (1983). The design of an optimum tree classifier, which will be

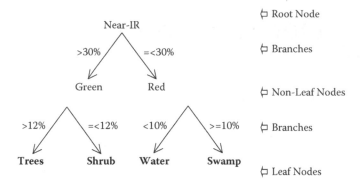

FIGURE 6.1   A simple hierarchical decision tree classifier using RGB reflectance as input.

reflected in both performance and computational efficiency, depends on the choice of the tree structure, the choice of features used in each terminal node, and the decision rules for performing the classification at each nonterminal node. Lee and Richards (1985) describe a classification strategy in which the classes are hierarchically separated in a piecewise linear fashion. In this approach, computational demands increase only linearly with the number of features used, and classification accuracy is claimed to be comparable to that of the maximum likelihood (ML) classifier. Kim and Landgrebe (1991) propose a hybrid approach to the design of decision tree classifiers. The rationale underlying their technique is that a class must simultaneously be of informational value and separable from other classes. As we know, supervised procedures, based upon training samples, can guarantee the former but not the latter; unsupervised procedures can guarantee the latter but not the former. Further, only terminal nodes must be both of separable and informational value. Nonterminal nodes are not required to be classes of informational value, but they must represent separable classes. A thorough review of the various methods used to develop decision tree classifiers is given in Safavian and Landgrebe (1991).

Interest in the use of automatic methods of designing and using decision tree classifiers (such as ID3 [Iterative Dichotomizer 3], C4.5, and CART [classification and regression tree] algorithms) has fast grown in recent years and the related performances are promising. This chapter illustrates methods of automatically constructing decision trees for classification. Indices for measuring the performance of a decision tree are first addressed. Several well-known schemes for forming decision trees are discussed in the following sections. Discussion of other interesting and relevant topics, such as methodologies for decision tree pruning, boosting, random forest classifiers, and applications in the remote sensing field, are also included.

6.1   FEATURE SELECTION MEASURES FOR TREE INDUCTION

As decision trees are iteratively built by recursively partitioning the data into leaf nodes, it follows that when generating a decision tree, the primary issue encountered by the user is to choose a suitable index to measure the performance of the decision tree (in terms of classification accuracy, for example). The two most frequently used indices in decision tree induction are known as *information gain* (Quinlan, 1979, 1993) and the *Gini impurity index* (Breiman et al., 1984), respectively. However, some decision tree induction algorithms may use statistical tests as feature selection criteria. For instance, one may use the Chi-square test to measure whether, for a certain subset of the samples, the attribute X is significantly coupled to the information class j. For each attribute X, the Chi-square test is performed on the pairs (X, j), and the attribute generating the smallest p-value resulting from the Chi-square test is selected as a split basis. This idea will be explained later in the context of the descriptions of the CHAID (CHi-squared Automatic Interaction Detector), CART (Classification and Regression Tree), and QUEST decision trees, respectively.

6.1.1 INFORMATION GAIN

Quinlan (1979, 1993) proposed a measure called *information gain* for tree induction based on Shannon's entropy measure (Shannon, 1948) used in information theory, which is expressed as $(-\log_2 \times$ probability) bits. Thus, eight equally probable messages each contain $-\log_2(\frac{1}{8}) = 3$ bits of information. The probability of a set of training data for a particular class is the relative frequency of the observations in that class (e.g., if training data class i contains 8 pixels and the total number of training data pixels is 100, then the probability of class i is 0.08). The concept of information theory is used to evaluate all possible tests for the subdivision of the tree, and the test that produces the greatest information gain is selected.

During the tree induction process, for a node t, the information gain for each attribute or feature (spectral band in the case of remote sensing) is calculated based on the split at node t on that attribute. The attribute with the greatest information gain is chosen for splitting at that node. In order to compute information gain, the entropy $I_E(t)$ for node t must first be calculated, which is expressed by

$$I_E\left(t\right) = -\sum_{j=1}^{m} f\left(t, j\right) \log_2 f\left(t, j\right) \tag{6.1}$$

where $f(t, j)$ is the proportion of training samples belonging to class j, $j \in \{1, 2, \ldots, m\}$, within node t. Here, m is the number of classes. That is, if node t contains N_t samples, then $f(t, j)$ is calculated by the following expression:

$$f\left(t, j\right) = \frac{1}{N_t} \sum_{i=1}^{N_t} \Gamma\left(y_i, j\right) \tag{6.2}$$

where

$$\Gamma\left(y_i, j\right) = \begin{cases} 1, & \text{if } y_i = j \\ 0, & \text{otherwise} \end{cases}. \tag{6.3}$$

In other words, $I_E(t)$ measures the information conveyed by the distribution of samples within node t. For instance, according to Equation (6.1), if the sample values contained in node t are (0, 1) then $I_E(t) = 0$. If the sample values are (0.3, 0.7), then $I_E(t) = 0.881$, and if the sample values are (0.5, 0.5), then $I_E(t) = 1$. Thus, the more uniform the sample distribution of $I_E(t)$ is, the greater the information content of node t will be (Shannon, 1948). Now, define attribute $X (= \{x_1, x_2, \ldots, x_r\})$. For a value x_i, suppose that there are n_i samples, and $f(t_{X(x_i)}, j)$ is the proportion of samples with the value x_i belonging to class j at node t. The entropy $I_E(t_{X(x_i)})$ associated with x_i for node t is computed by

$$I_E\left(t_{X(x_i)}\right) = -\sum_{j=1}^{m} f\left(t_{X(x_i)}, j\right) \log_2 f\left(t_{X(x_i)}, j\right). \tag{6.4}$$

Finally, the information gain associated with a split on attribute X is calculated by

$$Gain(t, X) = I_E(t) - \left(\frac{n_1}{N_t}\right) I_E\left(t_{X(x_1)}\right) - \left(\frac{n_2}{N_t}\right) I_E\left(t_{X(x_2)}\right) \cdots - \left(\frac{n_r}{N_t}\right) I_E\left(t_{X(x_r)}\right). \tag{6.5}$$

Equation (6.5) denotes the difference between the information needed to describe node t both excluding and including attribute X, i.e., the gain in information due to attribute X.

For further illustration, Table 6.1 shows a total of 15 samples at node t with the attributes *Brightness*, *Greenness*, and *Wetness* derived from the tasseled cap transform (Section 2.1.1), respectively, for identifying the class *Target*. The corresponding entropy $I_E(t)$ for node t is calculated by Equation (6.1) as

$$I_E(t) = -\left(\frac{4}{15}\right) \log_2 \left(\frac{4}{15}\right) - \left(\frac{11}{15}\right) \log_2 \left(\frac{11}{15}\right) = 0.837. \tag{6.6}$$

For the attribute *Wetness*, the entropy $I_E(t_{Wetness(high)})$ and $I_E(t_{Wetness(low)})$ can be calculated as follows according to Equation (6.4):

$$I_E\left(t_{Wetness(high)}\right) = -\left(\frac{3}{6}\right) \log_2 \left(\frac{3}{6}\right) - \left(\frac{3}{6}\right) \log_2 \left(\frac{3}{6}\right)$$

$$= 0.5 + 0.5 = 1$$

$$I_E\left(t_{Wetness(low)}\right) = -\left(\frac{1}{9}\right) \log_2 \left(\frac{1}{9}\right) - \left(\frac{8}{9}\right) \log_2 \left(\frac{8}{9}\right)$$

$$= 0.352 + 0.151 = 0.503. \tag{6.7}$$

TABLE 6.1

A Total of 15 Samples with Attributes *A* = *Brightness*, *B* = *Greenness*, and *C* = *Wetness*. See Text for Further Description

Sample No.	A: *Brightness*	B: *Greenness*	C: *Wetness*	Class: *Target*
1	High	High	Low	Yes
2	High	Low	High	Yes
3	Medium	High	Low	No
4	Low	Low	Low	No
5	Low	High	Low	No
6	Low	Low	High	No
7	Medium	High	High	No
8	High	Low	Low	No
9	High	High	Low	No
10	Low	Low	Low	No
11	High	High	High	No
12	Low	Low	High	Yes
13	Medium	High	Low	No
14	Medium	Low	High	Yes
15	Low	Low	Low	No

The information gain for *Wetness* (Equation [6.5]) is then calculated by

$$Gain\left(t, Wetness\right) = I_E\left(t\right) - \left(\frac{6}{15}\right)I_E\left(t_{Wetness(high)}\right) - \left(\frac{9}{15}\right)I_E\left(t_{Wetness(low)}\right) \tag{6.8}$$

$$= 0.837 - 0.4 - 0.302 = 0.135.$$

The gain for other the attributes, *Brightness* and *Greenness* are calculated in a similar way as

$$Gain\left(t, Brightness\right) = I_E\left(t\right) - \left(\frac{5}{15}\right)I_E\left(t_{Brightness(high)}\right)$$

$$- \left(\frac{4}{15}\right)I_E\left(t_{Brightness(medium)}\right) - \left(\frac{6}{15}\right)I_E\left(t_{Brightness(low)}\right) \tag{6.9}$$

$$= 0.837 - 0.324 - 0.216 - 0.26 = 0.037$$

$$Gain\left(t, Greenness\right) = I_E\left(t\right) - \left(\frac{7}{15}\right)I_E\left(t_{Greenness(high)}\right) - \left(\frac{8}{15}\right)I_E\left(t_{Greenness(low)}\right) \tag{6.10}$$

$$= 0.837 - 0.276 - 0.509 = 0.052.$$

Using the node-splitting rule described in the opening paragraphs of Section 6.1.1, the attribute *Wetness* shows the largest information gain and is therefore the first attribute selected for decision tree node splitting.

6.1.2 GINI IMPURITY INDEX

The Gini impurity index (named after Italian economist Corrado Gini, and originally used for measuring income inequality) measures the impurity of an input feature with respect to the classes (Breiman et al., 1984). Specifically, the Gini impurity index reaches its minimum (zero) when all attributes in the node fall into a single information class. The Gini index associated with attribute X (= $\{x_1, x_2, ..., x_r\}$) for node t is denoted by $I_G(t_{X(xi)})$ and is expressed as

$$I_G\left(t_{X(x_i)}\right) = 1 - \sum_{j=1}^{m} f\left(t_{X(x_i)}, j\right)^2 \tag{6.11}$$

where $f(t_{X(x_j)}, j)$ is the proportion of samples with the value x_i belonging to class j at node t as defined in Equation (6.4). The decision tree splitting criterion is based on choosing the attribute with the lowest Gini impurity index of the split. Let a node t be split into r children, n_i be the number of records at child i, and N_t be the total number of samples at node t. The *Gini impurity index of the split* at node t for attribute X is then computed by

$$Gini\left(t, X\right) = \left(\frac{n_1}{N_t}\right) I_G\left(t_{X(x_1)}\right) + \left(\frac{n_2}{N_t}\right) I_G\left(t_{X(x_2)}\right) ... + \left(\frac{n_r}{N_t}\right) I_G\left(t_{X(x_r)}\right); \tag{6.12}$$

Here, again for illustration purpose, Table 6.1 is used as an example. From Equation (6.11) the Gini values for attribute *Wetness* ($I_G(t_{Wetness(high)})$ and ($I_G(t_{Wetness(low)})$)) are calculated as follows:

$$I_G\left(t_{Wetness(high)}\right) = 1 - \left(\frac{3}{6}\right)^2 - \left(\frac{3}{6}\right)^2$$

$$= 1 - 0.25 - 0.25 = 0.5$$

$$I_G\left(t_{Wetness(low)}\right) = 1 - \left(\frac{1}{9}\right)^2 - \left(\frac{8}{9}\right)^2$$

$$\tag{6.13}$$

$$= 1 - 0.013 + 0.79 = 0.197.$$

The Gini impurity index of the split (Equations [6.12] and [6.13]) is then calculated by

$$Gini(t, Wetness) = \left(\frac{6}{15}\right) I_G\left(t_{Wetness(high)}\right) + \left(\frac{9}{15}\right) I_G\left(t_{Wetness(low)}\right)$$

(6.14)

$$= 0.2 + 0.118 = 0.318.$$

The Gini impurity for splitting attributes *Brightness* and *Greenness* can be calculated in a similar way as

$$Gini(t, Brightness) = \left(\frac{5}{15}\right) I_G\left(t_{Brightness(high)}\right) + \left(\frac{4}{15}\right) I_G\left(t_{Brightness(medium)}\right)$$

$$+ \left(\frac{6}{15}\right) I_G\left(t_{Brightness(low)}\right) = 0.16 + 0.1 + 0.111 = 0.371$$

$$Gini(t, Greenness) = \left(\frac{7}{15}\right) I_G\left(t_{Greenness(high)}\right) + \left(\frac{8}{15}\right) I_G\left(t_{Greenness(low)}\right)$$

(6.15)

$$= 0.25 + 0.114 = 0.364.$$

Based on the results of the above calculations, the attribute *Wetness* has the smallest Gini impurity index (0.318) and so it is the first attribute to be split at this node.

6.2 ID3, C4.5, AND SEE5.0 DECISION TREES

6.2.1 ID3

Quinlan (1979) proposes a decision induction method called ID3. He noted that the building of a decision tree requires a "divide and conquer" strategy that uses a recursive testing procedure with the aim of generating a small tree. ID3 uses information gain (Section 6.1) as a basis for tree induction. The ID3 algorithm is shown in Figure 6.2. Note that the algorithm in Figure 6.2, step 7, uses a technique to recursively self-call program ID3 to build the decision tree. Based on the example shown in Table 6.1, the procedure for using ID3 algorithm to generate a decision tree is shown in Figure 6.3. In Figure 6.3a, attribute *Wetness* is first chosen as candidate for splitting as it has the largest gain (see Equation [6.8] = 0.135) in comparison with the gains computed for attributes *Brightness* (Equation [6.9] = 0.037) and *Greenness* (Equation [6.10] = 0.052), respectively. Figure 6.3b shows the further tree induction under the circumstance that *Wetness* = high. The corresponding gain for *Brightness* and *Greenness* can be calculated by

---

Notations: 1. A: a set of attribute types used as predictors, originally, $A = \{A1,$
　　　　　　　$A2, ..., An\}$, where $A1$, $A2$, ..., An each may represent a spectral
　　　　　　　band in the case of remotely sensed imagery.
　　　　2. t: the node with corresponding attribute samples A contained
　　　　　　　inside.

ID3 (A, t)
　Begin
　　　1.　If t consists of samples all belonging to the same class, label the
　　　　　node with that class then exit the program;
　　　2.　If A is empty, label the node t as the most frequent of the class
　　　　　contained in the node t then exit the program;
　　　3.　Compute information gain for each attribute type using Equation
　　　　　(6.5) and let Ak be the attribute with largest gain among A;
　　　4.　Let $\{Ak(i)|i = 1, 2, .., r\}$ be the values of attribute Ak;
　　　5.　Let $\{tk(i)|i = 1, 2, .., r\}$ be the child-nodes of t consisting
　　　　　respectively of attribute samples according to $Ak(i)$;
　　　6.　Label node t as Ak and branches labeled $Ak(1)$, $Ak(2)$, .., $Ak(r)$ going
　　　　　respectively to the child-node $tk(1)$, $tk(2)$, ..., $tk(r)$;
　　　7.　For $i = 1$ to r
　　　　　　　ID3($A-\{Ak\}$, $tk(i)$);
　　　　End
　End

FIGURE 6.2　Decision tree induction using the ID3 algorithm.

$$Gain\left(t_{Wetness(high)}, Brightness\right) = I_E\left(t_{Wetness(high)}\right) - \left(\frac{2}{6}\right)I_E\left(t_{Wetness(high)\&Brightness(high)}\right)$$

$$-\left(\frac{2}{6}\right)I_E\left(t_{Wetness(high)\&Brightness(medium)}\right) - \left(\frac{2}{6}\right)I_E\left(t_{Wetness(high)\&Brightness(low)}\right)$$

$$= 1 - 0.333 - 0.333 - 0.333 = 0.001$$

$$Gain\left(t_{Wetness(high)}, Greenness\right) = I_E\left(t_{Wetness(high)}\right) - \left(\frac{2}{6}\right)I_E\left(t_{Wetness(high)\&Greenness(high)}\right)$$

$$-\left(\frac{4}{6}\right)I_E\left(t_{Wetness(high)\&Greenness(low)}\right) \tag{6.16}$$

$$= 1 - 0 - 0.54 = 0.46$$

Accordingly, the attribute *Greenness*, with a gain value of 0.46, is the attribute selected for further splitting from the parent node *Wetness* = high. For the circumstance *Wetness* = low, the corresponding gains are computed as

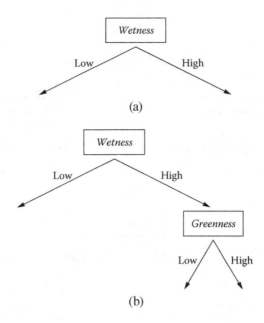

FIGURE 6.3 Tree induction process based on the samples in Table 6.1. (a) Attribute *Wetness* is firstly chosen as a splitting candidate. (b) The tree induction under the circumstance that *Wetness* = high and *Greenness* is selected for further splitting according to gain.

$$Gain\left(t_{Wetness(low)}, Brightness\right) = I_E\left(t_{Wetness(low)}\right) - \left(\frac{3}{9}\right)I_E\left(t_{Wetness(low)\&Brightness(high)}\right)$$

$$- \left(\frac{2}{9}\right)I_E\left(t_{Wetness(low)\&Brightness(medium)}\right) - \left(\frac{4}{9}\right)I_E\left(t_{Wetness(low)\&Brightness(low)}\right)$$

$$= 0.503 - 0.306 - 0 - 0 = 0.197$$

$$Gain\left(t_{Wetness(low)}, Greenness\right) = I_E\left(t_{Wetness(low)}\right) - \left(\frac{5}{9}\right)I_E\left(t_{Wetness(low)\&Greenness(high)}\right)$$

$$- \left(\frac{4}{9}\right)I_E\left(t_{Wetness(low)\&Greenness(low)}\right) \tag{6.17}$$

$$= 0.503 - 0.401 - 0 = 0.102$$

Based on these results, the attribute *Brightness* is selected for splitting in the case *Wetness* = low. The result of further decision tree induction for *Wetness* = low is shown in Figure 6.3c. Similarly, according to the ID3 algorithm, the final decision tree inductions for attribute *Greenness* and *Brightness* are shown in Figures 6.3d and 6.3e, respectively. The resulting tree is a 4-level decision tree. Note that nodes

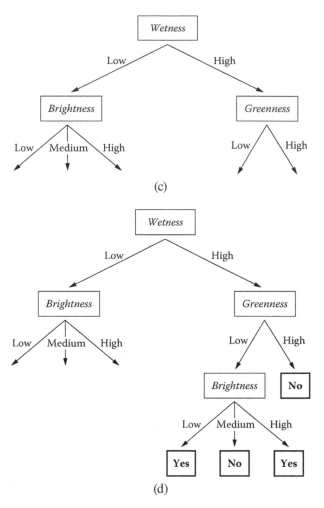

(c)

(d)

FIGURE 6.3 (continued) (c) When attribute *Wetness* = low, the attribute *Brightness* is chosen. (d) The left-hand side of the tree induction is complete.

numbered 11 and 12 will generate the classification ambiguity, i.e., both nodes choosing either *yes* or *no* will always have 50% classification errors. In such a situation, the user may use more reference data to help make an unequivocal decision. Alternatively, a probability can be used to provide for a fuzzy result (Chapter 5) instead of the originally crisp decision procedure. One should note that ID3 cannot handle numeric and continuous attribute values, is unable to deal with missing (or unknown) attributes, and does not provide users with a pruning facility. Quinlan (1993, 1996) therefore made a number of improvements to ID3, and the resulting improved decision tree version is called C4.5.

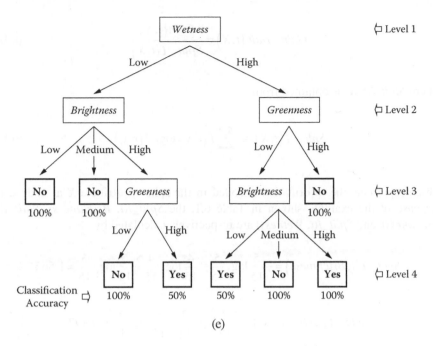

(e)

FIGURE 6.3   (continued) (e) Decision tree induction is finally finished according to the ID3 algorithm.

6.2.2   C4.5

C4.5 is an extension of the ID3 decision tree induction algorithm to account for the several issues not appropriately dealt with by ID3 as mentioned previously (Quinlan, 1986, 1993). The significant improvements include choosing an appropriate attribute selection measure, handling data with missing attributes, handling numeric and continuous attributes, and decision tree pruning. This section will discuss these characteristics, except that the tree pruning issue is left for discussion in Section 6.6.

C4.5 uses either information gain or a normalized version called the *gain ratio* (described in this section) to choose attributes as candidates for splitting. From Equation (6.5), it is seen that the information gain measure tends to favor attributes that have multiple values. For instance, in an extreme case, if attribute X holds a distinct value for each class, Equation (6.4) will always be 0, which in turn results in maximal gain in Equation (6.5), and thus this attribute becomes the first candidate for splitting. This feature makes the tree induction process more easily affected by noise. To compensate for this, Quinlan (1986) proposes an alternative index called the *gain ratio,* which can be regarded as a normalized version of the information gain measure. The gain ratio is defined as

$$Gain\_ratio(t,X) = \frac{Gain(t,X)}{Split\_I(t,X)} \tag{6.18}$$

where $Split\_I(t, X)$ is obtained from

$$Split\_I(t,X) = -\sum_{i=1}^{r} f(t,x_i) \log_2 f(t,x_i), \tag{6.19}$$

which represents the information encoded in the split of attribute X at node t. In the case of the example shown in Table 6.1, the $Split\_I(t, Brightness)$, $Split\_I(t, Greenness)$, and $Split\_I(t, Wetness)$ are respectively calculated by

$$Split\_I(t, Brightness) = -\frac{5}{15}\log_2\frac{5}{15} - \frac{4}{15}\log_2\frac{4}{15} - \frac{5}{15}\log_2\frac{5}{15} = 1.565$$

$$Split\_I(t, Greenness) = -\frac{7}{15}\log_2\frac{7}{15} - \frac{8}{15}\log_2\frac{8}{15} = 0.997$$

$$Split\_I(t, Wetness) = -\frac{6}{15}\log_2\frac{6}{15} - \frac{9}{15}\log_2\frac{9}{15} = 0.971. \tag{6.20}$$

Based on the information gain computed in Equation (6.8), (6.9), and (6.10), the corresponding gain ratio for each attribute is then computed by

$$Gain\_ratio(t, Brightness) = \frac{Gain(t, Brightness)}{Split\_I(t, Brightness)} = \frac{0.037}{1.565} = 0.024$$

$$Gain\_ratio(t, Greenness) = \frac{Gain(t, Greenness)}{Split\_I(t, Greenness)} = \frac{0.052}{0.997} = 0.052$$

$$Gain\_ratio(t, Wetness) = \frac{Gain(t, Wetness)}{Split\_I(t, Wetness)} = \frac{0.135}{0.971} = 0.139 \tag{6.21}$$

From these results, the attribute *Wetness* is the first candidate chosen for splitting as it has the highest gain ratio.

To deal with missing attributes, C4.5 uses a probability concept. In the example shown in Table 6.1, suppose that a new sample is inserted, for which *Wetness* = low, *Brightness* = high, and the value for *Greenness* is missing. One then traces the root node *Wetness* in the decision tree (see Figure 6.3e) first down left to node 3 *Brightness* and then to the right path to node 6 *Greenness*. At node 6, since the value of *Greenness* is missing, one has no idea whether the newly added sample should be classified as target (*yes*) or not-target (*no*). To solve this issue, one can find that if one set of unknown *Greenness* is high, the sample belongs to the target (*yes*) or not-target (*no*) will be 50% each (because node 12 in Figure 6.3e is 50% in accuracy). If one set *Greenness* to be low, the sample clearly belongs 100% to not-target (*no*) (node 13 in Figure 6.3e). Thus one can finally come to a conclusion for node 6 that under the condition where the value for *Greenness* is missing, the probability of the sample belonging to target (*yes*) is 0.333, and 0.666 for not-target (*no*), respectively. In a case where a crisp rather than a fuzzy decision is required, then based on the calculated probability values, the *no* alternative is selected.

To cope with attributes containing numeric data types (integers or real numbers, for instance) in addition to categorical data types, C4.5 uses the following procedures. The numeric data are first sorted in increasing or decreasing order. The algorithm then chooses one value, say x_i, from the data set and uses this value as a cutting point to partition the data, one part consisting of values including and greater than x_i, and the other part containing the values smaller than x_i. Gain ratios are computed for each partition. The procedures for choosing the cutting values and computing gain ratios are repeated for each numeric or continuous data item. The cutting value that produces partitions having the highest gain ratio is selected. Table 6.2 presents

TABLE 6.2

The 15 Samples Are the Same as Shown in Table 6.1 Except That Attribute B = *Greenness* Is Converted to Numeric Type

Sample No.	A: *Brightness*	B: *Greenness*	C: *Wetness*	Class: *Target*
1	High	12	Low	Yes
2	High	10	High	Yes
3	Medium	16	Low	No
4	Low	12	Low	No
5	Low	13	Low	No
6	Low	16	High	No
7	Medium	16	High	No
8	High	19	Low	No
9	High	12	Low	No
10	Low	12	Low	No
11	High	15	High	No
12	Low	12	High	Yes
13	Medium	12	Low	No
14	Medium	12	High	Yes
15	Low	18	Low	No

an example in which the attributes are mostly the same as Table 6.1 except that the data for attribute *Greenness* is turned into numeric values. One can use the same procedure introduced previously to calculate the gain or gain ratio. For *Greenness* one now has to choose cutting values one by one and to calculate the corresponding gain ratio. After exhaustive trials for *Greenness* values in Table 6.2, the best cutting value chosen is $x_{i.} = 13$, which partitions the feature *Greenness* into two categories, one with values less than 13 (a total of 8 records) and the second with values greater than or equal to 13 (7 records), giving a total of 15 samples. The resulting gain and gain ratio are computed as

$$Gain(t, Greenness) = I_E(t) - \left(\frac{8}{15}\right) I_E\left(t_{Greenness(<13)}\right) - \left(\frac{7}{15}\right) I_E\left(t_{Greenness(\geq 13)}\right)$$

$$= 0.837 - 0.533 - 0 = 0.304$$

$$Gain\_ratio(t, Greenness) = \frac{Gain(t, Greenness)}{Split\_I(t, Greenness)} = \frac{0.304}{0.997} = 0.302 \quad (6.22)$$

Either the gain or gain ratio computed from the attribute *Greenness* now is the largest in comparison with that computed for other attributes. Therefore the attribute *Greenness* is chosen as the first candidate for splitting. For further tree induction processes, one can iteratively implement the equation either for computing *information gain* (6.5) or *gain ratio* (6.18) to perform the tree induction. The final decision tree is shown in Figure 6.4.

6.2.3   SEE5.0

A further development of C4.5 called SEE5.0 can operate on several additional data types as well as those available in C4.5, including dates, times, time stamps, ordered discrete attributes, and case labels. In addition to dealing with missing values, SEE5.0 allows values to be flagged as not applicable and provides facilities for defining new attributes as functions of other attributes. SEE5.0 also introduces the important concept of *boosting*, which is a technique for combining multiple learning processes with appropriately weighted training class pixels to improve predictive accuracy. The weights given to incorrectly labeled (or unlabeled) pixels are increased relative to those given to correctly labeled pixels, and the tree-generating process is repeated. This cycle is repeated a number of times. Friedl et al. (1999) suggest that no more than 10 iterations are required, and their results show an improvement in accuracy of about 25% following boosting. We discuss this concept further in Section 6.7. Another characteristic of SEE5.0 is that it allows a separate cost to be defined for each predicted or actual class pair because in practical applications some classification errors are more serious than others. SEE5.0 then constructs classifiers to minimize expected misclassification costs rather than error rates.

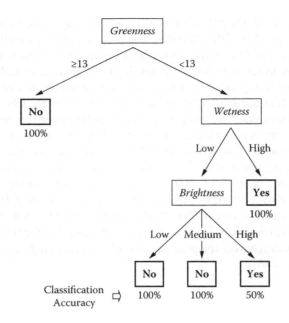

FIGURE 6.4   Tree induction using the values in Table 6.2 as inputs.

6.3   CHAID

CHAID (Kass, 1980; Chi-square Automatic Interaction Detector) is used to study the relationship between dependent variables (i.e., classes) and a series of predictor variables (e.g., spectral bands). CHAID selects a set of predictors and their interactions that optimally predict the dependent measure. Before using CHAID, one should know that the main purpose of the Chi-square test is to look at the relationship between two variables to determine the level of dependency between them. This is the primary characteristic of CHAID using the Chi-square test to determine which attribute is the most relevant to the class being predicted. This characteristic is quite different from the algorithms used in decision trees like the ID3 families and CART (Section 6.4), which use either information gain or the Gini index to choose the optimal splitting attribute. CHAID also uses Chi-square to judge if a decision tree should stop growing so as to avoid the issue of overfitting. Therefore, for a CHAID decision tree, there is no need to consider the issue of tree pruning (Section 6.6).

During the process of tree induction, CHAID carries out tests on each attribute to see whether splitting the sample of training pixels based on the attribute under test leads to a statistically significant discrimination in the dependent measure (i.e., the class). CHAID then builds a contingency table to store the results of these Chi-square tests for each attribute–class pair. The most relevant (i.e., the most significant) attribute is first selected and the samples are split in terms of a rule determined for that attribute (e.g., "if $x > b$ then node I"). After the first split, for each of the split

sample groups formed, CHAID then tests if the subgroup could be further significantly split using another of the attributes. The process continues until the end of the
tree induction process. One thus has a series of groups that are maximally different
from one another in terms of the attributes. In the general case, CHAID can only
deal with nominal (categorical) or ordinal (ordered categories ranked from small to
large) attributes. If an attribute is measured on a continuous scale, CHAID has to first
turn the continuous values into categorical values and then perform analysis of variance methods for optimal tree induction. For instance, if the attribute *Temperature* is
measured on a continuous scale from 0 to 100, CHAID may separate these continuous values into category *A*: 0 to 9, *B*: 10 to 19, …, etc. Such a technique for trans-
forming continuous values into categorical types is called *binning*. Since binning
expresses the original continuous attribute on a coarser scale, the CHAID tree may
lose accuracy to a certain extent. One should recognize that CHAID is analogous to
a forward stepwise regression analysis and thus contains all of the possible attendant
difficulties (such as ignoring the presence of outliers, and being restricted to *t*-ratio
criterion; Harrell, 2001).

6.4   CART

CART is the acronym for the Classification and Regression Tree (Breiman et al.,
1984). The main difference between CART and C4.5/SEE5.0 is that CART only
allows two branches (i.e., two children) to form at each spitting process, while C4.5/
SEE5.0 can generate different numbers of branches as required during the induction
process. In other words, the decision tree generated by the CART algorithm will
always be a binary tree.

As far the implementations of CART are concerned, when the classes under prediction are continuous in nature, one may use a *regression tree* to perform the prediction in terms of regression techniques. The information contained in each leaf
node will then be expressed as a continuous value. On the other hand, one may use
a *classification tree* to deal with categorical classes. It should be noted that the decision tree inducted by the CART algorithm generally tends to be overfitted. CART
then uses tests set to perform the decision tree validation and accordingly to prune
the tree (Section 6.6). Also, the measure that CART implements to perform tree
induction can be either based on univariate or multivariate criteria as detailed below
in this section.

For univariate tree induction criteria, the decision tree induction algorithms introduced in Section 6.2 (ID3 and descendent) and 6.3 (CHAID) all belong to the univariate induction domain in which the decision boundaries formed at each node of the
tree are defined by the outcome of a test applied to a single attribute that is evaluated
at each internal node. On the basis of the test outcome, the data are split into two or
more subsets. This process continues until a leaf node is reached, and the class label
associated with that leaf node is then assigned. CART may implement information
gain, gain ratio, or the Gini index to perform univariate decision tree induction.

One should note that univariate decision tree algorithms at each test use a
single feature only; it is restricted to a split through the feature space that is
orthogonal to the axis representing the selected feature as shown in Figure 6.5

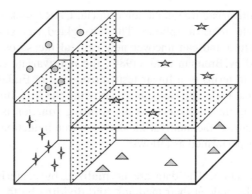

FIGURE 6.5 Example of decision boundaries formed by the univariate decision tree algorithm.

in which the decision boundaries are all orthogonal to the axes formed by attributes, so that this feature space is divided into a set of *hyper-rectangles*. The univariate decision tree is thus similar to the simple parallelepiped classifier in the way it partitions feature space. To overcome this restriction, CART allows the implementation of multivariate criteria to perform decision tree induction as detailed below.

Multivariate decision tree algorithms allow node splits extended to include linear combinations of features (Friedl and Brodley, 1997). A set of linear discriminant functions is estimated at each interior node of a multivariate decision tree, with the coefficients for the linear discriminant function at each interior node being estimated from the training data. The decision boundaries generated by a multivariate decision tree will therefore be more flexible than that formed by a univariate decision tree. Figure 6.6 displays an example of decision boundaries formed by a multivariate decision algorithm. In these cases, the multivariate decision boundaries are not necessarily orthogonal to the axes of the feature space, and should provide a more flexible fit to the boundaries of the classes.

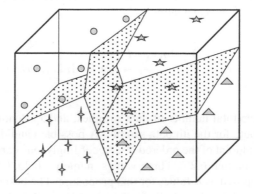

FIGURE 6.6 Decision boundaries formed by the multivariate decision tree algorithm.

When using multivariate tree induction criteria, CART looks for weighted averages of input attributes to use as splitters. These weighted averages can reveal important database structure and can uncover new critical measures. According to the algorithm proposed by Breiman et al. (1984), the multivariate case search at each node is performed in terms of a linear test (introduced in Equation [6.23]), which splits the data for that node into two subsets showing the least impurity. The CART algorithm for finding the coefficients of the splitting function is a stepwise procedure. Specifically, during the multivariate decision tree induction process at a given node, the process is performed as follows:

1. Data normalization: The data are normalized by centering all the data samples for the node at their medians, and dividing by their inter-quartile ranges.
2. Univariate split: Suppose there are a total of K attributes denoted as X_k, $k = 1, 2, ..., K$, and let T denote the splitting threshold. The best univariate split $X_k \leq T$ (for instance, X_k may be attribute *Greenness* and T may be chosen as value 13 as shown in Table 6.2) is found in terms of minimizing the Gini impurity index measure as described earlier.
3. Perform the linear test: This part of the process involves a number of cycles. A given split

$$\sum_{k=1}^{K} a_k X_k \leq T$$

is improved by successively considering splits of the form:

$$\sum_{k=1}^{K} a_k X_k - \delta\left(X_k + \lambda\right) \leq T \tag{6.23}$$

where a_k is the weighting coefficient to the corresponding attribute X_k, and δ is the parameter to be optimized by taking λ as the fixed value. The method for updating these coefficients is introduced below. For attribute X_1, one can obtain the best δ^* by calculating

$$\delta^{(i)} = \frac{\sum_{k=1}^{K} a_k x_{ik} - T}{x_{1k} + \lambda}, \text{ for } i = 1, 2, ..., N \tag{6.24}$$

where N is the total number of data samples, and x_{ik} denotes the value of the kth attribute for the ith data sample. A possible candidate value for δ^* can be the midpoint of the ordered $\delta^{(i)}$. Accordingly, one can obtain the best split from Equation (6.23). Three values, namely −0.25, 0.0, and 0.25 are normally proposed as λ values for this procedure. The best $\delta^{(i)}$ among these three values is then taken as the δ^* that is used to update a_k in

$$\sum\nolimits_{k=1}^{K} a_k X_k$$

by the replacement with a_k^{new}, where

$$a_k^{new} = \begin{cases} a_k & \text{for } k \neq 1 \\ a_1 - \delta^* & \text{otherwise} \end{cases} \tag{6.25}$$

The new threshold T^{new} is computed by $T^{new} = T + \delta^* \times \lambda$. This step is repeated for the other attributes $X_2, X_3, \ldots, X_K$. At the end of each cycle, one has completed the test

$$\sum\nolimits_{k=1}^{K} a_k X_k \leq T.$$

The next task is to improve the splitting threshold T by finding the best test, fixing a_k, and updating only the threshold T. The whole process can be terminated under the circumstance that the decrease in impurity from one cycle to the next is below a predetermined small level.

Readers should note that even though a multivariate decision tree may appear to be more flexible than the univariate tree, several factors may affect classification performance. These factors include the algorithms implemented to estimate the splitting rule at internal nodes, the various feature selection methods, and the use of different classification algorithms at different nodes (Friedl and Brodley, 1997).

6.5   QUEST

The word QUEST stands for Quick, Unbiased, Efficient, and Statistical Tree algorithm (Loh and Shih, 1997). The decision tree inducted by the QUEST algorithm is a binary tree. When the number of classes is more than two, QUEST will initially group the classes into two clusters to ensure binary splits. This is performed by applying a two-means clustering algorithm, which minimizes the within-cluster sum of squares to the original class means (see Chapter 2). Normally, two of the most extreme means are used as initial cluster centers. QUEST then uses quadratic discriminant analysis (QDA) to carry out the splitting process. Instead of combining the problem of split point selection with that of attribute selection (as adopted by most decision tree induction algorithms), QUEST deals with them separately. In what follows, the QUEST algorithm is therefore addressed in two parts, one dealing with split point selection and the other with attribute selection, respectively.

6.5.1   SPLIT POINT SELECTION

In the case where the attributes being dealt with are all numeric variables, let X be the attribute selected to perform a split at node t. If the number of classes in node t

is more than two, the two-means clustering algorithm is first applied and groups the classes into superclasses A and B. Let $\bar{x}(A)$ and s_A^2 denote the class mean and variance for class A, and let $\bar{x}(B)$ and s_B^2 denote the corresponding quantities for class B. QDA splits the attribute X into three intervals, namely, $(-\infty, d_1)$, (d_1, d_2), and (d_2, ∞), where d_1 and d_2 are the roots of the following equation:

$$p\left(A|t\right)s_A^{-1}\Phi\left[\left(x-\bar{x}\left(A\right)\right)/s_A\right]=p\left(B|t\right)s_B^{-1}\Phi\left[\left(x-\bar{x}\left(B\right)\right)/s_B\right]$$

(6.26)

where

$$p\left(j|t\right)=p\left(j,t\right)\bigg/\sum_i p\left(i,t\right)$$

denotes the estimated posterior probability that a sample belongs to class j in node t, and $\Phi(y) = (2\pi)^{-0.5}\exp(-y^2/2)$ denotes the standard normal density function. The probability $p(j, t)$ is defined as

$$p\left(j,t\right)=\pi\left(j\right)N_{j,t}/N_j$$

(6.27)

where $\pi(j)$ denotes the prior probability for class j (normally estimated in terms of sample proportion of class j), N_j is the number of samples belonging to class j, and $N_{j,t}$ is the number of samples belonging to class j within node t. QUEST uses only one of the two roots (the one that is closer to the mean of each class) as the split point to ensure a binary split. By taking logs on both sides of Equation (6.26), one can obtain the quadratic equation $ax^2 + bx + c = 0$, where:

$$a = s_A^2 - s_B^2$$

$$b = 2\left(\bar{x}_A \times s_B^2 - \bar{x}_B \times s_A^2\right)$$

$$c = \left(\bar{x}_B \times s_A^2\right)^2 - \left(\bar{x}_A \times s_B^2\right)^2 + 2s_A^2 \times s_B^2 \log\left[\left(p\left(A|t\right)\times s_B\right)\bigg/\left(p\left(B|t\right)\times s_A\right)\right].$$

(6.28)

If $a = 0$ and $\bar{x}_B \neq \bar{x}_B$, there is only one root (i.e., split point d) as given by

$$d = \frac{\left(\bar{x}_A + \bar{x}_B\right)}{2} - \left(\bar{x}_A - \bar{x}_B\right)^{-1}s_A^2 \log\left[p\left(A|t\right)\big/p\left(B|t\right)\right].$$

(6.29)

In the case of $a \neq 0$ and $b^2 - 4ac < 0$, the split point d is defined as $d = (\bar{x}_A + \bar{x}_B)/2$, otherwise, d is defined to be the root of

$$(2a)^{-1}\left(-b \pm \sqrt{b^2 - 4ac}\right).$$

The above derivations assume the attributes being dealt with are numeric. If attributes are however categorical variables, QUEST first uses an algorithm called CRIMCOORD to transform categorical variables into numeric ones before starting the process of tree induction. For the detailed derivation of categorical-to-continuous transformation, refer to Loh and Shih (1997).

6.5.2 ATTRIBUTE SELECTION

The previous section shows how a spit point is determined to split a selected attribute. In what follows, how to choose such an attribute is further discussed. The basic idea adopted by QUEST to perform attribute selection is statistical tests. For continuous attributes, a threshold value α is chosen and an analysis of variance (ANOVA) F-statistic is computed for every variable. If the largest F-statistic exceeds α, the corresponding attribute is selected to split the node. In the case of categorical attributes, a contingency table and Chi-square tests of independence between the class and the categorical attributes are computed. A contingency table is also known as a cross-tabulation, in which the rows denote the information classes, the columns denote the attributes, and each cell in the table then contains the tabulated counts for the corresponding attribute labeled to the corresponding information class. Chi-square tests of independence for a contingency table then test the null hypothesis that the information class and the attributes are independent. Specifically, let X_1, X_2, ..., X_{K_1} denote continuous attributes, X_{K_1+1}, X_{K_1+2}, ..., X_K are categorical attributes, and let $x_{ik}^{(j)}$ denote the value of the kth attribute for the ith sample in the jth class. The process for attribute selection can be addressed in terms of the following steps:

1. Choose a value $\alpha \in (0,1)$ to be a prespecified level of significance. Compute an ANOVA F-statistic F_k for each continuous attribute X_k, $k = 1$ to K_1. Let $F_{ka} = \max\{F_k : k = 1, 2, ..., K_1\}$ and $\alpha_1 = \Pr\{F_{(m_t-1),(N_t-m_t)} > F_{ka}\}$, where m_t is the number of classes present in node t, N_t is the number of samples in node t, and $F_{u,v}$ denotes the F-distribution with u and v degree of freedom.
2. Calculate the P-value β_k for the Chi-square test of independence between class and categorical attributes for categorical attribute X_k, $k = K_1 + 1$, $K_1 + 2$, ..., K as stored in the contingency table. The degrees of freedom are given by the product $(n_r - 1) \times (n_c - 1)$, where n_r and n_c are the numbers of rows and columns of the table with nonzero totals. Let $\alpha_2 = \beta_{kb}$, where $\beta_{kb} = \min\{\beta_k : k = K_1 + 1, K_1 + 2, ..., K\}$.
3. $k^* = k^a$ if $\alpha_1 \le \alpha_2$, otherwise define $k^* = k^b$. If $\min\{\alpha_1, \alpha_2\} < \alpha/K$, choose attribute X_{k^*} to split the node.
4. If $\min\{\alpha_1, \alpha_2\} \ge \alpha/K$, compute ANOVA F-statistics F_k^z, $k = 1$ to K_1, for the continuous attributes based on the absolute deviations

$$z_{ik}^{(j)} = \left| x_{ik}^{(j)} - \overline{x}_k^{(j)} \right|,$$

where

$$\overline{x}_k^{(j)} = N_{j,t}^{-1} \times \sum_{i=1}^{N_{j,t}} x_{ik}^{(j)}.$$

5. Let

$$F_{k^c}^{(z)} = \max\left\{ F_k^{(z)} : k = 1, 2, ..., K_1 \right\},$$

and

$$\alpha_3 = \Pr\left\{ F_{(m_t-1),(N_t-m_t)} > F_{k^c}^{(z)} \right\}.$$

If $\alpha_3 < \alpha/(K_1 + K)$, choose attribute X_{k^c} to split the node. Otherwise, choose attribute X_{k^*}.

The QUEST algorithm uses linear combination splits and is essentially a recursive linear discriminant analysis on all the attributes. The process is similar to CART. Note that categorical attributes have to be first transformed to CRIMCOORD. If the number of classes is greater than two, a prior grouping of the classes into two superclasses is also carried out. In terms of classification accuracy, variability of split points, and tree size, Loh and Shih (1997) show that QUEST reveals some advantages when univariate splits are used, and that the QUEST trees based on linear combination splits are usually smaller than those generated by other tree induction algorithms.

In order to present a more comprehensive view, Table 6.3 lists the comparisons of characteristics among the different tree induction algorithms introduced above.

6.6 TREE INDUCTION FROM ARTIFICIAL NEURAL NETWORKS

As noted in Chapter 3, artificial neural networks (ANN) may be more robust than statistical classification methods, but it is hard to find clear roadmaps to explanations of how and why the classification results have come to be. This is due to the fact that the ANN mapping process between the inputs and outputs is performed in a *black box*. However, it is important that the decisions generated by an ANN are understood, rather than just arbitrarily taking its outputs. For resolving this issue, one may implement rule extraction algorithms (e.g., Craven, 1996) to uncover the hidden knowledge from ANN.

Among the numerous symbolic learning algorithms, currently the methods being frequently applied to build decision trees from ANN may be roughly categorized into two types known as the *decomposition* (Duch et al., 2001) and *pedagogical* (Craven, 1996; Taha and Ghosh, 1999) approaches, respectively. The fundamental concept of the decomposition method is to perform a searching process for mapping

TABLE 6.3

Comparisons among Different Tree Induction Algorithms

Characteristics	QUEST	CART	CHAID	C4.5
Split types				
Univariate	•	•	•	•
Linear combinations	•	•		
Number of branches at each node				
Always two	•	•		
Two or more			•	•
Missing value methods				
Imputation	•			
Alternate/surrogate splits		•		
Missing value branch			•	
Probability weights				•
Tree size control				
Stopping rule			•	
Pre-pruning				•
Test-sample pruning	•	•		
Cross-validation pruning	•	•		
Other Characteristics				
Choice of misclassification costs	•	•	•	
Choice of class prior probabilities	•	•		
Choice of impurity functions	•	•		
Bagging		•		
Error estimation by cross-validation	•	•	•	

a well-trained ANN into a decision tree. Such kinds of methods could suffer from a serious computational loading issue because as the dimension of the input vector grows, the computational time increases exponentially and thus requires a pruning procedure to downsize the tree structure. The methods known as SUBSET (Towell and Shavlik, 1994) and C-MLP2LN (Duch et al., 2001) belong to this category. Pedagogical approaches, however, treat a trained ANN as a black box, and perform decision tree induction from the ANN by observing the relationship between both the ANN inputs and outputs. Typical methods within this family are referred to TREPAN (Craven, 1996) and Binarized Input-Output Rule Extraction (BIO-RE) (Taha and Ghosh, 1999). Although building decision trees from neural networks may be a useful approach, the field still requires more investigation.

6.7 PRUNING DECISION TREES

A decision tree developed in the manner described previously will generally be complex. Although in some circumstances the user may use predefined conditions (such as the maximal depth a decision tree can grow, or minimal number of samples a node should hold) to limit the decision tree induction process, in practical applications the resulting decision tree will most often have long and very uneven paths and so some *postpruning* methods will be required. Pruning of the decision tree is performed by using a leaf node to replace a subtree. The general rule to decide whether a subtree should be pruned or not is based on the observation that the expected error rate within the subtree can be reduced in terms of its replacement by a single leaf node. For instance, assume a two-leaf subtree is formed based on three training samples. The left leaf node contains one sample indicating that when the attribute *Wetness* = high, the pixel should be labeled as class *target*. The right leaf node contains two samples showing that when the attribute *Wetness* = low then the pixel should be labeled as class *non-target*. Later, during the test phase, assume only four test samples are delivered to this subtree domain in which three test samples with the attribute *Wetness* = high actually belong to the class *non-target* and one sample with the attribute *Wetness* = low belongs to the class *target*. According to the test result, the misclassification occurring is 4 (all four test samples are misclassified). However, one may consider replacing this subtree by a single leaf node labeled as majority class *non-target*. After this replacement, the classification errors can then be reduced from 4 to 2 (since only one training sample and one test sample belong to class *target*).

In order to perform decision tree pruning, one may require an independent data set called *pruning set* to support such a pruning process (Esposito et al., 1997). Figure 6.7 shows a typical data partition in supporting the formation of a decision tree at each stage. The data set is first divided into a training and a test set. The training set is further partitioned into two sets, one for decision tree growing during the training phase, the other (the pruning set) for use in the pruning process.

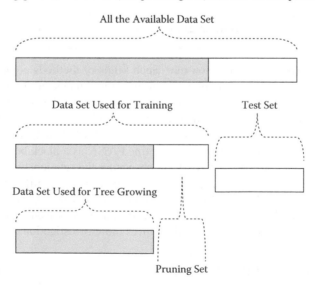

FIGURE 6.7 Typical data partitions for supporting decision tree induction at each stage.

There is a variety of descriptions of pruning methods available in the literature. For instance, Quinlan (1986, 1993) proposes methods called *reduced error pruning* (REP), *pessimistic error pruning* (PEP), and *error-based pruning* (EBP); Breiman et al. (1984) developed *cost complexity pruning* (CCP); Cestnik and Bratko (1991) present a Bayesian-based pruning method called *minimal error pruning* (MEP). These methods are described in the following sections.

6.7.1 REDUCED ERROR PRUNING (REP)

Reduced error pruning (REP) produces an optimal pruning of a decision tree. The resulting pruned tree will be the smallest tree among those with minimal class error with respect to a pruning set. REP is generally performed in two phases. In the first phase, a pruning set is input to the decision tree and the accuracies of individual nodes are estimated. Secondly, from a bottom–up direction (i.e., starts from leaf node), the subtrees, when removed without increasing the error, are pruned away. Specifically, one can express such a pruning process as follows. If S is a subtree rooted at node t, then the gain at t due to pruning subtree S, denoted as *pruning_gain$_{REP}$*(S,t), is given by the expression

$$pruning\_gain_{REP}(S,t) = error\left(S\right) - error\left(t\right) \qquad (6.30)$$

where error(v) denotes the number of misclassification samples occurred in v. The node with the largest *pruning gain* is pruned, and the process is repeated until only nodes with a negative *pruning gain* remain.

6.7.2 PESSIMISTIC ERROR PRUNING (PEP)

Pessimistic error pruning (PEP) does not require a *pruning set* (Figure 6.7). PEP estimates error directly from the training data. The error for leaf node t denoted as $Error_{PEP}(t)$ is estimated by

$$Error_{PEP}\left(t\right) = \frac{N\left(t\right) - n_j\left(t\right) + 0.5}{N\left(t\right)} \qquad (6.31)$$

where $N(t)$ is the number of samples within leaf node t, and $n_j(t)$ is the number of samples belonging to the most likely (i.e., majority) class j within node t. As a subtree S rooted at t is concerned, the error is computed by

$$Error_{PEP}\left(S\right) = \frac{\displaystyle\sum_{i \in \text{ leaf node of } S} \left(N\left(i\right) - n_j\left(i\right) + 0.5\right)}{\displaystyle\sum_{i \in \text{ leaf node of } S} N\left(i\right)}. \qquad (6.32)$$

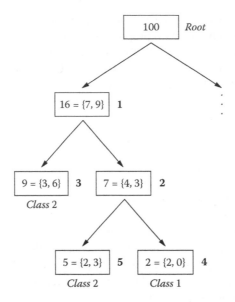

FIGURE 6.8 This example shows partial decision tree used for pruning experiments.

Equation (6.32) sums the errors at each leaf node within subtree S. The PEP then compares the error resulted from the subtree S and the error occurred at node t. If $Error_{PEP}(t) \le Error_{PEP}(S)$, then pruning is performed. Unlike most pruning methods, PEP runs in a top–down fashion, which tends to be faster since it only has to make one pass and looks at each node only once.

Figure 6.8 shows an example in which node 1 contains a total of 16 training samples with 7 samples belonging to class 1 and 9 samples belonging to class 2 (expressed as 16 = {7, 9}). Other nodes are described in a similar manner. For node 1, the $Error_{PEP}(node1)$ is

$$Error_{PEP}\left(node1\right) = \frac{7+0.5}{16} = \frac{7.5}{16} = 0.47 \, . \tag{6.33}$$

For the subtree S rooted at node 1, the $Error_{PEP}(S)$ is

$$Error_{PEP}\left(S\right) = \frac{3+0+2+0.5\times3}{16} = \frac{6.5}{16} = 0.4. \tag{6.34}$$

Since $Error_{PEP}(node1)$ is bigger than $Error_{PEP}(S)$, the subtree is not pruned.

6.7.3 Error-Based Pruning (EBP)

An improved version of PEP is called *error-based pruning* (EBP). EBP estimates the error by adding the so-called *one standard error pruning rule* to PEP. For a subtree S rooted at t, pruning is determined by the following equation

$$Error_{PEP}(t) \leq Error_{PEP}(S) + std\_error(S) \tag{6.35}$$

where std_error, the standard error, is calculated as

$$std\_error(S) = \sqrt{\frac{Error_{PEP}(S) \times (1 - Error_{PEP}(S))}{N(t)}}. \tag{6.36}$$

The corresponding standard error for subtree S rooted at node 1 is then calculated as

$$std\_error(S) = \sqrt{\frac{0.4(1-0.4)}{16}} = 0.122. \tag{6.37}$$

Since $Error_{PEP}(S) = 0.4$ (Equation [6.34] and $std\_error(S) = 0.122$ (Equation [6.37]), according to Equation (6.35), the sum of both being 0.522, which is greater than 0.47($Error_{PEP}$(node1) in Equation [6.33]), the result shows that based on EBP (with one standard error rule), the subtree should be pruned.

EBP has a special characteristic called *grafting*, which is a process whereby an internal node of a decision tree is removed and replaced by one of its subtrees. Figure 6.9 shows the difference between pruning (Figure 6.9a) and grafting (Figure 6.9b) using the decision tree shown in Figure 6.3e. In the case of decision tree pruning, Figure 6.9a shows that the subtree rooted at node 5 is removed and replaced by a leaf node, while in the case of grafting, Figure 6.9b shows that internal node 2 is removed and replaced by the subtree rooted at node 5.

6.7.4 COST COMPLEXITY PRUNING (CCP)

The *cost complexity pruning* (CCP) method, which is used in CART (Section 6.4), tries to find the best compromise between the errors estimated from a pruning set or cross-validation and the size of the decision tree. In other words, the aim of CCP is to minimize the error and complexity (equal to the number of leaves within a tree) of a decision tree. The total cost of a tree can be represented as

$$\text{Total cost} = \text{Error cost} + \text{Complexity cost}. \tag{6.38}$$

Let $R(t)$ denote the error cost at node t, $R(S)$ the error of subtree S, and α the complexity parameter (a measure of how much additional accuracy a split must add to the entire tree to warrant the additional complexity). Follows Equation (6.38), for a subtree S, the total cost of S denoted as Total_cost (S) is expressed as

$$\text{Total\_cost}(S) = R(S) + \alpha \times Num\_of\_leaves(S). \tag{6.39}$$

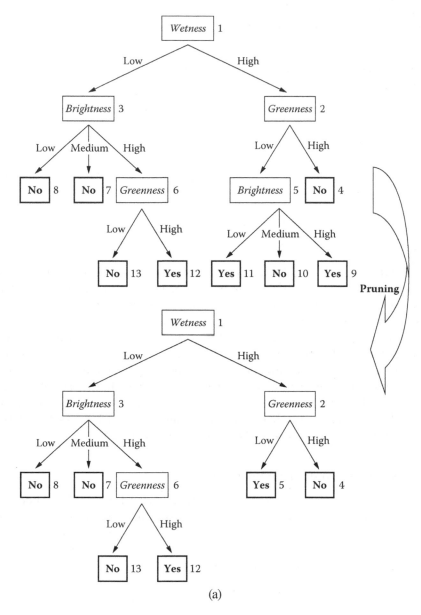

(a)

FIGURE 6.9   (a) Decision tree pruning in which the subtree rooted at node 5 is pruned and replaced by a leaf.

where *Num_of_leaves(S)* denotes the number of leaves within subtree *S*. If subtree *S* is rooted at node *t*, after *S* is pruned and replaced by *t*, the total cost of the leaf node *t* is expressed as

$$\text{Total\_cost}(t) = R(t) + \alpha. \tag{6.40}$$

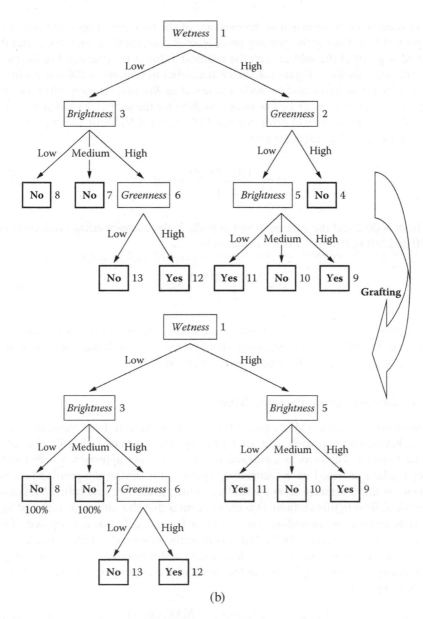

FIGURE 6.9 (continued) (b) The decision tree grafting process in which internal node 2 is removed and replaced by the subtree rooted at node 5.

When the costs of Equations (6.39) and (6.40) are equal, one obtains

$$\alpha = \frac{R(t) - R(S)}{Num\_of\_leaves(S) - 1} \qquad (6.41)$$

where α can then be regarded as the measure of the increase in apparent error rate per pruned leaf. During the pruning process, α is computed for each node, and the method is to select the subtree with the smallest value α for pruning. For instance, using the case shown in Figure 6.8, the total number of samples is 100 (shown in the root node) and the error cost for node 1 denoted as $R(node1)$ is then 7/100 (i.e., the proportion of error in node 1). The error cost $R(S)$ for the subtree rooted at node 1 is 5/100 (i.e., the proportion of error in node 3 [3 errors], 4 [error 0], and 5 [2 errors]). The value of α is then calculated by

$$\alpha = \frac{7/100 - 5/100}{3-1} = 0.01 . \tag{6.42}$$

For both node 2 and the subtree rooted at node 2, the corresponding error costs are 3/100 and 2/100. Parameter α is calculated by

$$\alpha = \frac{3/100 - 2/100}{2-1} = 0.01 . \tag{6.43}$$

Although both subtrees (rooted at node 1 and node 2, respectively) take on the same value of α, the best tree pruning policy should prune at node 1, because reduction in tree size is greater than if the tree is pruned at node 2.

6.7.5   MINIMAL ERROR PRUNING (MEP)

Minimal error pruning (MEP) also estimates error directly from training data. It uses a bottom–up approach, and after pruning, the estimated classification error can be reduced. The error is estimated using a Bayesian approach, which can be either Laplace or based on *m*-estimation. Specifically, during the pruning process, the error is first estimated from the subtree, which generates a so-called *backed-up error* (described in this section). One then assumes that this subtree is replaced by a leaf node and the corresponding *expected error* for that leaf node is computed. If the *expected error* is less than the *backed-up error*, the subtree should be pruned.

For a subtree being replaced by a leaf node i and labeled as class j (note that j is the majority class of samples within leaf node i), the corresponding *expected error* can be computed as

$$Expected\_error(i) = \frac{N + c - n_j - 1}{N + c} \tag{6.44}$$

where N is the number of samples assigned to leaf node i, c is the number of classes, and n_j is the number of samples belonging to the most likely (i.e., majority) class j within node i. Equation (6.44) is called the *Laplace error* estimate, which is based on the assumption that the probabilities that samples belonging to each class are uniformly distributed. It is apparent from Equation (6.44) that the degree of decision tree pruning is affected by the number of classes.

The expected error for leaf node i can also be calculated in terms of m-estimate, which is expressed by

$$Expected\_error(i) = \frac{N + m - prior(j) \times m - n_j}{N + m} \tag{6.45}$$

where $prior(j)$ is the prior probability of class j determined by the user, and m is a nonnegative parameter generally tuned in terms of experience. However, the general principle for determining value m (generally m can be as small as 1 and as large as N) is that if data contains little noise, one may choose a lower value of m, which will result in little pruning. On the other hand, if the data contains considerable noise, a higher value of m can be chosen, causing more pruning. In short, the choice of m depends on the quality of data. Also note that unlike the Laplace estimate, Equation (6.45) is not sensitive to the number of classes.

Now calculate the error upon node i, which is the father node of the original subtree containing s child nodes. The backed-up error for the subtree can then be calculated as

$$Backed\_up\_error(i) = \sum_{k=1}^{s} Expected\_error(k) \times P_k \tag{6.46}$$

where P_k is the percentage of samples (i.e., the number of samples divided by N) contained in child node k, and $Expected\_error(k)$ is calculated according to Equation (6.45).

Again, by using Figure 6.8 as an example, the expected error rate for node 2 (containing $7 = \{4, 3\}$, that is, 4 samples assigned to class 1 and 3 samples classified as class 2) according to Equation (6.44) is calculated as

$$Expected\_error(i) = \frac{7 + 2 - 4 - 1}{7 + 2} = 0.44. \tag{6.47}$$

For nodes 4 and node 5, the expected error rates are 0.25 and 0.375, respectively. The backed-up error rate for node 1 according to Equation (6.46) is

$$Backed\_up\_error(i) = 0.43 \times \frac{5}{7} + 0.25 \times \frac{2}{7} = 0.377. \tag{6.48}$$

Since the backed-up error (0.377) is smaller than the expected error (0.44), the subtree rooted at node 2 is thus not pruned. For the subtree rooted at node 1, the expected error and backed-up error are, respectively

$$Expected\_error(i) = \frac{16 + 2 - 9 - 1}{16 + 2} = 0.44 \tag{6.49}$$

$$Backed\_up\_error(i) = 0.377 \times \frac{7}{16} + 0.364 \frac{9}{16} = 0.370 \tag{6.50}$$

Since the backed-up error is greater than the expected error again, the subtree is not pruned.

6.8   BOOSTING AND RANDOM FOREST

6.8.1   Boosting

Boosting is an approach to improve the performance of an unsatisfactory classifier. The basis of the boosting technique is to assign weights to the individual elements of the training data set. The weights of the elements are initially set to be equal. One then compares the classifier's output and the known label of each element of the training data to determine which elements of the training data that have been classified incorrectly. These incorrectly classified training data elements are assigned a larger weight, and the classifier is run again. The increased weighting of these training data gives a heavier punishment to any misclassification and should force the classifier to focus on these cases. A method similar to boosting is described by Jackson and Landgrebe (2001). Quinlan (1996) concludes that about 10 iterations is the optimum number, and that little is gained by performing additional boosting runs. Pal and Mather (2003) find that in the case of dealing with remotely sensed data, 10 to 15 boosting iterations are sufficient to achieve an improvement in classification accuracy.

Among the boosting algorithms presented in the literature, this section focuses on the widely used Adaboost algorithm (Freund and Schapire, 1999) for its robustness in solving many of the practical difficulties of the earlier boosting algorithms.

The Adaboost algorithm takes a training set $(x_1, y_1), (x_2, y_2), ..., (x_n, y_n)$ as inputs, where x_i and y_i denote the input features and the corresponding information class, respectively. The classifier is called repeatedly by Adaboost in a series of iterations denoted as $k = 1, 2, ..., K$. The core of Adaboost is to maintain a set of weighting coefficients over the training set. The weight of training sample i on iteration k is denoted by $W_k(i)$. At the start of the training, all training samples are given equal weight, and as the training iterations proceed, the incorrectly classified samples are assigned a larger weighting coefficient to force the classifier to focus on such *hard* samples. Specifically, in a two-class case, the purpose of Adaboost is to let the classifier find a so-called *weak hypothesis* $h_k : x \rightarrow \{+1, -1\}$ appropriate for the weighting set $W(i)$. The goodness of a weak hypothesis is measured in terms of its error ε_k (i.e., the sum of weights of misclassified samples).

$$\varepsilon_k = \sum_{i:H_k(x_i) \neq y_i} W_k(i) \tag{6.51}$$

Adaboost then chooses a parameter α_k to measure the importance of each weak hypothesis, where α_k is represented by

$$\alpha_k = \frac{1}{2} \ln\left(\frac{1-\varepsilon_k}{\varepsilon_k}\right). \tag{6.52}$$

Note that the larger the value of ε_k, the smaller the value α_k will be. Also note that the process of selecting α_k and $h_k(x)$ can be interpreted as a single optimization step that minimizes the upper bound on the empirical error. Improvement of the bound is guaranteed, provided that $\varepsilon_k < \frac{1}{2}$. The weighting coefficient in the next $k + 1$ run is then updated by

$$W_{k+1}(i) = \frac{W_k(i)}{Z_k} \times \begin{cases} e^{-\alpha_k} \text{ if } h_k(x_i) = y_i \\ e^{\alpha_k} \text{ if } h_k(x_i) \neq y_i \end{cases}$$

$$= \frac{W_k(i) \times \exp\left(-\alpha_i h_k(i) y_i\right)}{Z_k}$$

(6.53)

where Z_K is a normalization factor, usually defined as

$$Z_k = 2\sqrt{\varepsilon_k(1 - \varepsilon_k)}$$

(6.54)

Each incorrectly classified training sample will be accompanied by an increase of the weight. The weight can be regarded as the upper bound on the error of a given training sample. The final hypothesis or decision for an input attribute x, denoted as $H(x)$, is then expressed as

$$H(x) = sign\left(\sum_{k=1}^{K} \alpha_k h_k(x)\right).$$

(6.55)

Figure 6.10 shows the results of using Adaboost for classifying binary samples. The 150 dot samples within the 2-dimensional space shown in Figure 6.10a are within the range [0.25, 0.75] on each axis, while 150 samples are also drawn from the white area. Results of forming classification boundaries for each iteration k (from 1 to 5) by Adaboost are displayed. The corresponding errors are shown in Figure 6.10b. It can be seen that Adaboost converges very quickly within around 10 iterations.

The case illustrated above is only concerned with binary classification. For the multiclass case, the most straightforward method is to transfer the multiclass issue into a larger binary problem. For instance, for each training sample x, the problem is to determine whether x belongs to class y or class z where class z comprises all classes except y (Schapire and Singer, 1998). There are other Adaboost-relating techniques that adopt similar concepts for dealing with the multiclass problem. Readers should refer to Freund and Schapire (1999) for details.

6.8.2 RANDOM FOREST

The *random forest classifier* consists of a combination of tree classifiers where each classifier is generated using a random vector sampled independently from the

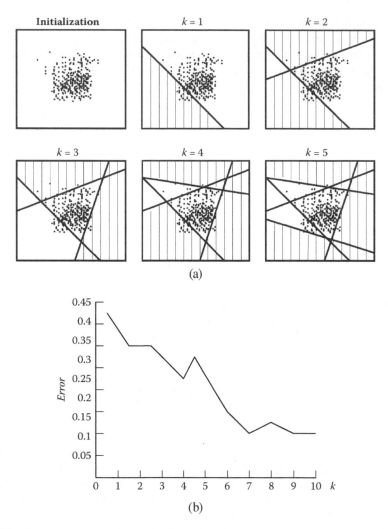

FIGURE 6.10   (a) Progress of classifications using the Adaboost technique. (b) Classification error at each iteration using Adaboost.

training set of input vectors, and each tree casts a unit vote for the most popular class in which to place a given input vector (Breiman, 2001). The technique for generating a random forest is generally a combination of the *bagging* and *random subspace* methods. Bagging (Breiman, 1996) is a technique to improve classification accuracy and avoid overfitting. Given a training set of size N, bagging generates a number of new training sets each of size n (where $n < N$) by randomly drawing samples with replacement from the original training set. Also suppose the data contains M attributes (e.g., spectral bands). For each node of the tree, m ($m < M$) attributes are randomly chosen to provide the base for calculating the best split at that node. Once the random forest is formed, each sample is classified to the class taking the most popular votes from all the tree predictors in the forest. Breiman (2001) concludes that the random forest technique holds several advantages: (a) its accuracy is as good as Adaboost and sometimes better, (b) it is relatively robust to outliers and noise, (c) it is faster than bagging or boosting, and (d) it shows useful internal estimates of error and the importance of the variable.

6.9 DECISION TREES IN REMOTELY SENSED DATA CLASSIFICATION

Using decision tree techniques to classify remotely sensed imagery has grown quickly in recent years. Hansen et al. (1996) apply decision trees to land cover classification, and a number of other authors have followed their lead (see, for instance, Muchoney et al., 2000; Friedl et al., 1999; Pal and Mather, 2003; 2006). Friedl and Brodley (1997) discuss many of these methods within the context of research with remotely sensed data. Simard et al. (2002) use a CART decision tree (Breiman et al., 1984) based on the Gini index to classify JERS-1 GRFM (Global Rain Forest Mapping project) and ERS-1 CAMP (Central Africa Mosaic Project) SAR (synthetic aperture radar) data (L-band and C-band) for large tropical coastal area mapping. The inputs to CART include ERS and ERS amplitude imagery both involving 100-m and 200-m resolutions, JERS/ERS amplitude ratio imagery in 100-m resolution, and texture features derived from the wavelet transform in 200-m and 400-m resolution. The results show that the decision tree successfully deals with the combined and complicated L-band and C-band data sets and an overall accuracy rate of around 84% is achieved. De Colstoun et al. (2003) examined the feasibility of using SEE5.0 with multitemporal Landsat Enhanced Thematic Mapper Plus (ETM+) data to map eleven land cover types at the area near Milford, PA. The boosting technique was implemented to improve classification performance, and the overall accuracy of 82% was achieved. De Colstoun and Walthall (2006) use SEE5.0 to evaluate the effects of the Bidirectional Reflectance Distribution Function (BRDF) on land cover classification accuracy at continental scale. The data set is the multidirectional global scale data from the first Polarization and Directionality of Earth Reflectances (POLDER) instrument, and fifteen land cover categories are evaluated. The results show that the decision tree method works well, and BRDF information contributes to the enhancement of classification accuracy.

A number of authors have conducted comparisons of decision trees and other classifiers. Gahegan and West (1998) find that the ML classifier did not perform as well as a decision tree in labeling the pixels of an image of a complex test area. The thematic map produced by ML shows severe fragmentation of land cover classes, possibly because the multivariate normal distribution provides an inadequate model. The use of a decision tree (C4.5) shows a significant improvement in performance. Muchoney et al. (2000) compare an artificial neural net (fuzzy ARTMAP), the ML classifier, and a decision tree in a study of vegetation and land cover mapping in Central America. They conclude that the decision tree produces superior results (see also Borak and Strahler [1999] for a similar comparison, in which a Fuzzy ARTMAP produces better results than a decision tree). Evans (1998) and Brown et al. (1993)

also describe comparisons among different classifiers. Pal (2005) presents the classification results obtained from random forest and support vector machines (SVMs) for Landsat Enhanced Thematic Mapper Plus (ETM+) data in the United Kingdom with 7 land cover classes. The study shows that the random forest classifier and SVMs perform equally with respect to the accuracy and training time concerned, and random forest requires less model parameters than do SVMs.

Other studies for providing detailed evaluations of decision tree and other classifiers can be found in Pal and Mather (2003, 2006). Pal and Mather (2003) evaluate the performance of two univariate decision algorithms, C4.5 and SEE5.0 (Quinlan, 1993, 1996), and one multivariate decision tree classifier, QUEST (Loh and Shih, 1997), for land cover classification. Two separate test and training data sets from two different geographical areas and two different sensors—multispectral Landsat ETM+ and hyperspectral DAIS—are used. Several factors considered to affect classification performance are investigated including the effects of variations in training data set size, the dimensionality of the feature space, together with the impact of boosting, attribute selection measures, and pruning. The classification accuracy achieved by decision trees is compared to results from back-propagating ANN and the ML classifiers. The results indicate that the performances of the univariate decision tree algorithms are acceptably good in comparison with that of other classifiers, except with high-dimensional data. The multivariate decision tree algorithm QUEST does not appear to perform any better than univariate decision trees. While boosting produces an increase in classification accuracy of between 3% and 6%, the use of attribute selection methods does not appear to be justified in terms of accuracy increases. However, neither the univariate decision tree nor the multivariate decision tree performed as well as the ANN or ML classifiers with high-dimensional data. In another study, Pal and Mather (2006) evaluate several classifiers including maximum likelihood, neural work, SVMs, and decision trees (SEE5.0) in the classification of Digital Airborne Imaging Spectrometer (DAIS) hyperspectral data (5-m resolution, and a total 72 bands within the wavelength range 0.4 to 12.5 μm) in a region of the province of La Mancha, Spain. The results show that the use of decision tree–based feature selection techniques (note that the decision tree classifier implicitly performs a feature selection procedure as the most effective feature is sequentially selected in each node-splitting stage) and the classification accuracy being achieved was nearly equivalent to the accuracy obtained by using raw data as inputs. This finding indicates that the decision tree approach can be effectively used for feature selection with hyperspectral data.

For demonstration purposes, experiments are made using four types of decision tree induction algorithms including C4.5 (Quinlan, 1993), CHAID (Kass, 1980), CART (Breiman et al., 1984), and QUEST (Loh and Shih, 1997), respectively. The CART algorithm is developed by Salford Systems company (2008), and CHAID and QUEST can be found in the Statistical Package for the Social Sciences (SPSS) package (2007). Also, for tree induction criteria, the maximum depth of trees is set to be under 20, and each leaf node has to contain at least 50 samples. A set of four-band Landsat MSS data (Michie et al., 1994) evaluated in Chapter 4 (in relation to support vector machines) is again tested here. It involves 36 attributes in total, made up of pixel values in the four spectral bands with 9 pixels within 3 × 3 windows. The class name and the corresponding number of training (total 4435) and test (total 2000) samples are shown in Table 6.1. For convenience, let N denote the pixel numbering within a 3×3 window, and B the band number. Specifically, within a window, $N1B1$ denotes the value of band 1 for the top-left corner pixel, $N3B2$ for band 2 at the top-right corner pixel, and the central pixel in band 4 will be marked as $N5B4$.

The information gain in the C4.5 algorithm is also used as an attribute selection criterion. The resulting decision tree is composed of 12 levels with 98 nodes in total among which 50 are leaf nodes. The root node chosen by C4.5 is $N5B1$ (≤ 78) (i.e., the value of window central pixel in band 1 divided into two branches with one branch ≤ 78, and the other >78). The average accuracy for learning samples is 85.6%, while it is 80.4% for test samples. CHAID uses a Chi-square test to perform tree induction. After the experiment, the resulting CHAID decision tree is made of 5 levels with 76 nodes (48 leaf nodes) in total. Only 18 attributes are significant for CHAID to grow the decision tree. The root node selected by CHAID is $N5B2$, i.e., the value of band 2 for the central pixel within 3×3 windows. Details of these attributes are shown in Table 6.4. The average accuracy obtained for training samples is 84.4%, while 80.8% accuracy is achieved for test samples.

CART includes univariate and multivariate types for growing decision trees. In the case of univariate split, the criterion for attribute selections is based on the Gini impurity index. The resulting decision tree is made of eight levels with 37 nodes in total (19 leaf nodes). The attribute chosen for the root node is $N5B2$ (≤ 45.5) (i.e., the value of the window central pixel in band 2 divided into two branches with one branch ≤ 45.5, and the other >45.5). The average accuracy for learning samples is 84.58%, and 82.7% for test samples. In the case of a multivariate split, the resulting decision tree is composed of 10 levels with a total of 23 nodes of which 12 are leaf nodes. All the attributes contribute to the classification. The root is made through the linear combination of $N3B4$ and $N5B2$ (-0.534 ($N3B4$) + 0.845($N5B2$)). The average accuracy for learning samples is 86.58%, and 83.35% for test samples. When QUEST is used for classification, the decision tree it generates has six levels with a total of 19 nodes, 10 of which are leaf nodes. The overall number of attributes

---

TABLE 6.4

Attributes Included by the CHAID and QUEST Decision Tree Algorithms to Test Satellite Imagery Classification

	Type	Attributes Being Included for Classification
(a)	CHAID	N5B2, N2B3, N5B3, N9B1, N1B1, N5B4, N2B4, N1B3, N5B1, N1B4, N9B3, N3B4, N7B2, N6B1, N8B1, N3B3, N7B3, N4B3.
		(18 attributes)
(b)	QUEST	N5B2, N6B2, N4B2, N2B2, N8B2, N9B2, N5B4, N5B3, N6B3, N6B4, N3B3, N4B3, N5B1, N6B1, N4B1, N8B1, N2B1, N9B1, N4B4, N8B4, N9B4, N7B1, N3B1, N7B4, N3B2, N3B4, N2B4.
		(27 attributes)

Note: The sequence of the attributes present is according to their significance level.

involved by QUEST for tree induction is 27 as shown in Table 6.4. The root node formed by CHAID is again *N5B*2, the same as that selected by CHAID and CART (univariate). Classification accuracies for training samples are 82.5% in average and 82.2% for test samples.

Based on the experiments previously discussed, it appears that the multivariate CART algorithm performs slightly better than others in terms of both training and testing accuracy. However, the reader should note that the experiments discussed earlier only show preliminary results, and that the accuracy might change when various criteria for tree growing or pruning and boosting are applied.

6.10 CONCLUDING REMARKS

Several decision tree induction algorithms that have been applied within the field of remotely sensed image classification are introduced in this chapter. The performance of these algorithms may depend on the choice of criteria to be used or on certain predefined indices (information gain or Gini index) or statistical test (Chi-square or *F*-test). Decision boundaries may be either univariate or multivariate. Comparison of the classification performance of these algorithms is provided. For reducing the overfitting issue generally seen in the process of decision tree induction, several pruning techniques are described in detail. In addition to these fundamental skills, boosting techniques or random forest methods may be applied with the aim of achieving robust learning. Above all, this chapter provides the knowledge necessary for readers to perform imagery classification by using decision trees. Due to their hierarchical characteristics, decision trees are very useful for data mining and knowledge discovery. This is a great advantage over artificial neural networks (though as mentioned in Section 6.5, some efforts have been devoted to extract rules from neural networks, however this approach has not proved popular in the field of remote sensing). Knowledge mining has received considerable attention in recent years, and developments in the use of decision trees in this application area should contribute to developments in remote sensing image classification.

7 Texture Quantization

Human beings use both spectral and spatial features to interpret visual information. Spectral features describe variations in tone over a grayscale image, while spatial features reflect the spatial distribution of these tonal variations, which contain two kinds of spatial relationship. One such relationship is tonal variation focused upon the object of interest, representing the structure of the object, while the other measures the broader-scale relationship between the object being analyzed and the remainder of the scene. Generally, the words texture and context are used to represent these two forms of spatial relationship. This chapter focuses on the problem of quantifying texture. The use of context-based methods is described in Chapter 8.

Spectral and textural features are interdependent for, as Haralick et al. (1973) noted, "texture and tone have an inextricable relationship to one another. Tone and texture are always present in an image, although one property can dominate the other at times" depending on the fineness or roughness of the surface of the object and on the image resolution relative to the surface roughness of the object. If tonal variation inside a limited range is relatively small, spectral information will dominate. For example, at a given resolution, an image may include areas of relatively constant tone, such as water bodies or expanses of concrete or tarmac. Conversely, if tonal variation is large and presents meaningful structures, then texture will be dominant, as in images of rocky areas, settlements, and some kinds of clouds.

Texture is an innate property of objects. It contains important information about the structural arrangement of surfaces. The use of texture in addition to spectral features for image classification might be expected to result in some level of accuracy improvement, depending on the spatial resolution of the sensor and the size of the homogeneous area being classified (Coburn and Roberts, 2004; Ouma et al., 2006). Where the spatial resolution of the image is fine relative to the scale of tonal variation, texture can be a valuable source of discriminating information. Conversely, where homogeneous regions in the image are small it may prove difficult to estimate texture, for texture is a property of an area rather than of a point. If the area over which texture is measured contains two or more different regions or categories, then texture measures may not prove to be useful.

The operational definition of texture features is difficult. The main texture recognition approaches can be categorized into four groups in terms of their different theoretical backgrounds. The first approach defines texture features that are derived from the Fourier power spectrum of the image under study via frequency domain filtering. Since different textures demonstrate different frequency patterns, it is reasonable to postulate that texture features are related to the distribution of spatial frequency components. The second approach is based on statistics that measure local properties that are thought to be related to texture, for example, the local mean or standard deviation. The third approach is the use of the joint gray-level probability density (Haralick, 1973). The final approach is based on modeling the image using assumptions, such as (1) the image being processed possesses fractal properties (Mandelbrot, 1977, 1982), or (2) it can be modeled using a random field model such as the multiplicative autoregressive random (MAR) field (Frankot and Chellapa, 1987).

This chapter presents a basic description of texture measures for image segmentation. Seven approaches based on different theoretical background are introduced. The first four methods are derived from multifractal theory; the fifth method is based on frequency domain filtering; the sixth method uses the gray-level co-occurrence matrix (GLCM), and the seventh texture quantization approach is based on the multiplicative autoregressive random field model.

7.1   FRACTAL DIMENSIONS

If an object possesses the *fractal* property, it can be thought of as being constructed of an infinite number of copies of itself, varying in scale from large to small. Two examples of fractal patterns are shown in Figure 7.1. The fractal dimension of an object is a measure of its complexity. The higher the fractal dimension, the more complex is the object shape or structure. For example, Figure 7.1a shows objects that have a fractal dimension of 1.58, while Figure 7.1b shows objects with a fractal dimension of 1.89. Fractal dimension can thus be applied as a measurement for texture quantization, as complex surfaces (such as the intensity surface represented by an image) have a rough texture, while simple (smooth) surfaces have a fine texture.

Real-world images (or objects) are not truly fractal; hence, the multifractal concept was introduced. The image is considered to be composed of a finite number of subsets, and its multifractal dimension considers each subset's property and is thus considered to be a more powerful measure of image texture (Tso, 1997). An introduction to the use of fractal and multifractal measures as texture descriptors is given in Section 7.1.1. Readers requiring a more detailed description of fractals are referred

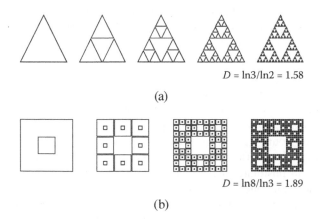

$D = \ln3/\ln2 = 1.58$

(a)

$D = \ln8/\ln3 = 1.89$

(b)

FIGURE 7.1   (a) The fractal dimension of the image is 1.58; (b) the fractal dimension of the image is 1.89.

to Agterberg and Cheng (1999), who introduce a special issue of *Computers and Geosciences* devoted to the topic of fractals and mulifractals; also see Mandelbrot (1977, 1982); and Feder (1988).

7.1.1 INTRODUCTION TO FRACTALS

Fractal geometry was first defined and explored by Mandelbrot (1977, 1982). Since then, this novel field has attracted the attention of a number of researchers. The idea of fractals is now widely used in image processing and image compression (Barnsley, 1993; Barnsley and Hurd, 1993). Mandelbrot used the term *fractal* to represent the irregular and fragmented nature of real-world objects. The word itself is derived from the Latin word *fractus* meaning "irregular segments."

A fractal is, by definition, a set for which the Hausdorff-Besicovich dimension is strictly larger than the topological dimension. In other words, the fractal dimension value of a complex object should be larger than our intuitive definition of its dimension (i.e., its Euclidean dimension).

The concept of the fractal dimension of an object can be explained by using the well-known example of a coastline whose length has to be measured on a map using a measuring instrument that has a specific step length, for example, a pair of dividers. The step length of the measuring instrument is denoted by l, and the coastline can be completely covered by n steps of the instrument. The length of the coastline (M) can be obtained by

$$M = nl^D \tag{7.1}$$

where D is the Euclidean dimension of the measuring instrument. Figure 7.2a displays a portion of coastline being measured. Figures 7.2b and 7.2c show the coastline as measured by two different measuring instruments, a stick and a box, whose Euclidean dimensions are equal to one and two, respectively.

The coastline measurement problem demonstrates another difficulty: The apparent length of the coastline changes as the step size of the measuring tool increases or decreases (i.e., it is not consistent). Moreover, whatever the step length l of the measuring instrument selected, coastline features whose size is smaller than l will

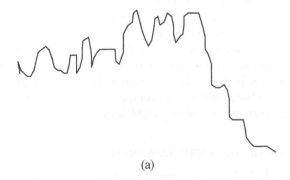

(a)

FIGURE 7.2 (a) A portion of coastline to be measured for fractal dimension.

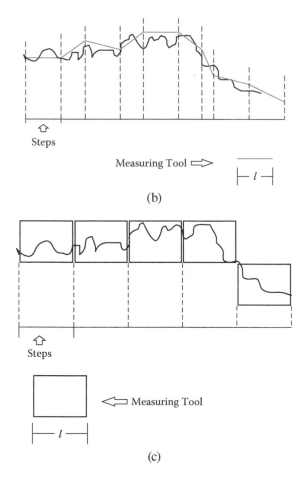

FIGURE 7.2 (continued) (b) Coastline measured using sticks; (c) coastline measured using boxes.

not be measured. It is clear that the apparent coastline length depends not only on the geometry of the coastline itself but also relates to the size of the measuring tool.

Mandelbrot was attracted by this measurement inconsistency problem. He found that the measurement inconsistency is due to the loss of fractional features (features whose size is smaller than the step size of the measuring instrument). In order to contain the contribution of these fractional features, Mandelbrot pointed out that we must generalize the notion of dimension to include these fractional dimensions. Once these "lost" fractional features are compensated for, the fractal dimension will yield a uniform measure at any measurement scale.

7.1.2 ESTIMATION OF THE FRACTAL DIMENSION

There are several methods of estimating fractal dimension. Mallat (1989) provides a detailed demonstration of fractal dimension estimation via the wavelet transform.

He pointed out that if the signal is truly fractal, then the fractal dimension will show some relation to the power spectrum and thus can be estimated. Peleg (1984) use the ε-blanket method to obtain an estimate of fractal dimension, which is a generalized version due to Mandelbrot (1982). Pentland (1984) fitted a model called fractional Brownian motion (FBM) to the image in order to measure its fractal dimension. The box-counting method is another way to obtain an estimate of the fractal dimension (Clarke, 1986; Gonzato, 1998; Keller and Chen, 1989; Sarkar and Chaudhuri, 1994; Voss, 1986). Since the methods used in the FBM and box-counting approaches proposed by Sarkar and Chaudhuri (1994) are straightforward and efficient, they are discussed in detail in the following sections.

7.1.2.1 Fractal Brownian Motion (FBM)

Pentland (1984) presents a method to estimate the fractal dimension of an object by fitting an FBM model to the frequency domain representation of the image intensity surface. A random function $f(t)$ is a fractional Brownian function if the following relationship holds for all t and Δt:

$$\Pr\left(\frac{f(t+\Delta t)-f(t)}{\|\Delta t\|^H} < y\right) = F(y) \tag{7.2}$$

$F(y)$ is a cumulative distribution function with the self-affine* property, which indicates that Equation (7.2) is still valid given affine transform of the function $f(t)$, and H is called the *Hurst coefficient*, which determines the distribution of $F(y)$. It should be noted that in Equation (7.2), t and Δt can be represented by vector quantities. Hence, this formula can be extended to higher Euclidean dimensions. The term $\|\Delta t\|$ denotes the length of vector Δt. In the case of an image, $\|\Delta t\|$ is the distance between two pixels of interest. The fractal dimension D is calculated by $D = 2 - H$. Equation (7.2) implies that, for every Δt,

$$E\left(\left|f(t+\Delta t)-f(t)\right|\right)\|\Delta t\|^{-H} = C \Rightarrow \frac{\ln E\left(\left|f(t+\Delta t)-f(t)\right|\right)}{\ln\|\Delta t\|} = H. \tag{7.3}$$

$E(|\Delta f_{\Delta t}|)$ is the expected value of the change in the intensity surface over Δt. C is a constant, equal to the mean of the random variable $|y|$ (Yokoya et al., 1989). One can therefore measure the quantities of $E(|\Delta f_{\Delta t}|)$ for various Δt, then use a least square fit in the log-log domain to obtain H (see Equation [7.3]) and thus determine the fractal dimension D. Based on the above description of the derivation of the parameter H, it is clear that for the same distribution function $F(y)$, a higher (or lower) value of H will

---

* In a Euclidean space of dimension E, a collection of positive real ratios $r = [r_1, r_2, ..., r_n]$ determines an affinity and transforms each point $[s_1, s_2, ..., s_n]$ into $[r_1 s_1, r_2 s_2, ..., r_n s_n]$, hence a set S is scaled by r into rS.

result in a lower (higher) value of dimension D, which indicates a relatively smooth (rough) image intensity surface. D is seen to be directly related to image texture.

7.1.2.2 Box-Counting Methods and Multifractal Dimension

Box-counting methods are based on Mandelbrot's self-similarity equation. Mandelbrot (1977, 1982) pointed out that one criterion of a surface being fractal is based on the self-similarity property. Consider a pattern A in two-dimensional space. The pattern A is defined to be self-similar if A is the union of N distinct (nonoverlapping) copies of itself, each of which is similar to A scaled down by a ratio r in each dimension. Thus, the fractal dimension D of A can be expressed by the following equations:

$$1 = Nr^D \text{ or } D = \frac{\ln N}{\ln(1/r)}. \tag{7.4}$$

Using this equation, the fractal dimensions of the patterns shown in Figure 7.1 can be computed. In Figure 7.1a, the scale r on each dimension is ½, the number of copies N is 3, and the fractal dimension D can be obtained as $\ln(3)/\ln(2) = 1.58$. The fractal dimension D for Figure 7.1b is computed in a similar manner as $ln(8)/ln(3) = 1.89$. The above formula is derived assuming a perfect self-similar nature. In practice, it is difficult to compute D using Equation (7.4) directly, so approximations to this relationship are employed.

First, the concept of image intensity surface is introduced. In general, an image is treated as a two-dimension matrix. At each intersection of column x and row y of the image matrix, we store the pixel value, denoted traditionally by the term DN. For instance, as shown in Figure 7.3a, the value of DN at the point $x = 2$, $y = 2$, is 16. If we interpret the DN axis as the third coordinate, z, the image can then be viewed as a three-dimensional intensity surface, as illustrated in Figure 7.3b.

Sarkar and Chaudhuri (1994) propose a relatively simple approach called *differential box counting* to estimate the fractal dimension. Consider an $M \times M$ image that has been divided into overlapping or abutted windows of size $L \times L$, where $M/2 \geq L \geq 3$, and L is an integer. For each window, there is a column of accumulated boxes, each box with a size of $L \times L \times L'$. If G denotes the gray-level range of the image (e.g., 256), L' is calculated from $\lfloor G/L' \rfloor = \lfloor M/L \rfloor$, where the symbol $\lfloor \ \rfloor$ indicates the "integer part of the argument." For instance, if $L = 3$, image side length $M = 13$, and gray level $G = 33$, then L' is computed as

$$\left\lfloor \frac{M}{L} \right\rfloor = \left\lfloor \frac{13}{3} \right\rfloor = 4 = \left\lfloor \frac{G}{L'} \right\rfloor = \left\lfloor \frac{33}{L'} \right\rfloor \Rightarrow L' = 8. \tag{7.5}$$

Assign the numbers 1, 2, ..., n in turn to each box in the column from bottom to top. Let the minimum and maximum gray levels of the pixels in the (i, j)th window fall in boxes numbered u and b, respectively. The number of boxes spanned on the (i, j)th window is

$$n_L(i, j) = u - b + 1. \tag{7.6}$$

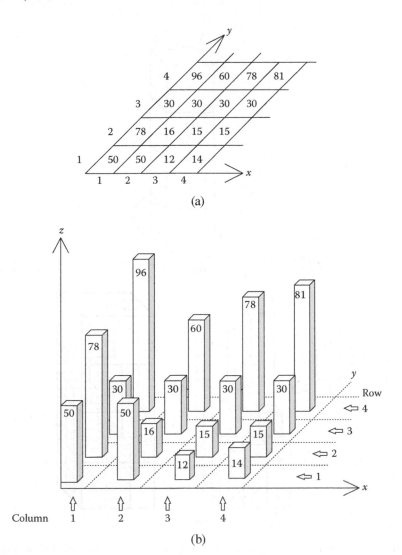

FIGURE 7.3 (a) A two-dimensional image space showing the DN values at corresponding coordinates (x, y); (b) an image viewed as a three-dimensional intensity surface.

For instance, in Figure 7.4a, the window size is set to $L = 3$, and each window thus contains nine pixels. If gray level G is 16, then $L' = 4$. We can then build up a column of boxes (see Figure 7.4a). Suppose that a column of boxes is centered on an image coordinate (3, 6), and the corresponding pixel values are as shown on the left-hand side of Figure 7.4b. After intersecting the pixel values with the column of boxes, we get $u = 4$, $b = 2$, and finally calculate $n_L(3, 6) = 4 - 2 + 1 = 3$ in terms of Equation (7.6) (see Figure 7.4b). The same calculation is performed for all windows, and the total number of boxes needed to cover the whole image with box size $L \times L \times L'$ is

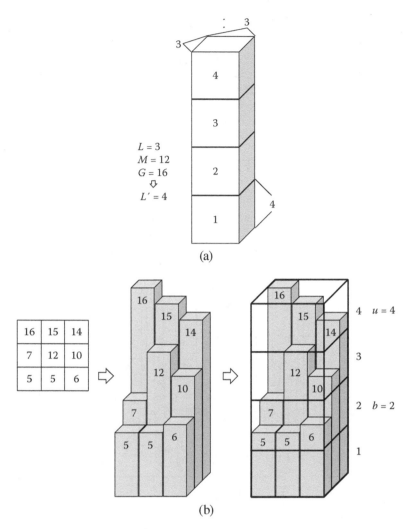

FIGURE 7.4 (a) Box counting approach with window size $L = 3$, box height $L' = 4$ (Equation [7.5]) and number of gray-level G = 16. A total of 4 boxes are generated. (b) The column of boxes are intersected with the pixel values; one gets $u = 4$, $b = 2$, and $n_L(3, 6) = 3$ according to Equation (7.6).

$$N_L = \sum_{i,j} n_L(i,j). \qquad (7.7)$$

According to Equation (7.5), the fractal dimension D is the negative of the slope of the least squares line with different value of L and N_L:

$$N_L \propto L^{-D} \Rightarrow \ln N_L = -D \ln L \qquad (7.8)$$

Begin
Do for $L = 2$ to $M/2$
    Define box size $L \times L \times L'$, where L' is calculated from (6.10).
    Partition the image into overlapping windows of size $L \times L$;
    Do for each window in the image
        Construct a column of boxes and assign numbers 1, 2, …, to the
            boxes from bottom to the top;
        Find the box numbers which intersect the maximal and the
            minimal pixel grey values within the window;
    EndDo
    Calculate N_L using (6.13);
EndDo
Estimate fractal dimension as the negative of the slope of the least
    squares linear fit to the data $\ln(L)$ and $\ln(N_L)$;
End

FIGURE 7.5 Algorithm using differential box counting proposed by Sarkar and Chaudhuri (1994) for computing the fractal dimension.

That is, D is the negative of the slope of the least squares line relating $\ln(L)$ and $-\ln(N_L)$. The algorithm for differential box counting is shown in Figure 7.5. Other approaches for calculating fractal dimension are described by Voss (1986), Keller et al. (1989), de Jong and Burrough (1995), and Gonzato (1998). Sun et al. (2006) provide a comprehensive review of fractal analysis in the remote sensing field. According to experiments made by Tso (1997), the methodology proposed by Sarkar and Chaudhuri (1994) for calculating fractal dimension is the most useful in terms of classification accuracy of texture imagery.

The algorithm shown in Figure 7.5 is used to generate unique measures of the fractal dimension for the entire image. However, for the purposes of image classification, we require an image showing a value of texture measured for the area surrounding each pixel, or at least for a number of subregions, rather than single overall fractal measurement. To generate a texture image using the fractal dimension approach, one can first divide the image into overlapped subimages. The fractal dimension can be derived separately for each subimage using one of the above algorithms, and the resulting fractal dimension is then assigned to the central pixel of the subimage (Figure 7.6a), thus building up a texture image. The pixel values in this image may then be mapped onto the interval [0, 255]. Note that the size of each subimage should not be too small, as a small subimage will limit the number of window sizes used in the fractal dimension estimation procedure, and the result will naturally be more biased. Conversely, if the subimage is too large it will exclude too many boundary pixels or overlap regions with different textures.

Figure 7.6b shows an HH-polarization SAR (synthetic aperture radar) image acquired by the Shuttle Imaging Radar C (SIR-C). The image is 256 × 256 pixels in size and shows part of the Red Sea Hills in Sudan. This area contains several lithological types, mainly volcano-sedimentary, intrusives, and granitic batholith,

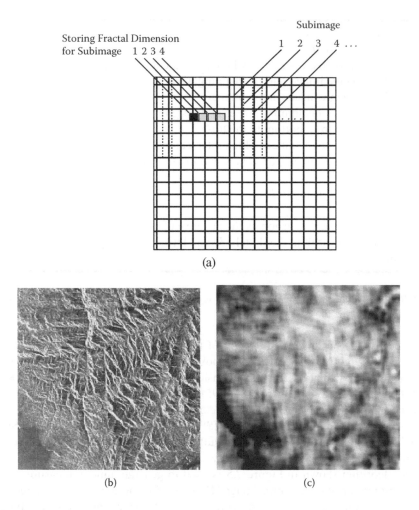

FIGURE 7.6 (a) The fractal dimension is calculated for each subimage and assigned to the central pixel. (b) An HH-polarization SAR image acquired by the Shuttle Imaging Radar C (SIR-C). (c) Result of fractal texture image calculated from Figure 7.6b using differential box counting method with a subimage size of 15 × 15.

which demonstrate different types of texture patterns (Mather et al., 1998). The intrusives rock type occupies the area around the lower-left corner of Figure 7.6b, and it is relatively smooth in appearance. Located to the right of the intrusives area is a region underlain by granitic rocks, termed granitic batholith. This area also appears smooth but the structure is different from that of the Intrusives group of rocks. The area around the middle of the image is underlain by volcano-sedimentary rocks, and shows a very rough texture pattern. This image is used as a basis for generating texture images in terms of fractal and other algorithms introduced in the following paragraphs. The fractal texture image derived from Figure 7.6b using the algorithm illustrated in Figure 7.5 with subimage size defined as 15 × 15 is shown in Figure 7.6c. Brighter areas denote higher fractal dimensions, indicating that the corresponding area is texturally rougher.

Note that the estimation of fractal dimension takes the contributions of all windows of an image and reflects the combined behavior of all windows. However, it is known that real-world images are not actually fractal. It is also the case that images with different texture patterns may have the same fractal dimension. Consequently, using fractal dimension to estimate the roughness of the image for segmentation purposes without regard to local characteristics may generate some confusion (Dubuc et al., 1989). To overcome this drawback, a function $D(q)$, called *the q moment generalized fractal dimension*, can be employed (Hentchel and Procaccia, 1983; Parisi and Frish, 1985; Tso, 1997):

$$D(q) = \frac{1}{q-1} \frac{\log \sum_{i,j} u_L{}^q(i,j)}{\log L} \quad \text{for } q \neq 1, \text{ and}$$

(7.9)

$$D(q) = \frac{\sum_{i,j} u_L(i,j) \log u_L(i,j)}{\log L} \quad \text{for } q = 1$$

where q is an integer, $u_L(i, j) = [n_L(i, j)/N_L]$, $n_L(i, j)$ and N_L are defined in Equation (7.6) and (7.7). Since $D(q)$ can vary with different values of q, $D(q)$ is thus called *multifractal dimension*.

7.2 FREQUENCY DOMAIN FILTERING

Frequency domain filtering provides a convenient way of extracting texture information. The most widely used frequency domain filter is a simple binary mask applied to the amplitude spectrum of the image, such that information covered by the mask remains unchanged while the remainder is suppressed. Although any kind of filter shape is possible, the square, wedge, and ring filters are generally used.

Several algorithms are available to transform image data from the spatial to the frequency domain. The best known of these is the Fourier transform. This algorithm generates a disc-like representation of image frequency information, with the low-frequency information located around the coordinate origin (the disc center) and the high-frequency information near the disc boundary. Analysis of the distribution of the spatial frequency representation of an image can provide measures that describe image texture.

7.2.1 FOURIER POWER SPECTRUM

The Fourier transform of a continuous function $f(x, y)$ of real variables (x, y) is defined by

$$F(u,v) \equiv \int\limits_{-\infty}^{\infty} \int e^{-2\pi\sqrt{-1}(ux+uy)} f(x,y) dx\, dy. \qquad (7.10)$$

For a discrete $N \times N$ digital image, the *discrete Fourier transform* (Brigham, 1974) is used. This is defined by

$$F(u,v) = \frac{1}{N^2} \sum_{x=0}^{N-1} \sum_{y=0}^{N-1} f(x,y) e^{-2\pi\sqrt{-1}(ux+vy)/N}, \ N-1 \geq u,\ v \geq 0. \qquad (7.11)$$

Both $F(u, v)$ and the magnitude function $|F(u, v)|$ are related by the expression

$$F(u,v) = |F(u,v)| e^{\sqrt{-1}\phi(u,v)}. \qquad (7.12)$$

The magnitude function $|F(u, v)|$ is called the *Fourier spectrum* of $f(x, y)$, and $\Phi(u, v)$ is the *phase spectrum*. The square of the amplitude spectrum is generally called the *Fourier power spectrum*, denoted by $P(u, v)$:

$$P(u,v) = |F(u,v)|^2. \qquad (7.13)$$

It is well known that the angular distribution of values in $P(u, v)$ is sensitive to the directionality of structure in $f(x, y)$. If $f(x, y)$ contains many lines or edges with some direction represented by angle θ, $P(u, v)$ will contain high values concentrated around the direction perpendicular to θ, i.e., $\theta + \pi/2$. If $f(x, y)$ shows a nondirectional structure, $P(u, v)$ will be nondirectional as well. Besides being sensitive to texture directions, the radial distribution of values in $P(u, v)$ is also known to be sensitive to the roughness of the structure in $f(x, y)$. A fine structure in $f(x, y)$ will result in high values distant from the origin of $P(u, v)$, while for a rough structure the high values in $P(u, v)$ are concentrated near the origin (Figure 7.8). Note that the coordinate origin of the amplitude spectrum is the center of the image, not the upper or lower left corner.

Based on these properties of $P(u, v)$, it is possible to design filters to extract radial and directional information concerning image structure from $P(u, v)$. Two filters that are often used are the wedge (W) and ring (R) filters. The W and R filters in the case of a discrete image surface can be expressed as

$$W_{\theta_1,\theta_2} = \sum_{\theta_1 \leq \tan^{-1}(v/u) < \theta_2} P(u,v) \qquad (7.14)$$

$$R_{r_1,r_2} = \sum_{r_1^2 \leq u^2+v^2 < r_2^2} P(u,v) \qquad (7.15)$$

where θ and r denote angle and radius, respectively. Both θ and r control the width of the filter's effective area. Equation (7.14) results in a wedge filter with a vertical alignment. If one wants to generate a wedge filter with a horizontal alignment, the value $\tan^{-1}(v/u)$ in Equation (7.14) is altered to $\tan^{-1}(u/v)$. A spatial interpretation of Equations (7.14) and (7.15) is shown in Figure 7.8. Note that in the calculation of the wedge filter, the value of $P(0, 0)$ is normally omitted because it is common to all wedges. One can therefore calculate the percentage of power in the area of the amplitude spectrum that is covered by the wedge filter by dividing the sum of the amplitudes in the area covered by the filter by the total power spectrum.

The Fourier power spectrum of the test image (Figure 7.6b) is shown in Figure 7.7a. It is clear that the higher values (indicated by bright areas) of the power spectrum have a diagonal (bottom-left to top-right) orientation. This is because the geological

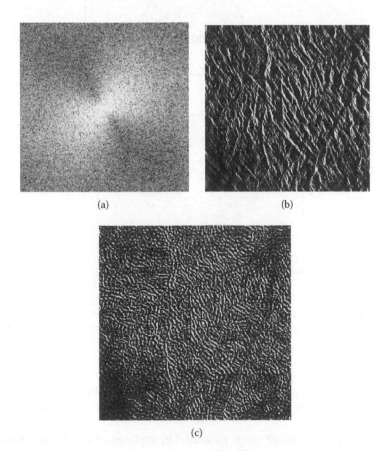

(a)   (b)

(c)

FIGURE 7.7   (a) Fourier power spectrum of Figure 7.6b. (b) Texture image generated by applying a ring filter within the frequency range [30, 100] on Figure 7.8a. (c) Image derived by a wedge filter with a horizontal direction with angles θ_1 and θ_2 defined by $-22.5°$ and $22.5°$, respectively.

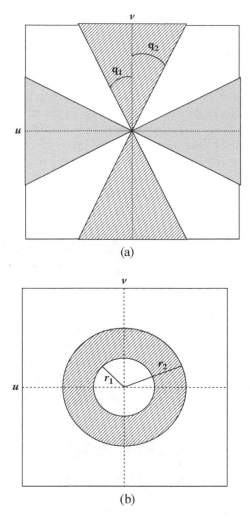

FIGURE 7.8   (a) A spatial interpretation of a frequency domain wedge filter according to Equation (7.14). (b) A spatial.interpretation of a frequency domain ring filter according to Equation (7.15).

patterns in the image have a NW–SE direction (recall that the direction of features on the power spectrum is perpendicular to the direction of image patterns in the spatial domain).

Figure 7.7b shows a texture image generated by applying a ring filter within the frequency range [30, 100] on Figure 7.7a. Note that the original image has become blurred because only low-frequency information is preserved. Figure 7.7c shows an image derived by a wedge filter with a horizontal direction with angles θ_1 and θ_2 defined by −22.5° and 22.5°, respectively, and the resulting patterns have a vertical orientation.

7.2.2   Wavelet Transform

It is recognized that most images, especially in the field of remote sensing, contain signals that are nonstationary (i.e., whose frequency response varies in time or over space). Thus, the Fourier transform introduced above may not be suitable for nonstationary signals since the transform is only capable of providing frequency information rather than a time- or space-frequency representation. An alternative way of dealing with such an issue can use the wavelet transform since this is capable of providing the time/space and frequency information simultaneously (i.e., giving a time- or space-frequency representation of the signal). This is the major advantage that the wavelet transform has in comparison with the Fourier transform.

The continuous wavelet transform decomposes the image on a base of elementary functions called *wavelets*. Such a base is formed by translations and dilations of a kernel function ψ called the *mother wavelet*. The wavelet transform displays the multiresolution characteristics of a signal or image by correlating it with different scales and shifts of a mother wavelet function. Formally, we define $a, b \in$ R, as the dilation step and translation step of translation and dilation processes, respectively. The base is then given by

$$\psi_{a,b} = D_a T_b\ \psi \tag{7.16}$$

where $T_b\psi(x) = \psi(x - b)$ and $D_a\psi(x) = |a|^{-1/2}\ \psi(x/a)$ are the translation and dilation operators, respectively (Mallat, 1989; Daubechies, 1992). There are numerous kinds of mother wavelets. However, for $\psi(x)$ to be characterized as a mother wavelet, the following admissibility condition must hold

$$\int_{-\infty}^{\infty} \frac{\left|\hat{\psi}(\omega)\right|^2}{|\omega|}\,d\omega < \infty \tag{7.17}$$

where $\hat{\psi}$ is the Fourier transform of ψ.

Once the wavelet has been defined, the continuous wavelet transform of a function $f(x)$ can be generated as

$$W_{\psi_{a,b}}f(x) = \frac{1}{\sqrt{a}} \int_{-\infty}^{\infty} f(x)\overline{\psi}\left(\frac{x-b}{a}\right)dx \tag{7.18}$$

where $\overline{\psi}$ denotes the complex conjugate of ψ (Daubechies, 1992). In practice, when the data being processed is restrained to a discrete image, ψ in a discrete form rather than continuous wavelet is applied. Discrete wavelet analysis also holds advantages that overcome the wavelet redundancy and computational issues encountered by the continuous wavelet (Burrus et al., 1998). In the case of processing an image, the wavelet transform is generally performed in a multiresolution sense, which requires a scaling function $\varphi(x)$ and the associated mother wavelet $\psi(x)$ in the construction of a complete basis.

The computation is based on the dyadic sequence (2^r), and the so-called two-scale relation (Sheng, 1996) holds for $\varphi(x)$ and $\psi(x)$ as

$$\varphi_{r,k}\left(2^r x\right)=\sum_k h_{r+1}\left(k\right)\varphi_{r+1,k}\left(2^{r+1}x-k\right) \tag{7.19}$$

$$\psi_{r,k}\left(2^r x\right)=\sum_k g_{r+1}\left(k\right)\varphi_{r+1,k}\left(2^{r+1}x-k\right) \tag{7.20}$$

where $h(k)$ and $g(k)$ are low pass and high pass filters, respectively. The above relation shows that the scaling and wavelet functions at a certain scale can be expressed in terms of translated scaling and wavelet functions at the next scale. The two filters are related to each other and are known as a *quadrature mirror filter* (Mallat and Zhong, 1992). A signal $f(x)$ can be expressed in terms of basis functions

$$f\left(x\right)=\sum_k \lambda_r\left(k\right)\varphi_{r,k}\left(2^r x-k\right) \tag{7.21}$$

Or, in turn,

$$f\left(x\right)=\sum_k \lambda_{r+1}\left(k\right)\varphi_{r+1,k}\left(2^{r+1}x-k\right)+\sum_k \delta_{r+1,k}\psi_{r+1,k}\left(2^{r+1}x-k\right) \tag{7.22}$$

where λ and δ are coefficients in the low pass and high pass domains, respectively. As the scaling function $\varphi(x)$ and the associated mother wavelet $\psi(x)$ are orthogonal, one can compute the coefficients λ and δ by taking the inner products:

$$\lambda_r\left(k\right)=\left\langle f\left(x\right),\varphi_{r,k}\left(x\right)\right\rangle \tag{7.23}$$

$$\delta_r\left(k\right)=\left\langle f\left(x\right),\psi_{r,k}\left(x\right)\right\rangle \tag{7.24}$$

and one can finally obtain the following equations:

$$\lambda_{r+1}\left(k\right)=\sum_n \lambda_r\left(n\right)h\left(n-2k\right) \tag{7.25}$$

$$\delta_{r+1}\left(k\right)=\sum_n \lambda_r\left(n\right)g\left(n-2k\right). \tag{7.26}$$

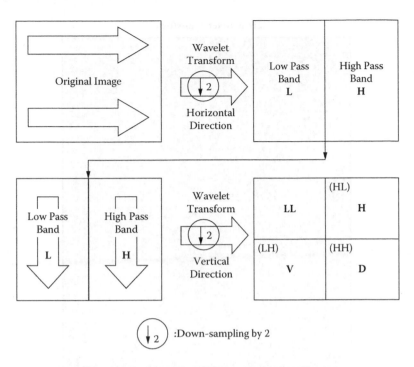

FIGURE 7.9   Example of wavelet transform for two-dimensional imagery.

The above formula can be regarded as a *down-sampling* (by 2) *convolution*. The discrete wavelet transform for two-dimensional imagery is similar. The transform is first performed along the row direction and then in the column direction. Figure 7.9 shows the process of the wavelet transform for a two-dimensional image. The process is first applied to each row to create an intermediate matrix of row coefficients. The intermediate matrix is then subject to a wavelet transform, this time in a column-wise sense. After the two-dimensional wavelet transform, the resulting image is made of four parts in which HL, LH, and HH bands are all high-frequency wavelet textures, and the LL band is the down-sized (1/4) version of the original image. The terms HL, LH, HH, and LL can be interpreted by recalling that the frequency represented by the row-wise transform increases from left to right, and for the vertical transform, from top to bottom of the image. The upper left quadrant, after one level of decomposition, is the LL image. The upper right quadrant represents the region in which low-frequency horizontal coefficients are combined with high-frequency vertical coefficients, while the lower left quadrant has high-frequency vertical coefficients and low-frequency horizontal coefficients. The bottom left quadrant combines high-frequency horizontal and vertical coefficients. See Mather (2004) for more details. Figure 7.10a presents an example of dealing with Figure 7.6b using the discrete wavelet transform process.

One useful texture measure derived from wavelets is called *entropy* and was introduced by Bian (2003). For a 2-dimensional wavelet-transformed image, let Wf_i denote

2D Wavelet Transform

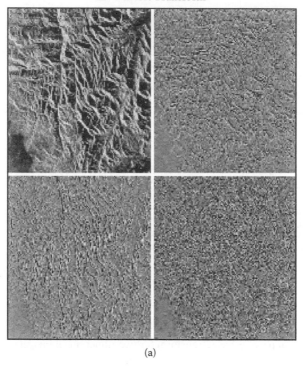

(a)

FIGURE 7.10   (a) Wavelet transform of Figure 7.6b.

the ith wavelet coefficient (in any of the HH, HL, or LH bands) within a window of n pixels centered at pixel k. The entropy measure for pixel k is then expressed as

$$Entropy_k = -\sum |p_i| \ln |p_i| \qquad (7.27)$$

where

$$|p_i| = \frac{|Wf_i|}{\sum |Wf_i|}. \qquad (7.28)$$

Figure 7.10b shows the wavelet entropy imagery computed from Figure 7.6b based on Equation (7.27). Mather (2004), Ouma et al. (2006) and Strand et al. (2006) provide descriptions of the use of wavelets in remotely sensed data processing. The wavelet transform method can also be used for a wide range of applications such as de-noising, compression, and multi-resolution representations. The interested reader may refer to Daubechies (1992), Mallat and Zhong (1992), and Burrus et al. (1998) for details.

Entropy Texture

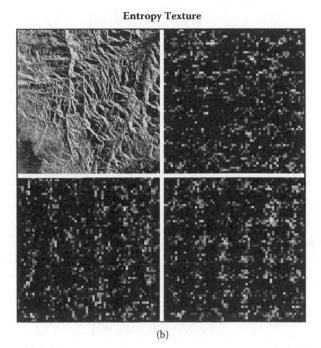

(b)

FIGURE 7.10   (continued) (b) Wavelet texture using entropy measure given by Equation (7.27).

7.3   GRAY-LEVEL CO-OCCURRENCE MATRIX (GLCM)

This section describes the texture feature extraction technique based on the gray-level co-occurrence matrix (GLCM), sometimes called the gray-tone spatial-dependency matrix. The principal concept of GLCM is that the texture information contained in an image is defined by the adjacency relationships that the gray tones in an image have to one another. In other words, it is assumed that the texture information is specified by values f_{ij} within the GLCM, where f_{ij} denotes the frequency of occurrence of two cells of gray tone i and j, respectively, separated by distance d with a specific direction on the image. Values of f_{ij} can be calculated for any feasible direction and distance d. Generally, only four directions corresponding to angles of $0°$, $45°$, $90°$, and $135°$ are used.

7.3.1   Introduction to the GLCM

Consider Figure 7.11a, which represents a 4×4 image with four gray levels. Figure 7.11b displays the general form of the corresponding GLCM. For example, the value contained in cell (2, 3) represents the number of times that gray levels 2 and 3 occur with a specific direction and distance, d. Figures 7.11c through 7.11f present the results for four directions given above with $d = 1$, while H, V, LD, and RD each denotes the cell number calculation for angles $0°$ (horizontal), $90°$ (vertical), $135°$ (left diagonal), and $45°$ (right diagonal), respectively, as indicated by the arrows.

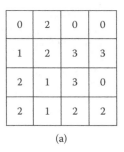

(a)

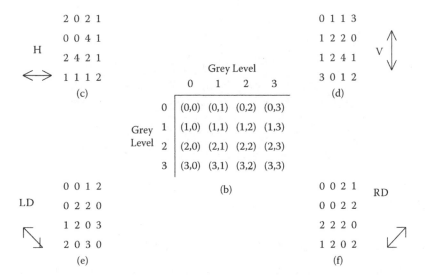

FIGURE 7.11   (a) A 4 × 4 image with four gray levels. (b) General form of GLCM with gray levels 0–3. The value in cell (i,j) stands for the number of times gray levels I and j occur with a specific direction and distance d. (s)–(f) present the results with respect to every pixel in image (a) for four directions with angles 0, 90, 135, and 45 degrees and d = 1.

Instead of using the frequency values in a GLCM directly, it is common practice to normalize them to the range [0, 1] to avoid scaling effects. The normalization procedure for each cell inside the matrix can be easily calculated. For the horizontal direction, with $d = 1$, there will be 2 × (number of columns − 1) pairs in each row. Consequently, the total number of nearest neighbor pairs can be obtained by the expression 2 × (number of columns − 1) × (number of rows). For instance, in the case of Figure 7.11a, the total number of pairs of horizontal direction is computed by 2 × (4 − 1) × 4 = 24. It can be easily verified that the sum of the entries in Figure 7.11c is 24. For the vertical direction, the same rule can be applied, providing a total of 2 × (number of rows − 1) × (number of columns) nearest neighbor pairs (see Figure 7.11d; the sum of the entries is 24). For the left-diagonal case with $d = 1$, there will be 2 × (number of columns − 1) pairs for each row except the last. This provides a total of 2 × (number of columns − 1) × (number of rows − 1) left-diagonal nearest neighbor pairs (Figure 7.11e; the sum of entries is computed as 2 × (4 − 1) × (4 − 1) = 18). For the right-diagonal direction, by symmetry, the number of nearest pairs is the same as the left diagonal (Figure 7.11f; the sum of entries is also 18). After the total number of pairs for each matrix has been obtained, each matrix is normalized by dividing each cell by the total number of pairs.

7.3.2 TEXTURE FEATURES DERIVED FROM THE GLCM

The principal ideas of texture measurement from GLCM are based on the following observations. If an image contains large homogeneous patches (i.e., it exhibits a coarse texture pattern), and the measurement distance d is relatively small in comparison with the texture structure, then neighboring gray-level pairs should have very similar values. Consequently, the joint neighboring pair distribution within the GLCM shows higher values concentrated around its principal diagonal (i.e., in the NW–SE direction, cells (i, j), $i = j$). Conversely, for a "busy" texture pattern (i.e., gray value varying considerably within a short range), if the measurement distance d is comparable to the scale of the texture structure, the gray level of points separated by d will be quite different. Therefore, high values in the GLCM are spread out away from the principal diagonal direction. In an extreme case, if all the pixel values are selected at random, the entries in the GLCM will show a uniform distribution. Once these relationships have been recognized, one can develop methods to quantify texture information contained in the GLCM.

To illustrate this idea, an example is shown in Figure 7.12 in which there are four gray levels. In Figure 7.12a the texture patterns are formed by several homogeneous patches (i.e., the texture pattern is coarse), while in Figure 7.12b the texture patterns are finer and smoother. The corresponding GLCMs are shown in Figures 7.12c and 7.12d. It is clear that the GLCM in Figure 7.12c contains large values around the principal diagonal direction; however, the values in Figure 7.12d show a relatively even spread.

Haralick et al. (1973) proposed a variety of texture measures based on the GLCM. These texture measures, called *textural features*, have been employed by a number of researchers (Franklin and Peddle, 1990; Sali and Wolfson, 1992), and the results have generally been successful. Four texture features, which are the most frequently used by researchers, are described here (Puissant et al., 2005). Other indices are given by Haralick (1973). In what follows, $p(i, j)$ denotes the $(i, j)^{th}$ entry in a normalized GLCM, and N_G denotes number of gray levels in the quantized image.

Angular second moment (ASM):

$$ASM = \sum_i \sum_j [p(i, j)]^2 . \qquad (7.29)$$

The measure of the *ASM* will output a higher value when co-occurrence frequency $p(i, j)$ is concentrated in few places in the GLCM, e.g., the main diagonal direction. If the $p(i, j)$ are close in value, then *ASM* will generate a small value. For instance, the *ASM* for the pattern shown in Figure 7.12a is 0.19, while for Figure 7.12b it is 0.06.

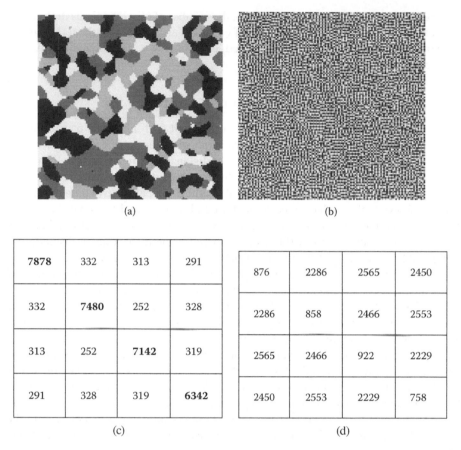

FIGURE 7.12   (a) Texture patterns are formed by patches with four gray levels. (b) Texture patterns are finer and smoother in comparison with Figure 7.12a. (c) Result of GLCM computed from Figure 7.12a. Most nonzero values are located around the principal diagonal direction. (d) Result of GLCM computed from Figure 7.12b. The values are more evenly spread than those in Figure 7.12c.

Contrast (Con):

$$Con = \sum_{n=0}^{N_G-1} n^2 \times \left[\sum_i \sum_{\substack{j \\ |i-j|=n}} p(i,j) \right]. \tag{7.30}$$

This measure assigns a higher weight to the $p(i, j)$ that are distant from the main diagonal of the GLCM. When the difference between neighboring pairs becomes large, the contrast increases. In this situation, co-occurrence frequencies $p(i, j)$ will have higher values away from the main diagonal of the GLCM, and consequently this equation will generate a higher value of *Con*. For instance, *Con* for Figure 7.12a is 0.37, while for Figure 7.12b, *Con* is 3.04.

Inverse difference moment (IDM):

$$IDM = \sum_i \sum_j \frac{1}{1+(i-j)^2} p(i,j). \tag{7.31}$$

This measure will generate higher values for an image containing large homogeneous patches because such an image generates high values on the main diagonal of the GLCM. It gives lower weight to those $p(i, j)$ that are located away from the main diagonal. For instance, the data shown in Figure 7.12a gives an *IDM* value of 0.92, while the value of *IDM* for the image in Figure 7.12b is 0.39.

Entropy (Ent):

$$Ent = -\sum_i \sum_j p(i,j) \log p(i,j). \tag{7.32}$$

The entropy measure outputs a higher value for a homogeneous distribution of $p(i, j)$, and lower otherwise. The value of *Ent* computed for Figure 7.12a is 1.85 while for Figure 7.12b it is 2.7.

7.4 MULTIPLICATIVE AUTOREGRESSIVE RANDOM FIELDS

The multiplicative autoregressive random field (MAR) model was proposed by Frankot and Chellapa (1987). It is used in modeling both the spatial correlation structures and the distribution of gray-level values in an image. Although the MAR model was originally used to model radar image data, the parameters of the model have been found to be highly correlated with the spatial distribution of the data, and consequently can be used as texture descriptors (Schistad et al., 1994).

7.4.1 MAR MODEL: DEFINITION

MAR is a natural extension of the Gaussian autoregressive random model with multiplicative behavior. Let an image $x(i, j)$ be represented by the following white-noise-driven multiplicative system:

$$x(i,j) = \prod_{[r=(m,n)] \in N} \left[x(i+m, j+n) \right]^{\theta_r} v(i,j). \tag{7.33}$$

N is the neighborhood set defining model support, $v(i, j)$ is a log-normal white-noise process referred to as the *driving process*, and θ_r, acts as an exponent weighting factor

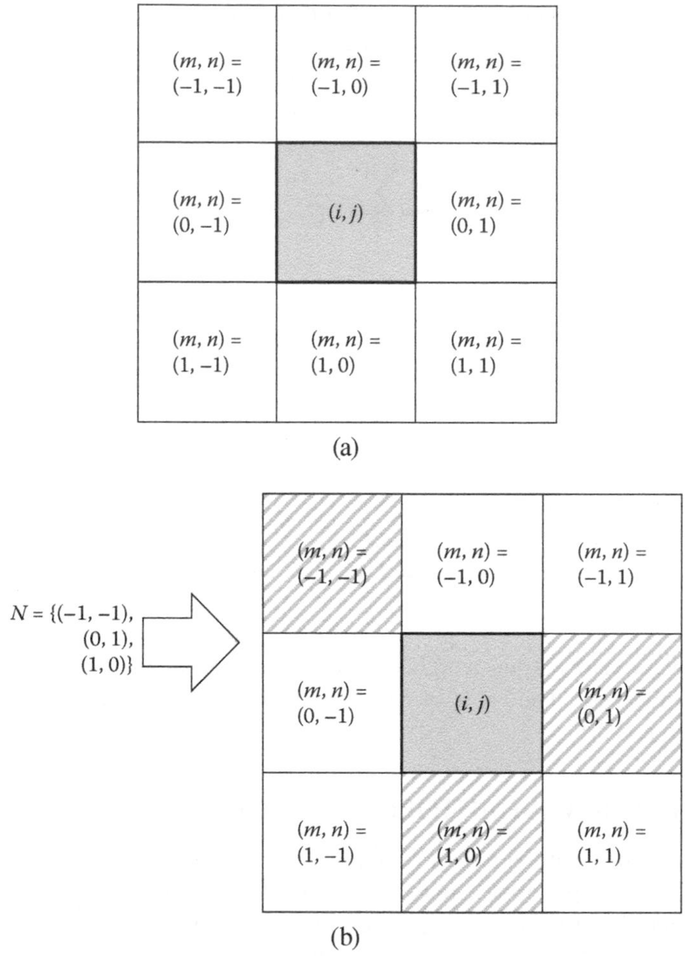

$(m, n) =$ $(-1, -1)$	$(m, n) =$ $(-1, 0)$	$(m, n) =$ $(-1, 1)$
$(m, n) =$ $(0, -1)$	(i, j)	$(m, n) =$ $(0, 1)$
$(m, n) =$ $(1, -1)$	$(m, n) =$ $(1, 0)$	$(m, n) =$ $(1, 1)$

(a)

$N = \{(-1, -1),$ $(0, 1),$ $(1, 0)\}$

$(m, n) =$ $(-1, -1)$	$(m, n) =$ $(-1, 0)$	$(m, n) =$ $(-1, 1)$
$(m, n) =$ $(0, -1)$	(i, j)	$(m, n) =$ $(0, 1)$
$(m, n) =$ $(1, -1)$	$(m, n) =$ $(1, 0)$	$(m, n) =$ $(1, 1)$

(b)

FIGURE 7.13 (a) Pixel (i, j) and its first- and second-order neighborhood. (b) Define neighborhood set N as $\{(-1, -1), (1, 0), (0, 1\}$. Only these three particular neighbors contribute to the central pixel (i, j) when applying the MAR model.

for neighborhood r. For instance, pixel (i, j) and its first- and second-order neighborhood is illustrated in Figure 7.13a. The concept of neighborhood is explained further in Chapter 8. If we define the neighborhood set N as $\{(-1, -1), (1, 0), (0, 1\}$, then only these three particular neighbors contribute to the central pixel (i, j), as illustrated in Figure 7.13b. According to Equation (7.33) the central pixel value $x(i, j)$ can be modeled as

$$x(i, j) = x(i-1, j-1)^{\theta_{(-1,-1)}} \times x(i+1, j)^{\theta_{(1,0)}} \times x(i, j+1)^{\theta_{(0,1)}} \times v(i, j). \qquad (7.34)$$

The random field $x(i, j)$ is said to obey a log-normal MAR model if $y(i, j) = \ln[x(i, j)]$ obeys the following Gaussian autoregressive random field model with $u(i, j) = \ln[v(i, j)]$:

$$y(i,j) = \sum_{[r=(m,n)]\in N} \theta_r y(i+m, j+n) + u(i,j) \qquad (7.35)$$

where $u(i,j)$ is zero-mean white Gaussian noise. The covariance of $u(i,j)$ is given by

$$Cov_u(r) = \begin{cases} \sigma_u^2, & r = (0,0) \\ 0, & r \neq (0,0) \end{cases} \qquad (7.36)$$

where σ_u denotes the variance of u.

7.4.2 ESTIMATION OF THE PARAMETERS OF THE MAR MODEL

Three parameters of the MAR model are generally estimated and used as texture descriptors. These are the neighborhood weighting parameter vector θ, the noise variance σ_u^2, and the mean value δ_y of the stationary random process y, respectively (Equation [7.35]). The estimation method employed here is that of least squares estimation (Kashyap and Chellappa, 1983). Given an image $x(i, j)$ of size $M \times M$, the least square parameter estimates, based on a log transform of $x(i,j)$, $\ln x(i,j) = y(i,j)$, are

$$\hat{\theta} = \frac{\sum_{i,j} z(i,j) \times (y(i,j) - \delta_y)}{\sum_{i,j} z(i,j) \times z^T(i,j)} \qquad (7.37)$$

with covariance

$$\sigma_u^2 = \frac{1}{M^2} \sum_{i,j} [y(i,j) - \delta_y - \theta^T \times z(i,j)]^2 \qquad (7.38)$$

and mean

$$\delta_y = \frac{1}{M^2} \sum_{i,j} y(i,j) \qquad (7.39)$$

where

$$z(i,j) = y(i+m, j+n) - \delta_y, r \in N. \qquad (7.40)$$

Because of variations in texture, different images will generally show different values of mean, noise covariance, and parameter θ, and so it is possible to use these three parameters to describe the texture pattern. An example is shown in Figure 7.14, in which images (a), (b), and (c) are constructed from Figure 7.6b by using the mean

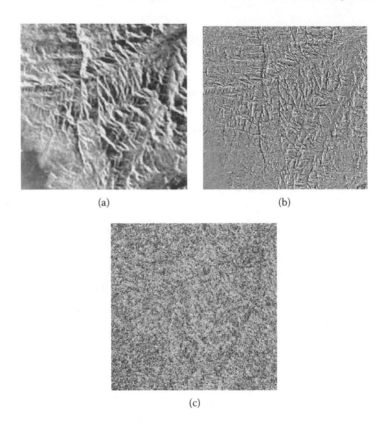

(a)   (b)

(c)

FIGURE 7.14   (a) MAR model texture formed by the mean of entries of vector θ (with neighborhood N defined as $\{(0, 1), \text{ and } (-1, -1)\}$ using Equation (7.37). (b) MAR model texture formed by variance σ_u^2 using Equation (7.38). (c) MAR model texture formed by mean value δ_y using Equation (7.39).

of entries of vector θ (with neighborhood N defined as $\{(0, 1), \text{ and } (-1, -1)\}$), variance σ_u^2, and mean value δ_y, respectively. The practical use of these methods is illustrated in Figure 7.15.

7.5   THE SEMIVARIOGRAM AND WINDOW SIZE DETERMINATION

Since textural features are scale dependent, the choice of window size has to be made carefully. If the window size is too small relative to the texture structure, the extracted texture feature will not accurately reflect the real texture property. In order to estimate the optimal window size, the semivariogram (Woodcock and Strahler, 1983; Woodcock et al., 1988a, 1988b; Curran, 1988; Curran and Atkinson, 1998; Zhang et al., 2004; Lloyd et al., 2004; Garrigues et al., 2006). The use of the semivariogram in texture estimation is described in the following text.

Given a transact across an image, where the digital number z of each pixel i, denoted by $z(i)$, is extracted at a predefined interval or lag h, then the relationship between pairs of pixels $z(i)$ and $z(i + h)$ can be expressed as the average variance

---

Begin
    Define the neighborhood set N (i.e., a set of values for pairs (m, n))
        for model support, where N is described in Figure 6.17;
    Divide the image into overlapping windows (e.g., $3 \times 3, 5 \times 5, \ldots$)
        centered on each pixel;
    Transform all pixel values into the lognormal domain;
    Do for each pixel (i, j)
        Calculate mean value δ_y from the window central on the (i, j)
           using Eq. (6.50);
        Calculate vector θ using Eq. (6.48) and (6.51);
        Using δ_y vector θ to calculate noise variance σ_u^2 using Eq. (6.49);
    EndDo
End

FIGURE 7.15 Algorithm for calculating MAR model texture coefficients.

of the differences between all such pairs. As the per-pixel variance is half this value, the semivariance S^2 for pixels at distance h apart is defined by

$$\gamma(h) = \frac{1}{2}\left[z(i) - z(i+h)\right]^2 \tag{7.41}$$

The average semivariance for lag h is

$$S^2(h) = \frac{1}{2k}\sum_i \left[z(i) - z(i+h)\right]^2 \tag{7.42}$$

$S^2(h)$ is an unbiased estimate of the average semivariance $\gamma(h)$ in the population and can be used as a measure of dissimilarity between spatially separate pixels (Woodcock and Strahler, 1983; Woodcock et al., 1988a,b).

In general, the semivariogram has a shape similar to that shown in Figure 7.16. The choice of a suitable window size h' should be made according to the criterion that at lag h', the semivariance has stopped increasing, or the rate of increase of the semivariance has fallen significantly. A suitable window size h' is indicated in Figure 7.16. We have computed the semivariogram for the test images (Figure 7.17) using a horizontal transect, and a window size of 16 was determined to be the most appropriate.

Atkinson and Lewis (2000) consider the use of the semivariogram for the estimation of image texture, and report that Carr and Miranda (1998) found that the use of the variogram produced classifications with greater accuracy than those based on the gray-level co-occurrence matrix (GLCM). Two methods of using the semivariogram are described by Atkinson and Lewis (2000). The first method uses the semivariogram values directly. Two windows are defined. The first is a larger window that contains the second, smaller, window. The larger window has dimensions of $l \times m$, and within this area a subwindow of size $r \times s$ is used to estimate semivariances for all

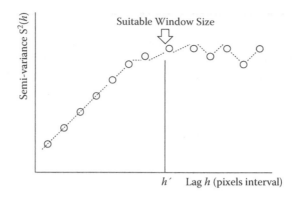

FIGURE 7.16   Determination of suitable window size h' where the corresponding semivariance $S^2(h)$ should show no further significant increase.

(a)   (b)

FIGURE 7.17   (a) A test image for evaluating different texture extraction approaches. From P. Brodatz. *Texture: A Photographic Album for Artists and Designers* (New York: Dover Books, 1966). With permission. (b) Test image faded by noise with *SNR* = 5.

possible lags (i.e., up to r-1 or s-1, whichever is the greater). The mean semivariance and the standard deviation of the semivariances for the subwindows are used as texture measures. Chica-Olmo and Abarca-Hérnandez (2000) describe an example of the use of this approach to texture, and Carr (1996) provides a Fortran computer program to compute the semivariogram-based measures of texture.

The second approach to semivariogram-based texture definition is to select a particular model to fit the semivariogram, and then to use selected parameters of the model as measures of texture. For example, Ramstein and Raffy (1989) use an exponential model for the semivariogram, and derive a parameter they call a' from this model. This parameter is used as a measure of texture. Atkinson and Lewis (2000) comment that the automatic model-fitting approach is unreliable, as the choice of model may not be appropriate for the region covered by the image. Further details of the use of geostatistical methods for the characterization of image texture are provided by Carr (1999) and van der Meer (1999b).

7.6   EXPERIMENTAL ANALYSIS

7.6.1   TEST IMAGE GENERATION

Brodatz's texture images (Brodatz, 1966) are widely used for texture analysis comparisons. Here we employ a combination of six Brodatz texture patterns (Figure 7.17) to test the performance of the different texture extraction approaches described in the preceding sections. Each subimage in Figure 7.17 is 128 × 128 pixels in size. From visual observation, the differences between the texture patterns within the image are clearly apparent. Experiments were also performed on a set of images subjected to an independent identical noise-fading environment. The original images were corrupted with additive noise denoted as

$$g(i, j) = x(i, j) + e(i, j) \tag{7.43}$$

where $x(i, j)$ is the original pixel value at location (i, j), $e(i, j)$ is the independent identical distribution (i.i.d.) Gaussian noise, and $g(i, j)$ is the resulting noise-fading image. The effect of noise fading is determined by the signal-to-noise ratio (SNR), defined as the ratio of average energy of the original image to the average energy of the random noise (Won and Derin, 1992):

$$SNR = \frac{\sum\limits_{i,j} x^2(i, j)}{\sum\limits_{i,j} e^2(i, j)}. \tag{7.44}$$

Let $v(i, j)$ be a normally distributed random variable at (i, j), then $e(i, j) = Kv(i, j)$, where

$$K = \sqrt{\frac{\sum\limits_{i,j} x^2(i, j)}{SNR \times \sum\limits_{i,j} v^2(i, j)}}. \tag{7.45}$$

Once the SNR is determined, Equation (7.45) is applied to obtain the values of K and then e. Equation (7.43) is then applied to generate the corresponding noise-fading image.

Four images were generated based on Figure 7.17a with different SNR values of 1, 5, 10, 100, and ∞ (the noise-free case). One of these images with SNR = 5 is shown in Figure 7.17b. Each image was divided into adjacent blocks of size 16 × 16. Texture features were then extracted for these blocks and subjected to clustering analysis. The results are shown in Section 7.6.3.

7.6.2 CHOICE OF TEXTURE FEATURES

Successful classification requires that texture features derived from each approach be carefully chosen. The selection of suitable texture features is generally image dependent. However, for some texture extraction approaches, the associated parameters can be determined through theoretical analysis while in other cases the texture features can only be extracted through trial-and-error methods.

7.6.2.1 Multifractal Dimension

From Section 7.1.2, it is clear that the range of parameter q is infinite (i.e., from $-\infty$ to $+\infty$), and thus will result in a considerable range of $D(q)$ values. However, it has been demonstrated by Tso (1997) that $D(q)$ changes considerably if q is approximately within the range $[-5, 5]$. In other words, there is no need to choose a value of q outside this range, as the resulting value of $D(q)$ will be approximately the same. Therefore, the values $q = 0$, $q = 3$, $q = -4$ were used in our experiments.

7.6.2.2 Fourier Power Spectrum

Two kinds of features extracted by applying W and R filters (Equations [7.38] and [7.39]) are used for the purposes of this example. The first is the proportion of the total power spectrum contained in the W filter, i.e., $[W_{\theta 1 \, \theta 2}/P(u, v)]$. Only horizontal and vertical directions within a predefined angular range of 45° (i.e., ±22.5°) were used. The second feature is the proportion of the power spectrum contained in the R filter, i.e., $[R_{r1,r2}/P(u, v)]$. Since both features are sensitive to orientation (i.e., $[W_{\theta 1, \theta 2}/P(u, v)]$) only, or to frequency (i.e., $[R_{r1,r2}/P(u, v)]$) only, it is better to use both filters together to perform texture segmentation. Note that there is no clear rule concerning the choice of a suitable range interval (r_1, r_2) for the R filter. The range was therefore decided by trial-and-error methods.

7.6.2.3 Wavelet Transform

There are many prototype mother wavelet functions that have been created for practical applications. Here, the selected mother wavelet is the Haar wavelet (Haar, 1910). The Haar wavelet is the simplest possible wavelet, which is expressed as

$$\frac{1}{\sqrt{2}}\begin{bmatrix} 1 & 1 \\ 1 & -1 \end{bmatrix}. \tag{7.46}$$

The advantage of the Haar wavelet is that the transform is equal to its inverse and is simple to compute and easier to understand. The disadvantage of the Haar wavelet is that it is not continuous and therefore not differentiable. Three texture features using entropy measures (cf. Equation [7.27]) of wavelet-transformed HL, LH, and HH imagery are used for segmentation purposes.

7.6.2.4 Gray-Level Co-Occurrence Matrix

Four texture features, ASM, Con, Ent, and IDM were derived from each subimage for the H, V, LD, RD directions with distance $d = 1$. In order to avoid the angular effect, in which the same texture pattern may spread out in different directions, the feature indices generated for each direction were not used separately. Instead, for each texture index, the sum of the four directions was used as input to the classifier.

7.6.2.5 Multiplicative Autoregressive Random Field

For each measured block, the corresponding neighbor set N used for model support was defined as $(1, 0)$, $(0, 1)$, and $(-1, -1)$. Three texture features σ_u^2, δ_y, and the mean value of entries of vector θ, were selected for the segmentation test.

7.6.3 SEGMENTATION RESULTS

The fuzzy c-means clustering algorithm (Chapter 5) was employed to cluster these texture features. The parameters chosen for the fuzzy c-means algorithm were $m = 2$ and number of clusters $n = 6$. The segmentation results are displayed in Figure 7.18.

The results of texture quantization methods, namely multifractal, GLCM, MAR, frequency domain filtering, and wavelet transform are shown in Figure 7.18. As indicated in the figure, multifractal does not seem to be very competitive with respect to other texture extraction approaches. GLCM produces the highest accuracy for the noise-free image. However, for noise-fading images, the MAR model seems to show more stable results. Note that texture classifications based on the MAR model use only three features as inputs (i.e., mean value of vector θ, covariance σ^2, and mean δ). These three features appear to characterize texture very well. An added feature of the MAR model is that the computational expense is low. Therefore, the MAR model can be regarded as the most useful tool for texture quantization.

The GLCM method generally needs many more computational resources than other methods. The computational requirement of the GLCM method is proportional to the number of gray levels used to represent the image. For instance, if the total number of

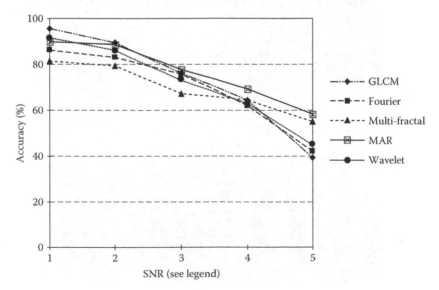

FIGURE 7.18 Segmentation accuracies obtained from different texture extraction approaches in dealing with different SNR images.

gray levels is 32, the calculation of each index for each pixel and on a specific direction requires a 32×32 matrix. However, if the number of gray levels is 256, the same process will need to operate on a 256×256 matrix, which increases the computational requirements very considerably. Therefore, a direct way of cutting the computational cost is to reduce image gray levels. A simple method known as the equal-probability quantizing algorithm (Haralick et al., 1973) can be used to achieve this end. Another method called the linked-list algorithm (Clausi and Jernigan, 1998), in which GLCM texture features are computed using linked lists, is also found to be successful in increasing the computational efficiency of the GLCM texture feature process.

Finally, with respect to the wavelet transform and Fourier frequency domain filtering, the wavelet transform shows good results when dealing with low-noise images. As the image is affected increasingly by noise, the performance of the wavelet transform drops off. For frequency domain filtering, the most successful texture segmentation result is derived from the combination of ring range [5, 7] and both horizontal and vertical wedges. One drawback, mentioned previously, is that the optimal ring range is hard to define, thus making segmentation results more unstable.

7.6.4 TEXTURE MEASURE OF REMOTE SENSING PATTERNS

The texture quantization approaches described above are next applied to real remotely sensed patterns. Figure 7.19 shows a test image of size 256×256, which is

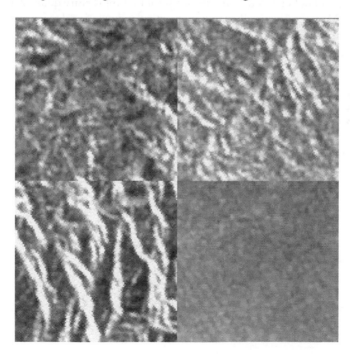

FIGURE 7.19   SIR-C SAR test image of size 256×256, made up of four 128×128 subimages extracted from different areas of the test image shown in Figure 7.6b.

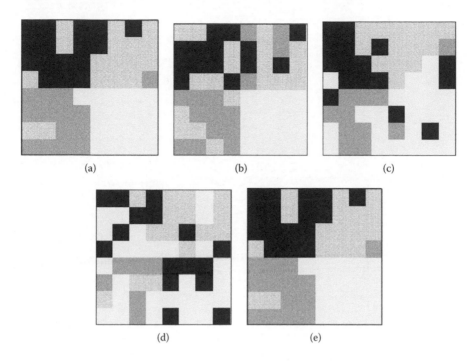

(a)   (b)   (c)

(d)   (e)

FIGURE 7.20 Segmentation result using (a) MAR model, (b) GLCM, (c) Fourier transform frequency filtering, (d) differential box counting approach, and (e) wavelet transform.

made up be four 128×128 subimages extracted from different areas of the SIR-C SAR (synthetic aperture radar) image shown in Figure 7.6b. The corresponding legends of these subimages (from upper left to lower right) are granitic, intrusives, volcano-sedimentary, and alluvium, respectively. The sub-images reveal different land surface structures. The MAR model, GLCM, frequency domain filtering, wavelet transform, and multifractal method with the same model parameters are again used to quantify these four texture patterns. Each subimage is further divided into 16 blocks, each 32×32 pixels in size. Classification results are shown in Figure 7.20. It is found that wavelet transform, MAR, and GLCM show relatively good texture discrimination ability compared with the other methods.

8 Modeling Context Using Markov Random Fields

Bayesian theory has had a long and profound influence on statistical modeling. Two key elements make up a Bayesian classification formula, namely, the *prior* and *conditional* probability density functions (p.d.f.). By combining these functions, a classification can be expressed in terms of maximum a posteriori (MAP) criteria (Chapter 2). In practice, there are difficulties in using MAP estimates. One of the problems is that prior information or information concerning the data distribution may not always available. As a result, alternative criteria must be used in place of MAP. For example, if knowledge of data distributions is available, but not prior information about the data under consideration, then the maximum likelihood (ML) criterion may be used. Conversely, if one has prior information but no knowledge about the data distribution, then the maximum entropy criterion can be employed (Jaynes, 1982).

The maximum likelihood criterion has been widely adopted in remotely sensed image classification, since in most classification experiments we generally use the normal (Gaussian) distribution to model the class-conditional p.d.f., while the prior p.d.f. is generally not used. It is likely that the classification result would be improved if (i) a reasonable assumption could be made in order to model the prior p.d.f., and (ii) the class-conditional p.d.f. could be incorporated in order to establish a MAP estimate. One assumption for modeling prior probability is context.

There has been an increasing interest in use of contextual information for modeling the prior p.d.f. (Derin and Elliott, 1987; Dubes and Jain, 1989; Jhung and Swain, 1996; Schistad et al., 1996; Magnussen et al., 2004; Tso and Olsen, 2005a). The concept is generally called the smoothness prior because the aim is to generate a smooth image classification pattern.

In the interpretation of visual information, context is very important. It may be derived from spectral, spatial, or even temporal attributes. The suitable use of context allows the elimination of possible ambiguities, the recovery of missing information, and the correction of errors (Li, 1995a; Magnussen et al., 2004). The use of context to model the prior p.d.f. in order to help in the interpretation of remotely sensed imagery is considered as a reasonable procedure, since a pixel classified as ocean is likely to be surrounded by pixels of the same class and is unlikely to have neighbors from categories such as pasture or forest. In other words, using the concept of context, pixels are not treated in isolation but are considered to have a relationship with their neighbors. Thus, the relationship between the pixel of interest and its neighbors is treated as being statistically dependent.

The Markov random field (MRF) is a useful tool for characterizing contextual information and has been widely used in image segmentation and image restoration

(Besag, 1974; Geman and Geman, 1984; Derin and Elliott, 1987; Dubes and Jain, 1989; Geman and Reynolds, 1991; Magnussen et al., 2004; Tso and Olsen, 2005a). The use of MRF models for linear feature detection has also achieved satisfactory results (Tupin et al., 1998). The practical use of the MRF relies on its relationship to the Gibbs random field (GRF), which provides a tractable way to apply the MRF to deal with context. Moreover, owing to the MRF's local property, the algorithm can be implemented in a highly parallel manner, which makes the MRF more attractive.

This chapter illustrates how MRF theory can be used to model the prior p.d.f. of contextually dependent patterns. Using both the prior p.d.f and class-conditional p.d.f., one can establish a MAP estimate. The fundamental concept of MRF is explained in Section 8.1. The construction of the energy function relating to remote sensing image classification and restoration is considered in Section 8.2. Parameter setting is another important issue in the application of MRF. A model is not complete if both the functional form of the model and its parameters are not fully specified. Methods for estimating model-associated parameters are therefore dealt with in Section 8.3. Classification algorithms based on the MAP-MRF framework and the experimental results derived from these algorithms are presented in the final sections of this chapter.

8.1 MARKOV RANDOM FIELDS AND GIBBS RANDOM FIELDS

Let a set of random variables $d = \{d_1, d_2, ..., d_m\}$ be defined on the set S containing m number of sites in which each random variable y_i ($1 \leq i \leq m$) takes a label from label set L. The family d is called a *random field*. The set S is equivalent to an image containing m pixels; d is a set of pixel DN values, and the label set L depends upon the application. The label set L is equivalent to a set of the user-defined information classes, e.g., L = {water, forest, pasture, or residential areas}, while in case of boundary detection, the label set L = {boundary, nonboundary}. There are many kinds of random field models describing ways of labeling the random variables. Here, we are concerned with two special types of random fields, the Markov random field and the Gibbs random field.

8.1.1 MARKOV RANDOM FIELDS

Given the definition of a random field specified earlier, we define the configuration w for the set S as $w = \{d_1 = w_1, d_2 = w_2, ..., d_m = w_m\}$, where $w_r \in$ L ($1 \leq r \leq m$). For convenience, we simplify the notation of w to $w = \{w_1, w_2, ..., w_m\}$. A random field, with respect to a neighborhood system is a Markov random field if its probability density function satisfies the following three properties:

1. Positivity: $P(w) > 0$ for all possible configurations of w,
2. Markovianity: $P(w_r|w_{S-r}) = P(w_r|w_{Nr})$, and
3. Homogeneity: $P(w_r|w_{Nr})$ is the same for all sites r.

$S - r$ is the set difference (i.e., all pixels in the set S excluding r), w_{S-r} denotes the set of labels at the sites in $S - r$, and Nr denotes the neighbors of site r.

The first property, that of positivity, can usually be satisfied in practice, and the joint probability $P(w)$ can be uniquely determined by local conditional properties as

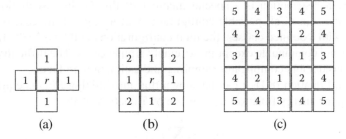

FIGURE 8.1 (a) The first-order neighbors of a pixel are the four pixels sharing a side with the given pixel. (b) The second-order neighbors are the four pixels having corner boundaries with the pixel of interest. (c) Higher-order (up to five) neighbors are extended in a similar manner.

long as the positivity property sustains (Besag, 1974). Markovianity indicates that labeling of a site r is only dependent on its neighboring sites. The property of homogeneity specifies the conditional probability for the label of a site r, given the labels of the neighboring pixels, regardless of the relative position of site r in S. An MRF may also incorporate other properties such as isotropy. Isotropy is the property of direction independence. That is, the neighboring sites surrounding a site r have the same contributing effect to the labeling of site r.

The usual neighborhood system used in image analysis defines the first-order neighbors of a pixel as the four pixels sharing a side with the given pixel, as shown in Figure 8.1a. Second-order neighbors are the four pixels having corner boundaries with the pixel of interest, as shown in Figure 8.1b. Higher-order neighbors can be extended in a similar manner. The neighborhood order up to five is shown in Figure 8.1c. More specifically, when the sites form a regular rectangular lattice, as do pixels in a two-dimensional image, the site $r = (i, j)$ has four nearest (first-order) neighbors, denoted by $N(i, j) = \{(I - 1, j), (I + 1, j), (i, j - 1), (i, j + 1)\}$.

8.1.2 GIBBS RANDOM FIELDS

A Gibbs random field (GRF) provides a global model for an image by specifying a p.d.f. in the following form:

$$P(w) = \frac{1}{Z} \exp\left[-\frac{U(w)}{T}\right]$$

(8.1)

where w is defined in Section 8.1.1, $U(w)$ is called an energy function, T is a constant termed *temperature* (which should be assumed to take the value 1 unless otherwise stated), and Z is called the partition function, which is expressed by

$$Z = \sum_{\text{all possible configurations of } w} \exp\left[-\frac{U(w)}{T}\right].$$

(8.2)

The Gaussian distribution is a special member of this Gibbs distribution family. Since Z is the sum of all possible configurations of w (e.g., in the case of a 256×256 image with 8 defined classes, the total configurations will be $8^{256 \times 256}$) hence Z is regarded as not computable except in extremely simple cases. The difficulty of computing Z considerably complicates sampling and estimation problems.

Based on Equation (8.1), it can be shown that maximizing $P(w)$ is equivalent to minimizing the energy function $U(w)$, given by Equation (8.3):

$$U(w) = \sum_{c \in C} V_c(w). \tag{8.3}$$

In this equation, C is known as a *clique* (defined below). $C = C_1 \cup C_2 \cup C_3 \cup ...,$ is the collection of all possible cliques, and $V_c(w)$ is called the potential function with respect to clique type C. A clique C is a subset in which all pairs of sites are mutual neighbors. It can be a single site, or a pair of neighboring sites, or a triple of neighboring sites, and so forth. Figure 8.2 shows clique types of the first and second order. The parameters associated with each clique type are also given. Figures 8.2a and 8.2b show those from the first-order neighborhood system; Figures 8.2c, 8.2d, and 8.2e are derived from the second-order neighborhood system. Clearly, as the order of the neighborhood system increases, the number of cliques grows rapidly, and so does the computational complexity. To help clarify the concept of the neighborhood system, Figure 8.3 shows cliques C_1, C_2, C_3, and C_4 relating to a pixel r of interest.

The energy function $U(w)$ in Equation (8.3) is more easily understood when it is expressed in the following expanding form:

$$U(w) = \sum_{\{r\} \in C_1} V_1(w_r) + \sum_{\{r,r'\} \in C_2} V_2(w_r, w_{r'}) + \sum_{\{r,r',r''\} \in C_3} V_3(w_r, w_{r'}, w_{r''}) + ... \tag{8.4}$$

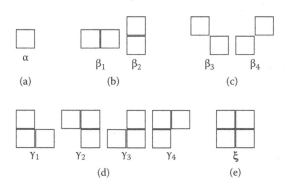

FIGURE 8.2 The clique types of the first- and second-order neighborhoods and the associated parameters. (a) single site, (b) horizontal and vertical neighbors, (c) diagonal neighbors. (d) triplets, and (e) four neighbors.

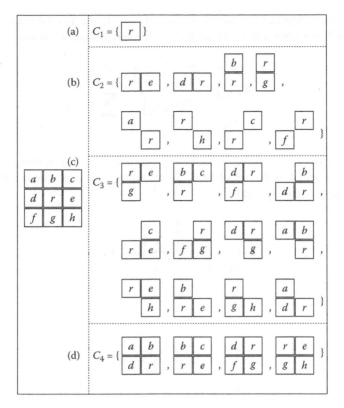

FIGURE 8.3 The cliques C_1, C_2, C_3, and C_4 relating to a pixel r of interest: (a) single site clique (C_1), (b) pair-site clique (C_2), (c) triple-site clique (C_3), and (d) quadruple-site clique (C_4).

C_1, C_2, C_3 and C_4 each represent a single-site clique (Figure 8.3a), a pair-site clique (Figures 8.3b and 8.3c), a triple-site clique (Figure 8.3d), and a quadruple-site clique (Figure 8.3e), respectively.

Schröder et al. (2000) provide a comprehensive account of Gibbs random field models and their applications in spatial data analysis.

8.1.3 MRF-GRF Equivalence

An MRF is defined in terms of local properties (i.e., the classification label assigned to a pixel is affected only by its neighbors), whereas a GRF describes the global properties of an image (i.e., the label given to a specific pixel is affected by the labels given to all other pixels) in terms of the joint distribution of classes for all pixels. The Hammersley-Clifford theorem describes the equivalence of GRF and MRF properties. The theorem states that a unique GRF exists for every MRF as long as the GRF is defined in terms of cliques on a neighborhood system. Several authors present the proof of this theorem (Besag, 1974; Moussouris, 1974; Kindermann and Snell, 1980). A proof that the GRF is an MRF is given below.

Let $P(w)$ be a Gibbs distribution on the set of sites (image pixels) S with respect to the neighborhood system N. Consider the conditional probability defined by

$$P\left(w_r \middle| w_{S-\{r\}}\right) = \frac{P\left(w_r, w_{S-\{r\}}\right)}{P\left(w_{S-\{r\}}\right)} = \frac{P(w)}{\sum_{w'_r \in L} P(w')} \tag{8.5}$$

where $w' = \{w_1, \ldots, w_{r-1}, w'_r, \ldots, w_m\}$ is any configuration that agrees with w at all sites except r, and L is the set of labels (user-defined classes). The combination of Equations (8.1) and (8.3) yields

$$P(w) = \frac{\exp\left[-\sum_{c \in C} V_c(w)\right]}{Z}$$

and, similarly,

$$P(w') = \frac{\exp\left[-\sum_{c \in C} V_c(w')\right]}{Z}. \tag{8.6}$$

Substituting Equation (8.6) into Equation (8.5) gives

$$P\left(w_r \middle| w_{S-\{r\}}\right) = \frac{\exp\left[-\sum_{c \in C} V_c(w)\right]}{\sum_{w'_r \in L} \exp\left[-\sum_{c \in C} V_c(w')\right]}. \tag{8.7}$$

If clique C is subdivided into two sets A and B with A consisting of pixels containing r and B being the pixels not containing r, then Equation (8.7) can be written as

$$P\left(w_i \middle| w_{S-\{i\}}\right) = \frac{\exp\left[-\sum_{c \in A} V_c(w)\right] \times \exp\left[-\sum_{c \in B} V_c(w)\right]}{\sum_{w'_r \in L} \left\{ \exp\left[-\sum_{c \in A} V_c(w')\right] \times \exp\left[-\sum_{c \in B} V_c(w')\right] \right\}}. \tag{8.8}$$

Since $V_c(w) = V_c(w')$ for any set of pixels c that does not contain pixel r, then the following terms can be deleted from both the numerator and denominator of Equation (8.8):

$$\exp\left[-\sum_{c \in B} V_c(w)\right]$$

and

$$\exp\left[-\sum_{c \in B} V_c(w')\right].$$

Hence, this probability depends only on the potential functions V_c of the pixels containing r, which gives the relationship

$$P\left(w_r \middle| w_{S-\{r\}}\right) = \frac{\exp\left[-\sum_{c \in A} V_c(w)\right]}{\sum_{w_r' \in L} \exp\left[-\sum_{c \in A} V_c(w')\right]}. \tag{8.9}$$

That is, the conditional probability $P(w|w_{S-\{r\}}) = P(w|w_{N_r})$ depends on the neighbors of site r. This proves that Gibbs random field is a Markov random field as far as the cliques are concerned.

The proof of MRF–GRF equivalence provides a simple way for dealing with MRF-based context. It also reduces the complexity of the MRF model, as we can specify the MRF model in terms of GRF model formation. That is, the joint probability $P(w)$ is specified through the energy function $U(w)$ via the *potential* $V_c(w)$, as shown in Equation (8.4).

8.1.4 Simplified Form of MRF

The choice of the form and the parameters of the function shown in Equation (8.4) for the energy function $U(w)$ is a major topic in MRF modeling. When all the parameters and the form of the energy are defined, the model is completely specified. The results of image segmentation are thus dependent both on the classification algorithm and on how well the form and parameters of the function are defined.

The most general form of energy function $U(w)$ is based on up to pairwise clique potential functions that allow the following expression:

$$U(w) = \sum_{r \in S, \{r\} \in C_1} V_1(w_r) + \sum_{r \in S, \{r,r'\} \in C_2} V_2(w_r, w_{r'}). \tag{8.10}$$

Equation (8.10) is a special case of Equation (8.4). When the label set L only contains two labels, L = {1, −1}, and if the potential function $V_1(w_r) = \alpha \times w_r$, $V_2(w_r, w_{r'}) = \beta \times w_r \times w_{r'}$, then the corresponding expression for the energy function $U(w)$ becomes

$$U(w) = \sum_{r \in S, \{r\} \in C_1} \alpha \times w_r + \sum_{r \in S, \{r, r'\} \in C_2} \beta \times w_r \times w_{r'} \tag{8.11}$$

where α and β are the potential parameters. Equation (8.11) is called an *auto-logistic model* (Besag, 1974) and has been used for texture modeling by Cross and Jain (1983). If the distribution of $P(w)$ is Gibbsian (i.e., can be modeled by Equation [8.1]), then the conditional probability $P(w_r|w_{Nr})$ for pixel r in an auto-logistic model is calculated as

$$P\left(w_r|w_{Nr}\right) = \frac{\exp-\left[\alpha w_r + \sum_{\{r,r'\} \in C_2} \beta w_r w_{r'}\right]}{\sum_{w_r \in \{0,1\}} \exp-\left[\alpha w_r + \sum_{\{r,r'\} \in C_2} \beta w_r w_{r'}\right]}$$

$$= \frac{\exp-\left[\alpha w_r + \sum_{\{r,r'\} \in C_2} \beta w_r w_{r'}\right]}{1 + \exp-\left[\alpha w_r + \sum_{\{r,r'\} \in C_2} \beta w_r w_{r'}\right]}. \tag{8.12}$$

When the neighborhood system defined in Equation (8.12) is reduced to a first-order system (i.e., using nearest neighbors as shown in Figure 8.1a), the auto-logistic model is called the Ising model, after Ising (1925), and is the best-known lattice system.

The auto-logistic model can be further generalized into the multi-level logistic (MLL) model. The MLL model is also called a generalized Ising model (Geman and Geman, 1984) and it possesses the following properties:

1. The number of labels in the label set L must be greater than 2
2. For a single clique C_1, the potential function $V_1(w_r)$ is label dependent, and is defined as $V_1(w_r) = \alpha_k$ if w_r = label k, $k \in L$.
3. For cliques containing more than one site (i.e., C_2, C_3, …) the potential function is defined as
   $V_c(w) = -\zeta_c$ if all sites in clique c have the same label,
   $V_c(w) = \zeta_c$ 0 otherwise

where ζ_c (> 0) is a real number called the potential parameter. In general, we apply different symbols to denote potential parameters for different clique types. The potential parameters corresponding to clique types from C_2 to C_4 are illustrated in Figure 8.2 (e.g., α for C_1, β for C_2, γ for C_3, and ξ for C_4).

A special case of the MLL model is defined when only pairwise cliques are active, that is, only parameter β is nonzero ($\beta > 0$). The potential can therefore be simplified to

$$V_2(w_r, w_{r'}) = -\beta \qquad \text{if sites on clique } \{r, r'\} = C_2 \text{ have the same label,}$$
$$V_2(w_r, w_{r'}) = \beta \text{ or } 0 \quad \text{otherwise} \tag{8.13}$$

When the model is *anisotropic* (i.e., rotation variant), the β coefficients may take on different values for different orientations as illustrated in Figures 8.2 (b) and (c) from β_1 to β_4. Owing to its simplicity, the pairwise MLL model has been widely used for modeling regions and texture (Geman and Geman, 1984; Derin and Elliott, 1987; Won and Derin, 1992). It should be noted that under the isotropic assumption (i.e., the same value β for all directions), bloblike regions are generated. As the value of β increases, regions that are more homogeneous will be favored. However, if the isotropy limitation does not exist, the result will show patterns that are more texture-like. This topic is discussed further in the following sections.

8.1.5 GENERATION OF TEXTURE PATTERNS USING MRF

In order to show how an image configuration is affected by its associated MRF parameters, an example specifying the effect of parameter selection is given below in which we adopted an MLL model with only pairwise cliques considered. The images are generated using Metropolis's algorithm (Metropolis et al., 1953), which is illustrated in Figure 8.4. This algorithm has been successfully tested by Dubes and Jain (1989) and results showed that 50 iterations (i.e., $n = 50$ in Figure 8.4, first line) are sufficient to achieve convergence. In interpreting Figure 8.4, note that the higher the energy, the lower the probability.

Texture patterns derived using this algorithm are shown in Figure 8.5. Note that the resulting image patterns are not affected solely by the magnitude of parameter β. The choice of the second term in Equation (8.13), that is, β or 0, will also contribute some variation. Figures 8.5a and 8.5c were constructed by using β in the case $w_r \neq w_{r'}$, while in Figures 8.5b and 8.5d, β is replaced by 0. It is clear that the resulting texture patterns do reveal some level of difference. These texture patterns can be used for testing the robustness of texture quantization methodology, as described in Chapter 7.

The procedure illustrated in Figure 8.4 allows the generation of a unique texture pattern for an image. If an image with multitextured patterns is required, then a hierarchical model can be applied. A hierarchical model generally involves two levels. The higher level uses an MLL model with an isotropic potential parameter (i.e.,

---

Begin
   Define potential parameters β, iteration n, and label set L;
   Initialize image $w^{(0)}$: for every pixel r by randomly selecting a label
      from L, then assigning to r;
   Do For iteration $i = 1$ to n do
      Do For each pixel r in the image
         Randomly select label $w_r^{(i)}$ from L;
         Calculate energy for both old label $U(w_r^{(i-1)})$, and new
            label $U(w_r^{(i)})$ at pixel r.
         Let $p = \min\{1, \exp(-U(w_r^{(i)})/\exp(-U(w_r^{(i-1)}))\}$;
         Substitute $w_r^{(i-1)}$ by $w_r^{(i)}$ with probability p;
      EndDo
   EndDo
End

FIGURE 8.4 Metropolis algorithm for generating a texture image. From Metropolis, N., A. W. Rosenbluth, M. N. Rosenbluth, A. H. Teller, and E. Teller. "Equations of State Calculations by Fast Computational Machine." *Journal of Chemical Physics* 21 (1953):1087–1091.

unique β) to generate a bloblike image, which is sometimes called a region process. At the lower level, each blob region is filled with different texture patterns generated by the MLL model using anisotropic potential parameters (i.e., different β for different orientations). The resulting image will therefore contain multitextured patterns. At the higher level, one can sometimes use a predefined (or digitized) image to replace the region process. At the lower level of the hierarchical model, one may also use independent noise instead of texture patterns to fill each blob region. Figure 8.6 illustrates the construction of multiple texture images using a hierarchical model.

   Several studies (Derin and Cole, 1986; Derin and Elliott, 1987; Won and Derin, 1992; Chen and Huang, 2001) use images generated by the hierarchical model to test the robustness of image segmentation algorithms. We can regard the higher-level image of a hierarchical model as representing the desired segmentation, which is contaminated by texture or noise generated at the lower level of the hierarchical model. If the classification is based on a Bayesian formula, one can adopt a smooth assumption to model prior p.d.f. (Section 8.1.4), while the conditional p.d.f. of the data given class label w can be derived from the texture or noise at low level. More details of MAP-MRF labeling are given in Section 8.4.

8.2   POSTERIOR ENERGY FOR IMAGE CLASSIFICATION

Assume that we have a set of observed feature vectors (e.g., pixel gray in a number of different spectral bands), which are denoted by d. Traditionally each pixel is labeled based on d alone without considering contextual information. Once the context is included as *a priori* information and modeled by means of MRF, the current trend is to use the Bayesian formulation to construct an objective function called *posterior energy* and then to perform labeling by minimizing this posterior energy. Recall that

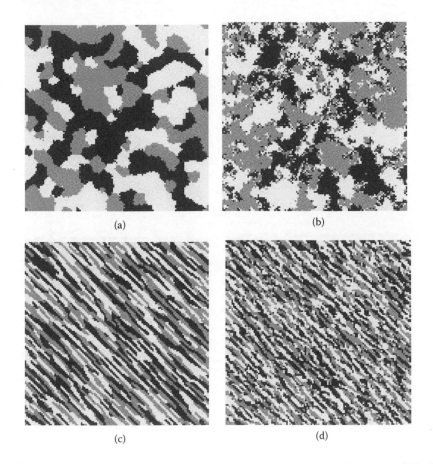

(a) (b)

(c) (d)

FIGURE 8.5 (a) and (c) were constructed using β in case of $w_r \ne w_{r'}$, while in (b) and (d) β is replaced by 0. The value of β is (0.5, 0.5, 0.5, 0.5) for (a) and (b) while $\beta = (1, 1, 1, -1)$ for (c) and (d).

the higher the energy $U(event)$, the lower is the probability $P(event)$, where *event* can be pixel values or class labels as far as an image is concerned.

Using Bayesian formulae, the posterior distribution $P(w_r|d_r)$ for label w_r given the observation d_r at pixel r is correlated with $P(d_r|w_r)P(w_r)$ (Chapter 2). According to Equation (8.1), $P(w_r) = (1/Z) \times \exp[-U(w_r)]$, and the $U(w_r)$ is therefore called the prior energy. Here, we use a smoothness assumption as prior information. If only cliques C_1 and C_2 are included in an MLL model (Section 8.1.4), then the prior energy (smoothness prior) for pixel r can be defined as

$$U(w_r) = \alpha_k + \sum_{\{r,r'\} \in C_2} \beta \times \delta(w_r, w_{r'}), \text{ if } w_r = \text{label } k, k \in L \tag{8.14}$$

where α_k is single-site clique potential parameter, which can be regarded as the first penalty term specially for label k. The larger the value of α_k, the less the probability

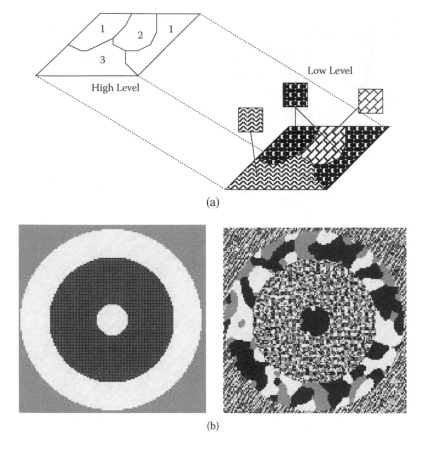

(a)

(b)

FIGURE 8.6 (a) Hierarchical model showing a region pattern at the higher level, while the texture patterns applied at lower level. (b) High level is replaced by a user-defined region image, which is then filled with texture at low level with $\beta = (2, 2, 2, 2)$ (in the white region), $(-1, 2, -1, 2)$ (in the gray region), and $(-1, -1, -1, -1)$ (in the black region).

that pixels in the image are assigned to label k. The term $\delta(a, b)$ is a step function defined as $\delta(a, b) = 1$ if $a \neq b$, and -1 otherwise, and β (>0) is a pairwise clique potential parameter. $\beta \times \delta(a, b)$ is the second penalty term. The larger the value of β, the stronger is the smoothing force. In other words, if w_r does not agree with its neighbors, higher prior energy (i.e., lower probability) will result. Note that in most applications, the probability of each class is considered to be the same, and thus α_k is set to 0, unless otherwise specified.

The conditional distribution of the observed data d_r (e.g., pixel values at site r) given the true label w_r ($w_r = $ class k) is often assumed to be Gaussian, and can be formulated as

$$P\left(d_r \middle| w_r\right) = \frac{1}{\sqrt{2\pi}^{\rho} \sqrt{\left|\Sigma_k\right|}} \exp\left[-U(d_r \middle| w_r)\right] \tag{8.15}$$

where ρ is the dimensionality of the feature space (e.g., the number of image bands), Σ_k is the class-conditional covariance matrix for class k and

$$U\left(d_r \middle| w_r\right) = \frac{1}{2}(d_r - u_k)\sum_k^{-1}(d_r - u_k) \tag{8.16}$$

is the class-conditional or likelihood energy, where u_k is the mean vector of class k.

By combining Equations (8.14) and (8.16), one obtains the posterior energy $U(w|d)$:

$$U\left(w \middle| d\right) = U\left(d \middle| w\right) + U(w)$$
$$= \sum_r \frac{1}{2}(d_r - u_k)\sum_k^{-1}(d_r - u_k) + \sum_r \sum_{\{r,r'\}\in C_2} \beta \times \delta(w_r, w_{r'}). \tag{8.17}$$

The MAP estimate, which maximizes posterior probability $P(w|d)$, and which is equivalent to minimizing the posterior energy, is defined by

$$\hat{w} = \arg\min_w U\left(w \middle| d\right). \tag{8.18}$$

The solution of Equation (8.18) requires a specially designed approach. An iterative algorithm is normally used because the labeling of each pixel has an effect on the labels to be assigned to its neighbors. Algorithms for the determination of energy minimization are considered in Section 8.5.

8.3   PARAMETER ESTIMATION

A probability model is not complete if the model-associated parameters are not fully specified, even if the functional form and the distribution are known. A good choice of parameters can successfully restore or segment a noise-disturbed image. Conversely, a poor selection of parameter values will usually generate poor results, as the following example demonstrates.

Two images are shown in Figure 8.7a. Each has two gray levels, 40 and 80, with pair-site parameters $\beta = \{-0.6, 0.6, 0.6, 0.6\}$ (Figure 8.7a, left-hand side) and $\beta = \{1, 1, 1, 1\}$ (Figure 8.7a, right-hand side), using the algorithm in Figure 8.4. After contamination by Gaussian noise with mean $\mu = 0$ and variance $\sigma 2 = 402$, the images shown in Figure 8.7b are derived. The results achieved using the known parameters (i.e., the values of μ, $\sigma 2$, and β that were used in generating the images) and incorrect estimates of β (i.e., $\{-2, -2, 2, 2\}$) are shown in Figures 8.7c and 8.7d, respectively.

In comparison with the study of probability function formulation, the question of deriving estimates of parameters is a relatively new research field. Methods of parameter estimation can be divided into supervised and unsupervised. Supervised methods are based on the assumption that the image only contains one type of MRF pattern (i.e., MRF texture), or that the image contains several MRF patterns, but we know exactly where the boundaries between those MRF patterns are located. In

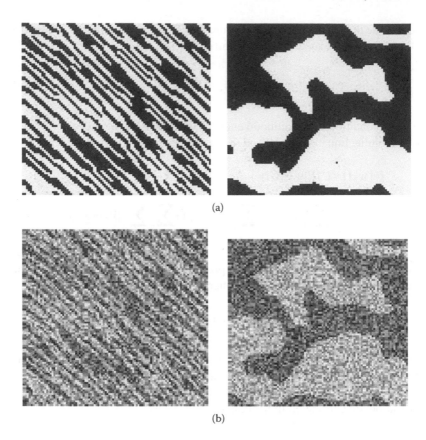

(a)

(b)

FIGURE 8.7   (a) Original image, (b) image faded by noise, (c) image classified by using the correct parameter values, and (d) classification using incorrect parameter values. See text for further details.

other words, the classes (MRF patterns) in the image are known. We can then use training data selected from each class to perform parameter estimation.

Unsupervised methods are designed to cope with the situation in which we do not have any *a priori* idea about how many MRF patterns are contained in the image. A general strategy is to perform image clustering and parameter estimation alternatively. Like traditional clustering algorithms, unsupervised methods are run in an iterative mode, but are more sophisticated. One supervised approach, the least squares fit (Derin and Elliott, 1987), is introduced below. For details of other supervised approaches, refer to Chandler (1987), Parisi (1988), and Saguib et al. (1998). For unsupervised estimation, refer to Dempster et al. (1977), Besag (1986), Wahba (1990), and Reeves (1992).

8.3.1 LEAST SQUARES FIT METHOD

The *least squares fit* (LSF) is a method that uses a regression technique to obtain estimates of Markovian parameters. The LSF method was proposed by Derin and

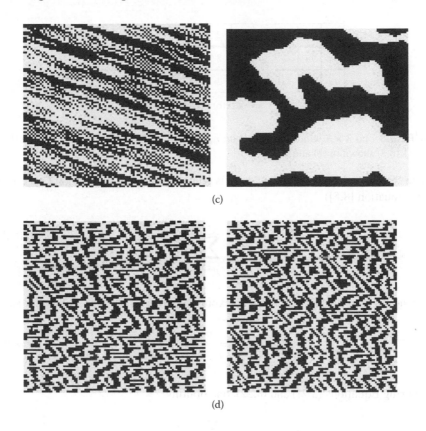

(c)

(d)

FIGURE 8.7 (continued)

Elliott (1987), who demonstrate that parameter estimation using LSF is more accurate than some other approaches (e.g., the method proposed by Besag, 1974).

For clarity, we consider a problem involving clique type $c = 2$. There will be four clique potential parameters (i.e., β_1, β_2, β_3, and β_4, as shown in Figure 8.2) to estimate. Define a vector θ, $\theta = [\beta_1, \beta_2, \beta_3, \beta_4]^T$, and consider a 3×3 window centered on the pixel r_0, as shown in Figure 8.8a. The potential function for pixel r_0 is defined as (cf. Equation [8.4])

$$U(w_{r0}, w_{Nr}) = V_{r0}(w_{r0}, w_{Nr}) = \theta \cdot \Psi(w_{r0}, w_{Nr}) \tag{8.19}$$

where

$$\Psi(w_{r0}, w_{Nr}) = [\psi(w_{r0}, w_{r4}) + \psi(w_{r0}, w_{r8}), \psi(w_{r0}, w_{r2}) + \psi(w_{r0}, w_{r6}),$$

$$\psi(w_{r0}, w_{r1}) + \psi(w_{r0}, w_{r5}), \psi(w_{r0}, w_{r3}) + \psi(w_{r0}, w_{r7})] \tag{8.20}$$

The function $\psi(a, b)$ is Boolean and is defined as $\psi(a, b) = -1$ if $a = b$, $\psi(a, b) = 1$ otherwise. For example, in Figure 8.8b, the function $\Psi(w_{r0}, w_{Nr}) = [2, -2, 0, 0]$, while in Figure 8.8c, $\Psi(w_{r0}, w_{Nr})$ is $[0, 0, -2, 2]$.

r_1	r_2	r_3
r_8	r_0	r_4
r_7	r_6	r_5

3	2	3
1	2	1
2	2	2

1	1	2
3	1	1
2	2	1

(a)   (b)   (c)

FIGURE 8.8   (a) 3×3 window centered on r_0, and two possible configurations with three labels (1 to 3) shown in (b) and (c).

The conditional probability $P(w_{r0}|w_{Nr})$ is related to the energy function $U(w_{r0}, w_N r)$ by (cf. Equation [8.5])

$$P\left(w_{r0}\Big|w_{Nr0}\right) = \frac{e^{-U\left(w_{r0}, w_{Nr0}\right)}}{\displaystyle\sum_{w_{r0} \in L} e^{-U\left(w_{r0}, w_{Nr0}\right)}} . \tag{8.21}$$

The conditional probability can also be expressed in terms of joint probability as

$$P\left(w_{r0}\Big|w_{Nr0}\right) = \frac{P(w_{r0}, w_{Nr0})}{P(w_{Nr0})} . \tag{8.22}$$

Comparing Equations (8.21) and (8.22), one obtains

$$\frac{e^{-U\left(w_{r0}, w_{Nr0}\right)}}{P(w_{r0}, w_{Nr0})} = \frac{\displaystyle\sum_{w_{r0} \in L} e^{-U\left(w_{r0}, w_{Nr0}\right)}}{P(w_{Nr0})} . \tag{8.23}$$

Note that the right-hand side of Equation (8.23) is independent of w_{r0}. One can extend the idea to show that the left-hand side is also independent of w_{r0}. Thus, if one defines w'_{r0} as another label on site r_0, according to Equation (8.23), the following relation can be created:

$$\frac{e^{-U\left(w_{r0}, w_{Nr0}\right)}}{P(w_{r0}, w_{Nr0})} = \frac{e^{-U\left(w'_{r0}, w_{Nr0}\right)}}{P(w'_{r0}, w_{Nr0})} . \tag{8.24}$$

By using the natural logarithm operator, Equation (8.24) is reduced to

$$U(w_{r0}, w_{Nr}) - U(w'_{r0}, w_{Nr}) = \ln P(w'_{r0}, w_{Nr}) - \ln P(w_{r0}, w_{Nr})$$

or, equivalently

$$\theta \cdot [\Psi(w_{r0}, w_{Nr}) - \Psi(w_{r0}, w_{Nr})] = \ln P(w'_{r0}, w_{Nr}) - \ln P(w_{r0}, w_{Nr}) . \tag{8.25}$$

For any $w_{r0} \in L$, the $\Psi(w_{r0}, w_{Nr})$ can be obtained by using Equation (8.20), and the joint probability $P(w_{r0}, w_{Nr})$ can be estimated by using a histogram technique. Assume there are a total of y windows of size 3×3 in an image, and a particular configuration of (w_{r0}, w_{Nr0}) occurs x times. Then

$$P(w_{r0}, w_{Nr0}) = \frac{x}{y}. \tag{8.26}$$

The relationship between the histogram technique and the ML estimation is examined by Gurelli and Onural (1994). Using Equations (8.25) and (8.26), one can obtain numerous equations (parameterized by θ) by counting different configuration for different class labels. For instance, if the number of configurations in Figure 8.8b occurs 50 times, and for Figure 8.8c it occurs 100 times, and if there are total 500 windows in an image, one can generate an equation such as

$$\theta \times [(2, -2, 0, 0) - (0, 0, -2, 2)] = \ln(100/500) - \ln(50/500)$$

$$\Rightarrow 2\beta_1 - 2\beta_2 + 2\beta_3 - 2\beta_4 = 0.693. \tag{8.27}$$

Once those equations are established, the solution for parameter vector θ can be solved by a least squares technique (refer to Mather (1976) for a comprehensive description of this approach). In order to reduce estimation bias, Derin and Elliott (1987) suggest that one should discard the case of $x = 0$.

Following from the above definitions, it is shown that the number of equations grows considerably as the number of labels L increases. An alternative approach, which reduces the amount of computation, is the *logit model fit method* (Dubes and Jain, 1989), which is a simple modification of least squares estimation. Let H be the collection of all possible configurations of site r_0 and its eight neighbors. Thus, the number of possible configurations in H denoted by $|H|$ for a two-label image is equal to 2^8. Define a relation $\approx$ on H by $h_i \approx h_j$, if $h_i, h_j \in H$ and $\Psi(w_{r0}, h_i) = \Psi(w_{r0}, h_j)$ (cf. Equation [8.20]). Relation $\approx$ partitions H into 81 disjoint classes because each component of the 4-couple vector Ψ can take on three values $\{-2, 0, 2\}$ (cf. Equation [8.20]), i.e., $3^4 = 81$. Parameter estimation can therefore use the same procedure based on the above equations. For instance, if there are other windows containing the pixel values as shown in Figure 8.9 then, based on the logit model fit method (Dubes and Jain, 1989), both window configurations will be treated as the same group as that shown in Figure 8.8b because they output the same vector $\Psi(w_{r0}, w_{Nr}) = [-2, 2, 0, 0]$.

8.3.2 Results of Parameter Estimations

Three test images, each of size 256×256 pixels, shown in the left-hand column of Figure 8.10, are subjected to parameter estimation experiments using the logit model fit method described in Section 8.3.1. The specified and estimated parameters are shown in Table 8.1. The images generated by these estimated parameters are shown in the right-hand column of Figure 8.10. As seen from Table 8.1,

2	1	3
2	1	2
2	1	1

(a)

1	1	2
3	1	3
3	1	3

(b)

FIGURE 8.9 The logit model fit method from Dubes, R. C., and A. K. Jain. 1989. "Random Field Models in Image Analysis." *Journal of Applied Statistics* 16 (1989):131–164. The two different configurations can be treated as the same group as Figure 8.8b because they each output $\Psi(w_{r0}, w_{Nr}) = [-2, 2, 0, 0]$.

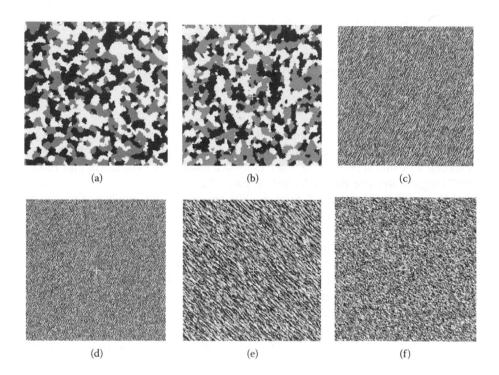

(a)   (b)   (c)

(d)   (e)   (f)

FIGURE 8.10 Comparisons of images using specified and estimated parameters. Refer to Table 8.1 for details of the parameter values used to generate these images.

although the estimates are not numerically good in rows 2 and 3, the images generated by the specified and estimated parameters yield a visually similar realization (Figure 8.10).

The results of parameter estimation are mainly affected by two factors, namely, the number of equations being used and the image size. As described in Section 8.3.1, the results of parameter estimation tend to fluctuate considerably when different numbers of equations are used. We carried out a number of experiments and concluded that the appropriate number of equations is case dependent. If too few or

TABLE 8.1
Specified and Estimated Parameters

Figure Number	Description	Number of Equations	β_1	β_2	β_3	β_4
8.10(a)	Specified-		0.40	0.40	0.40	0.40
8.10(b)	Estimated-	2803	0.25	0.4	0.32	0.37
8.10(c)	Specified-		−1.00	1.00	−1.16	1.00
8.10(d)	Estimated-	5954	−0.99	0.46	−1.13	0.48
8.10(e)	Specified-		0.50	0.50	0.50	−0.50
8.10(f)	Estimated-	7355	0.35	0.28	0.34	−0.36

too many equations are used, unacceptable estimates will result. There is a lack of clear rules for coping with this issue although Derin and Elliott (1987) suggest one should choose the equations containing large probabilities; how large the probability should be is not clear, and is obviously case dependent.

The second issue concerns image size. If the image size is not sufficiently large, then probabilities derived using histogram based techniques (cf. Equation [8.26]) will be biased and thus yield unsatisfactory parameter estimates. In order to generate reliable probability estimates using the histogram technique, we suggest that the image size should not be smaller than 256×256 pixels.

8.4 MAP-MRF CLASSIFICATION ALGORITHMS

Once the posterior energy and the associated parameters have been determined, the next step is to determine a solution. As noted previously, based on the Bayesian formulation, the criterion for pixel labeling is to find the MAP estimate. This is equivalent to a minimum-risk solution with a $0 - 1$ loss function. The MAP approach is also equivalent to a minimum-energy solution in terms of MRF modeling (see Equation [8.18]). A more detailed description of Bayes theory is given by Walpole (1982).

If the energy function is strictly convex (i.e., bowl-shaped, with one minimum point), a MAP-MRF solution can be obtained using a basic search approach, such as a gradient decent technique, because there is only one minimum in the solution space. However, for a nonconvex energy function, there may be many local minima. Therefore, in order to obtain a truly MAP estimate (i.e., to find a global minimum of the function shown in Equation [8.18]), one has to search all local minima over the entire solution space, and also show that there are no more local minima. It is evident that such a search process can be very lengthy.

Three algorithms, known as simulated annealing (Metropolis et al., 1953; Geman and Geman, 1984), iterated conditional modes (ICM; Besag, 1986) and maximizer of posterior marginals (MPM; Marroquin et al., 1987) have been proposed in the literature. These methods aim to approximate MRF-MAP estimates. All three algorithms are iterative in nature. There is also an alternative way to apply the MRF contextual concept based on only one trial. Readers may refer to Haslett (1985) and Pickard (1980) for details. More detailed descriptions of the MRF-MAP framework are contained in Geman and Geman (1984), Dubes and Jain (1989), Szeliski (1989), Geman and Gidas (1991), and Wei and Gertner (2003).

8.4.1 ITERATED CONDITIONAL MODES

Iterated conditional modes (ICM) is a local optimization method proposed by Besag (1986). This algorithm will converge to a local minimum of the energy function. The idea of ICM is based on two assumptions, the first of which is that the observation components $d_1, d_2, ..., d_m$ (m is the number of pixels) are class-conditional independent, and each d_r has the same known conditional density function $P(d_r|w_r)$ dependent only on w_r (i.e., the label on the pixel r). Thus, the following equation holds

$$P(d|w) = \prod_r P(d_r|w_r). \tag{8.28}$$

The second assumption is that w depends on the labels in the local neighborhood (i.e., the image has Markovian properties). Based upon these two assumptions, and using Bayes theorem, we get

$$P(w|d) \propto \prod_r P(d_r|w_r) \times P(w_r|w_{Nr}). \tag{8.29}$$

The MAP-MRF estimation is now transformed to maximize Equation (8.29), which can be done by maximizing $P(d_r|w_r) \times P(w_r|w_{Nr})$ on each pixel r locally.

The algorithm for ICM is shown in Figure 8.11. Note that the ICM algorithm can converge to a local minimal very quickly. Our experience suggests that 10 iterations are sufficient to allow the procedure to converge.

Begin
    Determine number of iterations N, the neighborhood system
      (usually only clique C_2 is used), and parameters (e.g., β_1 to β_4
      in Figure 8.2) for energy function;

    (1) Initialize image of configuration w for each pixel by choosing w, as
        the class at site r that minimises the non-contextual energy term

$$U(d_r|w_r) = \frac{1}{2}(d_r - u_k)\sum_k^{-1}(d_r - u_k) \text{ as given in Eq. (8.16)};$$

    (2) For each pixel r, compute $U(w_r'|d_r) = U(d_r|w_r') + U(w_r'|w_{Nr}')$ as
        shown in Eq. (8.17) for every class w_r', $w_r' \in \iota$, and upate w_r by the
        w_r' that minimize $\{U(w_r'|d_r), \forall w_r' \in \iota\}$.

    (3) Repeat (2) N times.
End

FIGURE 8.11   Iterated conditional modes algorithm.

8.4.2 SIMULATED ANNEALING

Simulated annealing (SA) is a type of stochastic relaxation algorithm. It was first proposed by Metropolis et al. (1953) to simulate the behavior of a system containing a large number of particles in thermal equilibrium. Geman and Geman (1984) applied a similar idea to image segmentation. The SA algorithm designed by Geman and Geman (1984), called the *Gibbs sampler*, generates a new label w_r' for each pixel r based on conditional probability $P(w_r'|w_{S-\{r\}})$ instead of energy change as used by Metropolis et al. (1953), and can be regarded as a "hot bath" version of Metropolis's algorithm (i.e., a temperature parameter is added).

The algorithm is shown in Figure 8.12. It is started at a high temperature T. After the Metropolis algorithm converges to equilibrium at the current temperature T, the value of T is decreased according to some carefully defined criterion called a cooling schedule. The process is repeated until the system becomes frozen (i.e., $T \rightarrow 0$). It is shown that a high temperature T can increase the probability of w_r being replaced by new class w_r' (because $\Delta < 0$, large T indicates large $\exp(\Delta/T)$), even though the energy $U(w_r'|d_r)$ of the new class w_r' is higher (that is, probability is lower) than that of class w_r. As the system cools down and temperature T decreases, only small increases of U are accepted. Near the freezing point ($T \rightarrow 0$), no increase of U can be accepted. The relationship between temperature T and $\exp(\Delta/T)$ is shown in Figure 8.13.

Begin

Determine neighborhood system (usually only clique C_2 is used), and parameters (e.g., β_1 to β_4 in Figure 8.2) for energy function;

(1) Choose an initial temperature T.

(2) Initialize image of configuration w for each pixel by choosing w, as the class at site r that minimizes the non-contextual energy term

$$U\left(d_r|w_r\right) = \frac{1}{2}(d_r - u_k)\sum_k{}^{-1}\left(d_r - u_k\right) \text{ as given in Eq. (8.16);}$$

(3) For all pixels r, perturb w_r by label w_r', which is randomly selected from the label set ι. Define energy difference Δ as

$$\Delta = U(w_r|d_r) - U(w_r'|d_r)$$

Both energy functions are calculated in terms of Equation (8.17).
If $\Delta > 0$ replace w_r by w_r',
else if $\exp(\Delta/T) \geq$ random [0,1] replace w_r by w_r';

(4) Repeat (3) N_{inner} times.

(5) Let $T = f(T)$ where f is a decreasing function.

(6) Repeat (3)–(5) until frozen, $T \rightarrow 0$;

End

FIGURE 8.12 Simulated annealing algorithm. See text for details of parameter assignments.

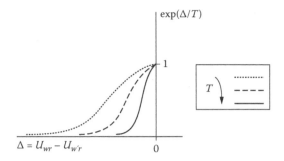

FIGURE 8.13   The probability for class replacement. As T is decreasing, only small increases of energy can be accepted.

The central idea of the SA algorithm is equivalent to the introduction of noise into the system to shake the search process away from a local minimum. The idea is similar to a process in metallurgy in which a small region of a metal structure is heated until it is pliable enough to be reconstructed into the desired shape. The metal is then cooled very slowly to make sure that it is given enough time to respond. Changes in temperature must be very small until the metal has hardened.

Geman and Geman (1984) present proofs of two convergence theorems. The first concerns the convergence of Metropolis's algorithm: if every configuration w is visited infinitely often, the distribution of generated configurations is guaranteed to converge to the Gibbs distribution. The second theorem concerns the temperature T of SA. It states that if the decreasing sequence of temperatures satisfies the following condition:

$$\lim_{k \to \infty} T_k = 0, \text{ and } T_k \geq \frac{m \times \Omega}{\ln(1+k)} \tag{8.30}$$

where

$$\Omega = \max_{w} U(w) - \min_{w} U(w) \tag{8.31}$$

and m is number of lattices (i.e., pixels), then convergence can be guaranteed. However, decreasing temperature T in terms of Equation (8.30) is too slow for practical applications. One may apply another temperature cooling function f defined as (Dubes and Jain, 1989)

$$T_{k+1} = f(T_k) = \frac{\ln(1+k)}{\ln(2+k)} T_k . \tag{8.32}$$

The initial value of temperature T_0 is usually set to a value of 2 or 3.

The parameter N_{inner} in Figure 8.12 requires further discussion. Generally, the value of N_{inner} can be set as small as 10 m or as high as 200 m, depending on how many configurations are intended to be tried at each temperature.

8.4.3   MAXIMIZER OF POSTERIOR MARGINALS

Maximizer of posterior marginals (MPM) is an alternative optimization algorithm (Marroquin et al., 1987). It is based on minimizing a criterion called the loss function describing the segmentation error. The MPM algorithm is similar to SA but without the application of any cooling schedule. MPM will therefore be much faster than the SA algorithm in terms of computation time. MPM requires that the posterior distribution of w given observations d be an MRF. The pixel-labeling scheme that minimizes segmentation error can be shown to maximize the marginal posterior distribution, so the new label w'_r for pixel r is chosen based on the comparison of all possible labels to satisfy

$$P\left(w'_r\middle|d\right) \geq P\left(w_r\middle|d\right), \text{ for } \forall \ w_r \in L. \tag{8.33}$$

The practical application of the MPM algorithm relies on an important assumption: that a Markov chain exists over w^m states where w^m is the number of possible configurations, w is the number of classes or gray levels, and m is the number of pixels. Once this Markov chain has reached a steady state, the marginal posterior probability can be approximated by counting the number of times each label is present at each pixel in a series of configurations. This approximation can be expressed by

$$P\left(w_r = b\middle|d\right) = \frac{1}{n-k} \sum_{i=k+1}^{n} \delta(w_r - b) \tag{8.34}$$

where b represents the possible label belonging to label set L. Function $\delta(a - b)$ is defined as

$$\delta(a-b) - 1 \qquad \text{if } a-b=0,$$
$$-0 \qquad \text{otherwise,} \tag{8.35}$$

which is equivalent to counting the times that pixel r is labeled as b during the process interval $(n - k)$.

The parameters k and n are chosen heuristically; k is the number of iterations needed for the Markov chain to reach a steady state and n is selected for accurate estimation at acceptable computational cost. For example, in the case of a two-label classification problem, if one chooses $k = 10$, $n = 110$, and if pixel r is labeled 1 on 20 occasions and labeled 2 on 80 occasions, then the probability $P(w_r = \text{Label } 1|d_r)$ will be $20/(n - k) = 0.2$, and $P(w_r = \text{Label } 2|d_r) = 0.8$. The pixel r will then be classified

Begin

　　Determine neighborhood system (usually only clique C_2 is used),
　　　and parameters (e.g., β_1 to β_4 in Figure 8.2) for energy function;

(1) Choose temperature T, recording interval k, and n in Equation (8.34);

(2) Initialize image of configuration w for each pixel by choosing w, as
　　the class at site r that minimizes the non-contextual energy term

$$U\left(d_r \middle| w_r\right) = \frac{1}{2}\left(d_r - u_k\right) \sum\nolimits_k^{-1} \left(d_r - u_k\right) \text{ as given in Eq. (8.16)};$$

(3) For all pixels r, perturb w_r by label w_r', which is randomly
　　selected from the label set $\mathcal{L}$. Define energy difference Δ as

$$\Delta = U(w_r|d_r) - U(w_r'|d_r)$$

　　Both energy functions are calculated in terms of Equation (8.17).
　　If $\Delta > 0$ replace w_r by w_r',
　　else if $\exp(\Delta/T) \geq$ random [0,1] replace w_r by w_r';

(4) Repeat (3) N times, and save configuration from w^{k+1} to w^n;

(5) For each pixel r, compute probability $P(w_r = b|d_r)$, for $\forall b \in \mathcal{L}$,
　　using Equations (8.34) and (8.35);

(6) For each pixel r, choose $w_r = \max\{P(w_r|d_r)$, $\forall w_r \in \mathcal{L}$, obtained by
　　step (5).

End

FIGURE 8.14　Maximizer of posterior marginals classification algorithm.

as label 2 according to Equation (8.33). It can be inferred that the computational cost of MPM is larger than that of ICM, but is much less than that required by SA. The algorithm for MPM estimation is given in Figure 8.14.

8.5　EXPERIMENTAL RESULTS

In this section, the results of classification experiments using the algorithms introduced in Section 8.4 are described. The test data set is the same as that used in Chapter 3 and described in Section 3.6.2.

　　The results of the classification experiments are shown in Figure 8.15 and corresponding confusion matrices are shown in Table 8.2. For comparison, the result produced by the maximum likelihood method is shown in Figure 8.15a. The ML method achieves an overall classification accuracy of 55.93% (Kappa = 0.460). It is clear that ML does not provide a very clean result. The addition of contextual information results in images that are much more patchlike. Figure 8.15b is the classification result output by the ICM algorithm, which achieved an overall classification accuracy of 63.78% (Kappa 0.561). In comparison with the result obtained by the

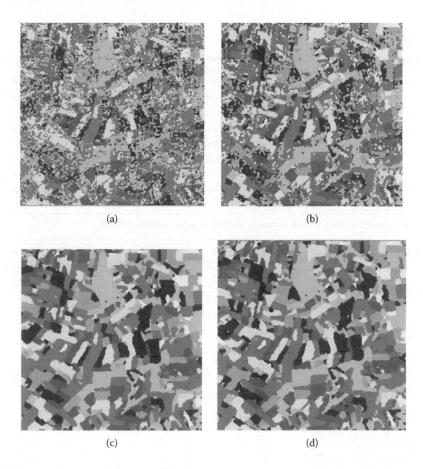

(a)   (b)

(c)   (d)

FIGURE 8.15   Classification results produced by (a) maximum likelihood, (b) ICM algorithm, (c) SA algorithm, and (d) MPM algorithm.

maximum likelihood algorithm, the ICM result shows an improvement of around 8% in terms of overall classification accuracy. The result generated by the SA algorithm is shown in Figure 8.15c. The initial value of parameter T was set to 3, and N_{nner} = 100 (see Figure 8.12 for a description of the algorithm). The SA algorithm achieved an overall accuracy of 68.21% (Kappa = 0.610). Compared to the maximum likelihood and ICM results, SA shows an improvement of around 13% and 5%, respectively. Figure 8.15d shows the results of the MPM algorithm, with $T = 1$, $k = 50$, and $n = 200$ (as described in Figure 8.14). The MPM algorithm achieved an overall classification overall accuracy of 67.91% (Kappa = 0.614), which is comparable to the result obtained by the SA algorithm.

For the ICM, MPM, and SA algorithms, only pair-site neighborhood systems (i.e., clique $c = 2$) were considered, and we used the isotropic assumption (i.e., β = {1.5, 1.5, 1.5, 1.5}). The results shown in Figures 8.15b through 8.15d provide a more patchlike classification, and also show an improvement in accuracy. It is also apparent that the results from SA and MPM show larger patch sizes than the results of

TABLE 8.2
Confusion Matrices Obtained Using the ML, ICM, SA, and MPM Algorithms

No.	1	2	3	4	5	6	7	u.a%.
1	**8738**	6160	872	498	1740	1327	83	45.0
2	692	**12178**	2382	414	630	742	118	70.9
3	582	10671	**4886**	681	1420	965	19	25.4
4	279	760	133	**10088**	4175	384	59	63.5
5	961	1629	298	2840	**19670**	988	221	73.9
6	302	2152	1181	689	1502	**5875**	158	49.5
7	314	488	22	168	1005	60	**1118**	35.2
p.a.%	73.6	35.7	50.0	65.6	65.3	56.8	62.9	**62553**

Note: Maximum Likelihood: Total accuracy: 55.20%. Kappa coefficient: 0.460.

No.	1	2	3	4	5	6	7	u.a%.
1	**9441**	4096	428	448	2353	599	83	54.1
2	350	**15331**	1804	376	341	317	13	82.7
3	575	10332	**6065**	512	951	609	2	31.8
4	56	303	106	**10614**	2583	120	2	75.0
5	1014	1729	215	2835	**21714**	1042	168	75.6
6	126	1528	1084	508	1836	**7641**	43	59.8
7	306	719	72	85	364	13	**1465**	48.4
p.a.%	79.5	45.0	62.0	69.0	72.0	73.8	82.4	**72271**

Note: ICM algorithm: Total accuracy: 63.78%. Kappa coefficient: 0.561.

No.	1	2	3	4	5	6	7	u.a%.
1	**10139**	3714	123	462	2547	407	871	58.0
2	122	**16676**	1089	317	425	282	14	88.1
3	468	9740	**7125**	553	886	548	1	36.8
4	48	264	119	**10979**	1418	185	2	84.3
5	858	1829	222	2593	**22919**	932	174	77.6
6	26	1264	1070	463	1769	**7982**	23	63.3
7	207	551	26	11	178	5	**1475**	60.1
p.a.%	85.4	48.9	72.9	71.3	76.0	77.1	83.0	**77295**

Note: SA algorithm: Total accuracy: 68.21%. Kappa coefficient: 0.614.

No.	1	2	3	4	5	6	7	u.a%.
1	**9938**	3730	221	4451	2415	421	76	57.6
2	315	**16188**	1207	382	310	268	12	86.6
3	417	10313	**6891**	564	996	350	1	35.2
4	44	252	129	**11122**	1429	158	2	84.6
5	911	1823	224	2368	**23229**	1032	187	78.0
6	38	1142	1095	477	1616	**8105**	19	64.8
7	205	590	7	14	147	7	**1479**	60.3
p.a.%	83.7	47.5	70.5	72.3	77.0	78.3	83.2	**76952**

Note: MPM algorithm: Total accuracy: 67.91%. Kappa coefficient: 0.610.

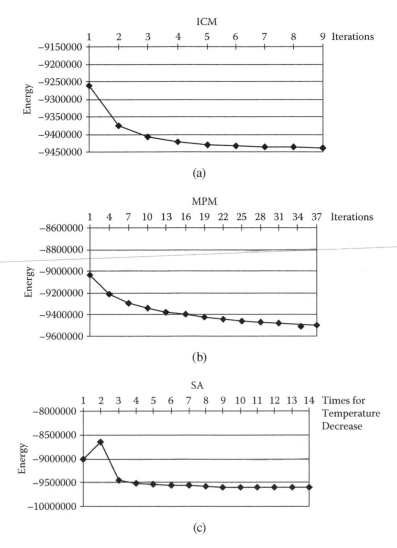

FIGURE 8.16 (a) The change in overall image energy at each iteration of ICM algorithm, (b) energy change in terms of MPM algorithm, and (c) energy change using SA algorithm.

the ICM algorithm. A larger patch size means that a lower overall energy value is detected. The change of energy following the progress of each algorithm is shown in Figure 8.16. The lowest energy and the highest classification accuracy were achieved by using SA algorithm. However, the SA algorithm is very time-consuming due to its cooling schedule. The MPM algorithm seems to be a good alternative.

9 Multisource Classification

Earth-observing satellites currently in operation or planned for the near future carry sensors that operate in the visible, infrared, and microwave regions of the spectrum. In addition, the widespread availability of GIS means that digital spatial data has become more accessible. Hence, greater attention is now being given to the use of multisource data in remote sensing image classification. Such data may consist of images produced by different sensor systems or digital spatial data. The assumption is made that classification accuracy should improve if additional features are incorporated. Such features may be derived from the image data set itself, or from two or more co-registered image sets from different sensors, or from associated geographical information such as surface elevation, soil type, or drainage pattern. This assumption is generally reasonable, because the greater the amount of relevant information that is included in a classification the greater the probability that interclass confusion will be reduced. Thus, it can be foreseen that the development of multisource classification methodologies will become increasingly important. It should, of course, also be remembered that the use of highly correlated features or features that show no variation between the classes of interest can obscure rather than illuminate the problem. One should always bear in mind that weighting is generally applied to input features. If features are standardized to a $0 - 1$ range, for example, then variation in feature a is considered to be equivalent to variation in feature b. Noting that variation can be considered to be equivalent to information, it is clearly not helpful to include surface elevation as a feature if it varies only slightly over the region of interest, and if these small variations are not related to the spatial boundaries of the classes.

A further consideration is the varying reliability and completeness of different data sources. It is necessary, therefore, to take the reliability or uncertainty of each data source into account when classification of multisource data is attempted. Source reliability factors are important parameters that determine how strongly a given source contributes to the multisource consensus pool. In this sense, reliability factors are equivalent to source weighting parameters. If these weighting parameters are not chosen properly, then the multisource consensus will give disappointing results although it is based on a theoretically robust mechanism.

In this chapter the principal approaches for dealing with multisource classification are addressed. Before introducing classification schemes relating to multisource data, techniques for image fusion are described. Image fusion is a formal framework for the bringing together of imagery originating from different sources. Several classification methods, based on the stacked-layer procedure, incorporation of topographic data, the extension of Bayesian theory (in which a further derivation to incorporate the Markov random field [MRF] concept to perform multisource MRF-MAP is also illustrated), and evidential reasoning, respectively, are then discussed. We also

283

introduce possible options for source weighting factor assignment, and experimental results are presented in the final section.

9.1 IMAGE FUSION

Image fusion is a process of combining images obtained by sensors of different wavelengths simultaneously viewing the same scene, to form a composite image with improved image content that makes it easier for the user to detect, recognize, and identify targets (Wald, 1999). Specifically, as far as spatial and spectral resolution are concerned, high spectral resolution can generally be helpful in discriminating land use/cover types, while high spatial resolution may hold advantages in identifying terrain features and structures. The user may therefore choose to fuse images from different sources to obtain complementary spectral and spatial characteristics to improve fused image contents, and further, when subjected to classification, to seek the possibility of improvement in classification accuracy. According to the literature, the most commonly implemented techniques for image fusion are based on principal component analysis (PCA), the intensity-hue-saturation (IHS) transform, Brovey's method, and the wavelet transform. The reader should note that before performing fusion activities, the images have to be preprocessed so as to obtain consistent image size and geolocation. For instance, in the case of fusing LANDSAT Thematic Mapper (TM) with panchromatic (PAN) imagery, the TM imagery is resampled at the PAN image size, and then the two image sets are co-registered before the fusion algorithm is used.

9.1.1 IMAGE FUSION METHODS

In the case of PCA-based fusion, the low resolution, multispectral imagery is first subjected to principal component analysis in order to obtain the first principal component (PC1) with maximum variance. PC1 is then replaced by the high-resolution image (a panchromatic image, for instance), which is stretched to give it the same variance and average as PC1. This combination was then transformed back into the RGB (red, green, blue) domain by inverse PCA transformation. The resulting image is subjected to a rescaling procedure (0 to 255) to obtain the final fused output.

For IHS-based fusion, the low resolution, multispectral imagery is subjected to an intensity-hue-saturation transformation leading to the extraction of three components of the original image. The intensity component representing the surface roughness or the spatial content of the image is then replaced by the high-resolution image channel, which is previously contrast-stretched (in terms of adjusting mean and variance) to match the original intensity component. A reverse IHS transformation is then applied to obtain the final fused image.

In the case of Brovey's fusion, the low-resolution imagery is normalized by dividing all individual digital number (DN) of the selected band with the sum of the corresponding DN values in all bands and multiplied by the high-resolution band, to incorporate the intensity or spatial component into the output. The mathematical expression of this process is perhaps clearer:

$$R_{new} = \frac{R_0 \times P}{\left(R_0 + G_0 + B_0\right)/3}$$

$$G_{new} = \frac{G_0 \times P}{\left(R_0 + G_0 + B_0\right)/3} \qquad (9.1)$$

$$B_{new} = \frac{B_0 \times P}{\left(R_0 + G_0 + B_0\right)/3}$$

where R_0, G_0, and B_0, are, respectively, the original DN value of the selected bands in the RGB domain, and P is a higher spatial resolution image for intensity modulation.

In recent years, several studies (Núñez et al., 1999; Aiazzi et al., 2002; Ulfarsson et al., 2003; Chen et al., 2006, González-Audícana et al., 2005) have proposed image fusion procedures using multiresolution analysis based on the discrete wavelet transform (cf. Chapter 7). In particular, both Mallat's (Mallat and Zhong, 1992) and the so-called *à trous* algorithms are the most popular approaches (González-Audícana et al., 2005). Mallat's algorithm, as introduced in Chapter 7 (Equation [7.22]) is an orthogonal, dyadic, nonsymmetric, decimated, nonredundant algorithm, while the *à trous* algorithm is nonorthogonal, shift invariant, dyadic, symmetric, undecimated, and redundant in nature. According to Mallat's algorithm (Mallat and Zhong, 1992), after a wavelet transform, the resulting image is made of four parts in which HL, LH, and HH bands are all high-frequency wavelet textures, and the LL band is the downsized (1/4) version of the original image (referred to in Section 7.2.2, "Wavelet Transform," in Chapter 7). However, in the case of the *à trous* algorithm, the image decomposition scheme uses a parallelepiped style. In each level of the *à trous* transform, the resulting image has a coarser spatial resolution by the same number of pixels as the original image (i.e., the image size always remains the same with only resolution degraded). Specifically, assume an image is of resolution 2^j, using the *à trous* transform at the nth level, the resulting image will be 2^{j+n} in resolution. For instance, if the original image resolution is of 2^2 (= 4) m, then after one pass with the *à trous* transform, the resolution will become 2^{2+1} (= 8) m. The *à trous* image is computed using scaling functions. The spatial detail lost between transformed level 2^{j+n} and 2^{j+n+1} is collected in just one wavelet coefficient image denoted as w^{j+n+1}, which involves horizontal, vertical, and diagonal spatial details at resolution 2^{j+n+1}. For the practical implementation of the *à trous* algorithm for an image, two-dimensional convolution masks associated with the scaling functions are used. The following are the two most popular scaling functions, namely, a 3×3 linear interpolation function

$$\begin{bmatrix} 1/16 & 1/8 & 1/16 \\ 1/8 & 1/4 & 1/8 \\ 1/16 & 1/8 & 1/16 \end{bmatrix} \qquad (9.2)$$

and a B3-spline function

$$\begin{bmatrix} 1/256 & 1/64 & 3/128 & 1/64 & 1/256 \\ 1/64 & 1/16 & 3/32 & 1/16 & 1/64 \\ 3/128 & 3/32 & 9/64 & 3/32 & 3/128 \\ 1/64 & 1/16 & 3/32 & 1/16 & 1/64 \\ 1/256 & 1/64 & 3/128 & 1/64 & 1/256 \end{bmatrix}. \tag{9.3}$$

In the case of fusion of multispectral and panchromatic images, the process for integrating the spatial detail of the panchromatic image into the multispectral images can be done in two steps. In the first step, a wavelet transform is applied to both the multispectral and panchromatic images to separate the images into a degraded image and wavelet coefficient images (HH, HL, and LH, for instance). In the second step, the wavelet coefficients of the panchromatic image are substituted for those of the multispectral image. An inverse wavelet transform is then performed.

It should be noted that either substitutive or additive methods can be applied to perform image fusion. The additive method is implemented by adding the panchromatic wavelet coefficients to the wavelet coefficients of the multispectral image bands rather than substituting one image band for another. Both methods may contribute different fusion qualities. González-Audícana et al. (2005) show that if Mallat's (Mallat and Zhong, 1992) algorithm is applied, the quality of the fused images via substitutive method is similar to that of fused images obtained by the additive approach. However, if the *à trous* algorithm is implemented, then the image quality obtained by the substitutive method is less ideal than that of fused images generated by the additive approach.

Figure 9.1 shows an example of the use of Mallat's wavelet transform to perform image fusion. Once both high-resolution images (e.g., panchromatic, denoted P in Figure 9.1) and low-resolution images (e.g., multispectral, denoted S in Figure 9.1) have been subjected to the wavelet transform, the wavelet coefficients (Figure 9.1 WP1, WP2, and WP3) of the high-resolution image are combined with feature T. An inverse wavelet transform is then performed, where feature T can be an approximation

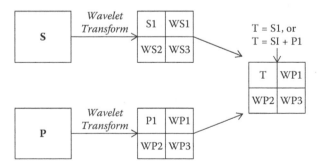

FIGURE 9.1 Image fusion in terms of the wavelet transform using Mallat's algorithm. See text for details.

image S1 of original low-resolution image S (if the substitution method is adopted) or T = S1 + P1 (if the additive method is used).

The wavelet-based image fusion method can also be incorporated with the PCA and IHS methods introduced above. When a multispectral image has been transformed into the IHS domain, a wavelet transform is then applied to both the intensity (I) image and panchromatic image, respectively. The wavelet coefficients of the panchromatic image are then substituted (in either a substitutive or additive way) into the intensity image in place of the wavelet coefficients for the intensity image, and an inverse IHS transform is carried out. Where transformation into the PCA domain is used, the fusion step is similar to that just described for the IHS method except that the wavelet coefficients of the panchromatic image are substituted for the first principal component (i.e., PC1) instead of the intensity image.

9.1.2 ASSESSMENT OF FUSED IMAGE QUALITY IN THE SPECTRAL DOMAIN

Assessment of fused image quality in the spectral domain is initially based on visual inspection. If classification is the major concern, then the classification accuracy would also be one of the indicators used to measure the fusion quality. Alternatively, one may use some or all of the following four quantitative indicators for measuring differences in spectral information between each band of the fused image and of the original image:

1. The correlation coefficient between the original and the fused images. A result closer to 1 is preferred (Wald et al., 1997, 2002).
2. Standard deviation of the difference image between both original and fused images. A low value of the standard deviation is preferred (Wald et al., 1997).
3. The ERGAS (Erreur Relative Globale Adimensionnelle de Synthèse) index, also called relative dimensionless global error (Wald, 2000) formulated as

$$ERGAS = 100 \frac{h}{l} \sqrt{\frac{1}{N} \sum_{i=1}^{n} \left(RMSE^2 \left(B_i \right) \big/ M_i^2 \right)} \tag{9.4}$$

where h is the resolution of the panchromatic image, l the resolution of the multispectral image, N the number of spectral bands, B_i the difference image between the original and the fused images for band i, M_i the mean radiance of each spectral band, and $RMSE$ the root mean square error, which is computed as

$$RMSE^2 \left(B_i \right) = \bar{B}_i^2 + \sigma \tag{9.5}$$

where $\bar{B}_i$ is the mean of B_i, and σ the standard deviation. Lower ERGAS index values are preferred.

4. The Image Quality Index, Q (Wang and Bovik, 2002):

$$Q = \frac{4\upsilon_{AB}\overline{A}\,\overline{B}}{\left(\sigma_A^2 + \sigma_B^2\right)\left(\overline{A}^2 + \overline{B}^2\right)} \qquad (9.6)$$

where A and B denote the original and fused images, respectively; $\overline{A}$ and $\overline{B}$ are the means of A and B; σ is the corresponding standard deviation; and υ_{AB} is the covariance of A and B.

9.1.3 PERFORMANCE OVERVIEW OF FUSION METHODS

Prasad et al. (2001) use IHS, PCA, and Brovey methods for fusing IRS LISS-III and PAN data covering the forest area around Hardwar, India, and the results are evaluated in terms of the success in interpretation of forest features. It is observed that PCA outperforms the IHS and Brovey methods. González-Audícana et al. (2005) perform image fusion based on the wavelet technique for IKONOS multispectral and panchromatic images. The images were acquired in October 2000 and cover the irrigated agricultural area of Mendavia (Navarre), in northern Spain. The main crops around the area in the year 2000 were corn, alfalfa, and grapes. Both the à trous algorithm and Mallat's algorithm are applied, and the results in terms of several spectral quality measures, show that the à trous algorithm generally performs better than Mallat's algorithm. González-Audícana et al. (2005) also indicate that due to the decimation process of Mallat's algorithm strongly oriented in the horizontal and vertical directions, the resulting merged images present, visually, a lower spatial quality than those obtained using the à trous algorithm.

Colditz et al. (2006) use PCA, IHS, Brovey, and wavelet methods for image fusion, and the fused results are evaluated by means of classification accuracy. The study imagery is Landsat 7 ETM+ acquired on 15 August 2001, and located around the Würzburg area in central Germany. The results show that the fusion images produced by wavelet and PCA methods achieve better classification results than that generated by using the IHS and Brovey approaches. Teggi et al. (2003) evaluate the performance of the wavelet transform, PCA, and IHS methods for fusing Indian IRS-1C-PAN sensors. The fusion quality is analyzed in terms of correlation, standard deviation (Wald et al., 1997), and classification accuracy. The results suggest that the wavelet transform outperforms both the PCA and IHS methods.

9.2 MULTISOURCE CLASSIFICATION USING THE STACKED-VECTOR METHOD

The most straightforward approach to deal with a multisource classification problem is simply to extend the dimension of the data vectors to include each source. This approach is known as the *stacked-vector* or *augmented-vector* method. For example, if one has a six-band TM image and a three-band SPOT image, then nine bands can

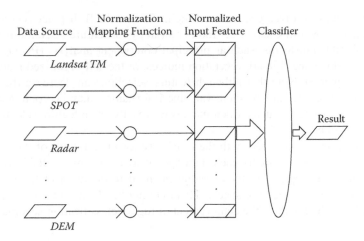

FIGURE 9.2 The stacked-vector classification process.

be used together as inputs to the classifier. Alternatively, one may first implement the fusion process to the available imagery to obtain a higher quality of image content in both the spectral and spatial domains, and then implement a stacked-vector method (in such a way that the dimension of data stacked can also be reduced). Although this method is easy to apply, several issues require attention.

The first issue concerns the scale of measurement of each source. Since different data sources are likely to have different measurement scales, it is generally recommended that all data be mapped into the same scale. Such a process is called normalization (Figure 9.2). For instance, if data from the optical spectrum are used, one may convert them into ground reflectance units. If radar images are used, one may convert these data into backscatter coefficients (σ^0, Chapter 1). However, where a variety of data sources is used, then the choice of a data normalization scheme may be logically difficult, and the classification process may be dominated by data sources that display a larger scale of variation. For instance, if the first data source in a two-source case has a range of [0, 32], and the second has a range of [0, 16384], then the second data source is more likely to dominate the classification process due to the effects of the measurement scale. Some classification procedures require explicit normalization of the input data (for example, a feed-forward neural network). Other methods, such as maximum likelihood (ML), use a hidden form of normalization. ML uses estimates of the variance–covariance matrices (S_i) of the classes in order to generate class membership probabilities. The calculation of S_i involves the subtraction of the class mean of each feature from the feature measurements.

A second issue is that the stacked-vector method may not be practical in terms of computational cost when the number of data vectors is large. For example, if one uses the statistical Gaussian maximum likelihood method for classifying N dimensional variables, the resulting computational cost is proportional to N^2. A way to reduce this problem is to use a feature selection procedure, which involves selecting a subset of data sources. A suitable subset can be selected by using some distance measure, such as the divergence index (Singh, 1984) or the B-distance (Haralick and Fu, 1983), as shown in Chapter 2. Another strategy for feature selection is through the use of data transforms, such as principal component analysis (PCA), tasseled cap (Crist and Cicone, 1984a), vegetation indices, or the self-organized feature map (SOM) (Chapter 3). Note that although feature selection can decrease the computational cost of classification, some valuable information may be contained in the discarded data sources. Such a dilemma always results when feature selection techniques are applied.

A third issue is concerned with the reliability (or uncertainty) of the data. The stacked-vector approach treats each data source as being fully reliable (i.e., each source contributes equally to the classifier in the determination of the location of decision boundaries in feature space). This may not be always the case in practice, and could be a drawback to the achievement of higher classification accuracy.

9.3   THE EXTENSION OF BAYESIAN CLASSIFICATION THEORY

Interest in techniques of multisource classification based on the extension of Bayesian theory was triggered by the studies of Lee et al. (1987) and Benediktsson et al. (1990), and later extended by Zhu and Tateishi (2006) to deal with multisource, multitemporal data for land cover mapping. The model proposed by Benediktsson et al. (1990) is a refinement of the method of Lee et al. (1987), in which a more complete statistical mechanism was derived. These two models are introduced next, together with a derivation using the results obtained by Benediktsson et al. (1990) to incorporate the Markov random field (MRF) concept in order to achieve multisource MRF-MAP classification.

9.3.1   AN OVERVIEW

The extension of Bayesian classification theory was compared with the evidential reasoning approach based on the Dempster-Shafer theory by Lee et al. (1987). Both showed satisfactory results for performing multisource data classification, although the method based on the extension of Bayesian classification theory performed slightly better than the evidential reasoning method (Lee et al., 1987). Benediktsson et al. (1990) adopted extended Bayesian classification theory to classify multisource remotely sensed data. They also compared these results with those achieved by a neural network stacked-vector approach. Both approaches reveal different advantages and disadvantages. More recently, Schistad et al. (1994) tested an extended Bayesian classification approach for classifying Landsat TM and synthetic aperture radar (SAR) images, and their results were better than those achieved by conventional single-source classification. A similar experiment was conducted by Kim and Swain (1995), who used the evidential reasoning approach to manage multisource data and achieve an acceptably good classification result.

Both the extension of Bayesian classification theory and evidential reasoning methods regard each data source as fully independent. Hence, one has to generate probability (or evidence) measures for each information class using each source separately, and then obtain the classification result in terms of probability (or

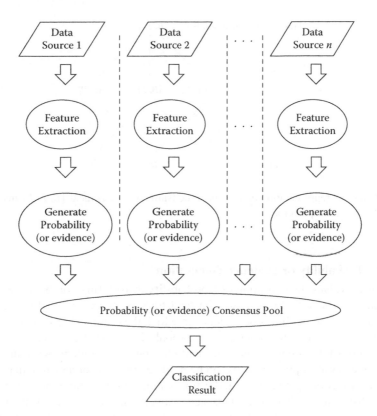

FIGURE 9.3 General steps in multisource classification based on evidential reasoning and extension of Bayesian theory.

evidence) consensus. The general steps are shown in Figure 9.3 and are explained as follows.

9.3.1.1 Feature Extraction

This is the preparation stage of the classification process, the aim of which is to determine what kind of input features (e.g., pixel gray values, texture measures) should be used in the classification process. The suitable selection of input features can enhance classification accuracy. Certainly, the choice of input features not only depends on the way in which the image was formed (e.g., optical or microwave sensor), but is also determined by the scale relationship between the ground objects of interest and image resolution. If, for example, image resolution is around 30 m, then in the case of lithological classification, textural information may contribute more significantly to interclass variation than tonal information. In the case of classifying crop types in an agricultural area, textural and tonal information may make no significant difference (Tso, 1997) because of the scale and the size of the object being classified (refer to Chapter 7 for a discussion of texture). Where radar images are used, textural features may be more useful than image tonal information (Ulaby et al., 1986b).

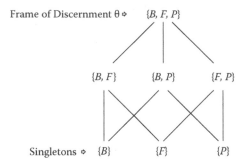

FIGURE 9.4 A representation of subsets of the frame of discernment $\{B, F, P\}$ made up of bare soil, forest, and pasture.

9.3.1.2   Probability or Evidence Generation

This stage involves the definition of a probability density function or some alternative methodology (e.g., by the use of a feed-forward neural network) to generate class-associated probabilities or evidence in which the user has high confidence. If a statistical model is used for generating the probability or statistical evidence, then the choice of probability density function should be source dependent (Kim and Swain, 1995). For example, optical images can be modeled in terms of a Gaussian probability density function (p.d.f), while radar images may more suitably be modeled as a Gamma distribution (Chapter 1). This step is an alternative way of normalization of the data scale, as each different source is mapped onto the probability domain.

9.3.1.3   Multisource Consensus

In this final stage, all the probability (or evidence) measures are combined together in order to generate a classified image (Figure 9.4). Where the evidential reasoning method is used for evidence consensus, the data sources are considered in a pairwise fashion, while if the consensus mechanism is based on the extension of Bayesian classification theory, all of probability measures are combined simultaneously.

9.3.2   Bayesian Multisource Classification Mechanism

The model described in this section is based on Lee et al. (1987). In the general multisource case, a set of observations from n ($n > 1$) different sources is used. Let $x_i, i \in [1, n]$, denote the measure of a specific pixel from measurement source i. The aim is to derive the probability of the pixel belonging to each of c information classes $\omega_j, j \in [1, c]$. It is possible to get some information about the prior probability of class ω_j, denoted by $P(\omega_j)$, i.e., the probability that an observation will be a member of class ω_j. A useful model for prior probability is context, as described in Chapter 8. The approach using contextual assumption to perform multisource classification is considered later.

According to Bayesian probability theory, the law of conditional probabilities states that

$$P\left(\omega_i\middle|X\right)=P\left(\omega_j\middle|x_1,\,x_2,\,...,\,x_n\right)=\frac{P\left(x_1,\,x_2,\,...,\,x_n\middle|\omega_j\right)\times P(\omega_j)}{P(x_1,\,x_2,\,...,\,x_n)} \qquad (9.7)$$

where $P(\omega_j|x_1, x_2, ..., x_n)$ is known as the *conditional* or *posterior* probability that ω_j is the correct class, given the observed data vector $(x_1, x_2, ..., x_n)$. $P(x_1, x_2, ..., x_n|\omega_j)$ is the p.d.f. associated with the measured data $(x_1, x_2, ..., x_n)$ given that $(x_1, x_2, ..., x_n)$ are members of class ω_j. $P(x_1, x_2, ..., x_n)$ is the p.d.f. of data $(x_1, x_2, ..., x_n)$. Assuming class-conditional independence among the sources, one obtains $P(x_1, x_2, ..., x_n|\omega_j) = P(x_1|\omega_j) \times P(x_2|\omega_j) \times ... \times P(x_n|\omega_j)$, and Equation (9.7) becomes

$$P\left(\omega_j\middle|x_1,\,x_2,\,...,\,x_n\right)=\frac{P\left(x_1\middle|\omega_j\right)\times P\left(x_2\middle|\omega_j\right)\times...\times P\left(x_n\middle|\omega_j\right)\times P(\omega_j)}{P(x_1,\,x_2,\,...,\,x_n)} \qquad (9.8)$$

Again, following the law of conditional probabilities,

$$P\left(x_i\middle|\omega_j\right)\times P(w_j)=P\left(w_j\middle|x_i\right)\times P(x_i)$$

$$\Rightarrow P\left(x_i\middle|\omega_j\right)=\frac{P\left(w_j\middle|x_i\right)\times P(x_i)}{P(w_j)}. \qquad (9.9)$$

If Equation (9.9) is substituted into Equation (9.8), one obtains

$$P\left(\omega_j\middle|x_1,\,x_2,\,...,\,x_n\right)=P\left(\omega_j\middle|x_1\right)\times P\left(\omega_j\middle|x_2\right)\times...\times P\left(\omega_j\middle|x_n\right)\times P(\omega_j)^{1-n}$$

$$\times\frac{P(x_1)\times P(x_2)\times...\times P(x_n)}{P(x_1,\,x_2,\,...,\,x_n)} \qquad (9.10)$$

If the intersource independence assumption has been made such that $P(x_1) \times P(x_2) \times ... \times P(x_1) = P(x_1, x_2, ..., x_n)$, then Equation (9.10) results in

$$P\left(\omega_j\middle|x_1,\,x_2,\,...,\,x_n\right)\propto P\left(\omega_j\middle|x_1\right)\times P\left(\omega_j\middle|x_2\right)\times...\times P\left(\omega_j\middle|x_n\right)\times P(\omega_j)^{1-n} \qquad (9.11)$$

Equation (9.11) is only valid under the circumstance that the data sources are mutually independent. The practical value of this assumption is considered further in Section 9.3.5.

When processing multisource information it is natural to consider the inclusion of a reliability (or uncertainty) measure for each different data source. In other words, one can assign different weights to different sources in order to achieve a more satisfactory multisource consensus. One simple way to adjust the contribution for each data source is to append exponents to the source-specific posterior probabilities (Lee et al., 1987). This modification to Equation (9.11) gives

$$P\left(\omega_j \middle| x_1, x_2, \ldots, x_n\right) \propto P\left(\omega_j \middle| x_1\right)^{\alpha_1} \times P\left(\omega_j \middle| x_2\right)^{\alpha_2} \times \ldots \times P\left(\omega_j \middle| x_n\right)^{\alpha_n} \times P(\omega_j)^{1-n} \quad (9.12)$$

where $\alpha_i \in [0,1]$ is the source-specific weighting parameter, which allows one to adjust the contribution for the ith source, and a measure of the sensitivity of $P(\omega_j|x_1, x_2, \ldots, x_n)$ to changes of the ith posterior probability, which reduces to following expression:

$$\frac{\delta P\left(\omega_j \middle| x_1, x_2, \ldots, x_n\right)}{P\left(\omega_j \middle| x_1, x_2, \ldots, x_n\right)} \propto \alpha_i \frac{\delta P\left(\omega_j \middle| x_i\right)}{P\left(\omega_j \middle| x_i\right)}. \quad (9.13)$$

Equation (9.13) shows that the difference in the posterior probability $P(\omega_j|x_i)$ for source i leads to a difference of α_i in $P(\omega_j|x_1, x_2, \ldots, x_n)$. The weighting parameter α_i thus represents the level of sensitivity of the data source i.

Given the above definition of α_i it is clear that if α_i is equal to the value 1 then the data source i is fully reliable. As α_i tends to 0, the source i is considered to be less reliable. Note that in the case $\alpha_i = 0$, one should not include any contribution from source i. Equation (9.12) cannot provide a satisfactory mechanism to cope with this situation. For instance, if there is a total of five data sources, and only the weighting parameters α_1 and α_2 for the first and second data sources are nonzero, the posterior probability derived from Equation (9.12) will use $P(w_j)^{-4}$ (where the exponent -4 is obtained from $1 - n$, and the number of data sources $n = 5$ in this case) as prior probability. A refined Bayesian multisource classification mechanism was proposed by Benediktsson et al. (1990) to deal with this difficulty.

9.3.3 A REFINED MULTISOURCE BAYESIAN MODEL

Consider that one more source is being added to the classification analysis. Equation (9.11) then becomes (Benediktsson et al., 1990)

$$P\left(\omega_j \middle| x_1, x_2, \ldots, x_n, x_{n+1}\right) \propto P\left(\omega_j \middle| x_1\right) \times P\left(\omega_j \middle| x_2\right) \times \ldots \times P\left(\omega_j \middle| x_{n+1}\right) \times P(\omega_j)^{-n} \quad (9.14)$$

By dividing Equation (9.14) by Equation (9.11), the probability ratio is seen to be

$$P\left(\omega_j \middle| x_{n+1}\right) \middle/ P(\omega_j) \quad (9.15)$$

This is the contribution of the $(n + 1)^{th}$ source, and therefore Equation (9.11) can be modified as follows:

$$P\left(\omega_j\middle|x_1, x_2, \ldots, x_n\right)\propto\left[P\left(\omega_j\middle|x_1\right)\middle/P(\omega_j)\right]\times\left[P\left(\omega_j\middle|x_2\right)\middle/P(\omega_j)\right]\times\ldots$$
$$\times\left[P\left(\omega_j\middle|x_n\right)\middle/P(\omega_j)\right]\times P(\omega_j) \tag{9.16}$$

For reliability measurements, we can again use the idea of appending exponents to each data source, and Equation (9.17) can be rewritten as

$$P\left(\omega_j\middle|x_1, x_2, \ldots, x_n\right)\propto\left[P\left(\omega_j\middle|x_1\right)\middle/P(\omega_j)\right]^{\alpha_1}\times\left[P\left(\omega_j\middle|x_2\right)\middle/P(\omega_j)\right]^{\alpha_2}\times\ldots$$
$$\times\left[P\left(\omega_j\middle|x_n\right)\middle/P(\omega_j)\right]^{\alpha_n}\times P(\omega_j) \tag{9.17}$$

Clearly, Equation (9.17) provides a more satisfactory relationship for dealing with multisource fusion problems in comparison with Equation (9.12). The reason is that when a source is found to be fully unreliable (i.e., $\alpha_i = 0$), both posterior and associated prior probabilities should be rejected simultaneously as shown in Equation (9.17). However, it is worthwhile to note that if prior probability is not used (which is equivalent to regarding $P(\omega_j)$ as being uniformly distributed for all information class j), then Equations (9.12) and (9.17) are equivalent.

The sensitivity measure as given by Equation (9.13) is converted to the form

$$\frac{\delta P\left(\omega_j\middle|x_1, x_2, \ldots, x_n\right)}{P\left(\omega_j\middle|x_1, x_2, \ldots, x_n\right)} = \alpha_i\frac{\delta P\left(\omega_j\middle|x_i\right)\middle/P(\omega_j)}{P\left(\omega_j\middle|x_i\right)\middle/P(\omega_j)}. \tag{9.18}$$

Another way to understand the behavior of α_i more clearly is to represent Equation (9.17) in logarithmic form (Benediktsson et al., 1990):

$$\ln P\left(\omega_j\middle|x_1, x_2, \ldots, x_n\right)\propto\ln P(\omega_j)+\sum_i\alpha_i\ln\left[P\left(\omega_j\middle|x_i\right)\middle/P(\omega_j)\right] \tag{9.19}$$

It is now apparent that α_i is a coefficient that corresponds to the weighting parameter associated with data source i.

9.3.4 MULTISOURCE CLASSIFICATION USING THE MARKOV RANDOM FIELD

In most remote sensing image classification experiments, the prior probability $P(w)$ is rarely used because of the problem of properly modeling the prior probability. However, as demonstrated in Chapter 8, the smooth contextual concept can be used to model prior probability using smooth prior modeling in terms of Markov random fields (MRF) to achieve an MRF-MAP classification. The concept of MRF can also be applied to multisource classification.

The method for deriving an MRF-MAP estimate, as described in Section 8.1, requires the specification of a posterior energy function. The posterior energy shown in Equation (8.17) is valid only for classifying a single data source. In the case of multisource classification, one has to construct a multisource posterior energy function. Equation (9.19) can be used to achieve this aim. According to Bayesian probability theory, it can be shown that $[P(w_j|x_i)/P(w_j)] \propto P(x_i|w_j)$, so that the right-hand side of Equation (9.19) can be expressed as

$$\ln P\left(\omega_j \middle| x_1, x_2, \ldots, x_n\right) \propto \ln P(\omega_j) + \sum_i \alpha_i \ln P\left(x_i \middle| w_j\right)$$

or, equivalently,

$$U\left(\omega_j \middle| x_1, x_2, \ldots, x_n\right) \propto U(\omega_j) + \sum_i \alpha_i U\left(x_i \middle| w_j\right). \tag{9.20}$$

$U(w_j|x_1, x_2, \ldots, x_n)$ denotes the multisource posterior energy, and $U(w_j)$ and $U(x_i|w_j)$ are the prior energy and the class-conditional energy. Each of these terms is defined in Chapter 8 (Equations [8.14] and [8.16], respectively). Using this definition of multisource posterior energy, one can use the classification algorithms described in Chapter 8 using one of the methods outlined in Figures 8.11, 8.12, or 8.14 to perform multisource MRF-MAP estimation (note that one has to substitute the multisource posterior energy in place of the original single-source posterior energy in the algorithm being applied). Schistad et al., (1996) use a similar equation in multisource classification.

From the foregoing description, it can be seen that not only will the posterior probability of each source influence the result of multisource consensus, but the weighting parameter α_i will also contribute to the final decision. The problem is: how to specify a reliability measure in order to obtain a more reliable multisource fusion result. Note also that in the single-source MRF-MAP estimate one has to choose a suitable potential parameter (Chapter 8) for the energy functions in order to obtain a good classification result. Here, in the case of multisource MRF-MAP estimation, both the potential parameter and the source-associated weighting factors have to be determined. The parameter estimation issue in multisource classification is considered in Section 9.5.

9.3.5 Assumption of Intersource Independence

The assumption of intersource independence means that the joint probability distributions for the measurements of sources are mutually independent, and the final joint probability function can be expressed as the product of each class-conditional probability function. Although this might not always be the case, the validity of the assumption is hard to determine.

In practical situations, multiple data sources usually contain complex but unknown interactions. For example, in the case of multisource classification, the data set may consist of optical and radar images and a digital terrain model. If there is no reliable information concerning the relationship between surface spectral reflectance (or backscattering) and terrain parameters (such as ground slope, surface shape, etc.), then lack of knowledge of these interactions will force us to ignore any intersource relationships, and therefore to treat these data sources as independent variables. However, rather than being a shortcoming in the data consensus analysis, the intersource independence assumption does provide an easy way to perform classification using multiple data sources.

9.4 EVIDENTIAL REASONING

The mathematical theory of evidence is a field in which a number of data sources can be combined to generate a joint inference concerning pixel labeling. The theory was first developed by Dempster in the 1960s and later extended by Shafer (1979), who provided details of the development of evidential theory, which has therefore become known as the Dempster-Shafer (D-S) theory of evidence. Garvey et al. (1981) discuss applications using the D-S theory of aggregating evidential knowledge. Barnett (1981) mentions some of the computational issues involved in order to reduce the computational requirements of the method. Gordon and Shortliffe (1985) and Shafer and Logan (1987) propose modified approaches that are mainly based on a hierarchical evidence space. The D-S theory has also been related to the field of artificial intelligence systems (Gouvernet et al., 1980; Friedman, 1994; Strat, 1984) with promising results.

The evidential reasoning approach also provides a valuable theoretical basis for dealing with the remotely sensed multisource classification problem. The approach has been tested in several studies (e.g., Lee et al., 1987; Srinivasan and Richards, 1990; Wilkinson and Megier, 1990; Peddle and Franklin, 1992; Kim and Swain, 1995; Franklin et al., 2002). The main issues relating to the evidential reasoning method are the need to generate the so-called *mass of evidence* and to measure the uncertainty (which determines the weight) of each data source. The basic concepts of D-S theory are described next.

9.4.1 CONCEPT DEVELOPMENT

One important aspect of the generality of the D-S theory is avoidance of the Bayesian restriction that commitment to belief of a hypothesis implies commitment of the remaining belief to its opponents, i.e., belief in event w is equivalent to $P(w)$ so that belief in the concept *Not w* is equal to $1 - P(w)$. In many real applications, evidence partially in favor of one particular hypothesis is not necessarily evidence against alternative hypotheses. In D-S theory, complete knowledge is represented as unity (i.e., the sum of all labeling possibilities is equal to one), which includes the case of uncertainty. In other words, the measure of belief assigned to each hypothesis may be less than one. If there is some evidence in favor of hypothesis w, then the remaining belief (which is equivalent to the value of uncertainty) will be assigned to the whole hypothesis space, not simply to its opponents. A simple example may illustrate this idea.

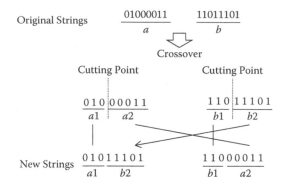

FIGURE 9.5 Method of crossover used by GA to generate new search strings.

Suppose that an analyst is trying to classify an image that involves labeling pixels as belonging to one of three classes $\{B, F, P\}$, where B denotes bare soil, F denotes forest, and P denotes pasture. In D-S theory, this set $\{F, B, P\}$ is generally called a *frame of discernment*, and is denoted by the symbol θ. The number of all possible subsets of the frame of discernment θ is equal to $2^{|\theta|}$, where $|\theta|$ denotes the number of one-element subsets (called *singletons*). In our case, there are a total of three singletons in $|\theta|$, and therefore the total number of subsets of θ is $2^3 = 8$, as shown in Figure 9.4. Note that the empty set { } is one of these subsets, but is not displayed in Figure 9.4.

D-S theory uses a number within the range of [0, 1] to indicate the degree of belief in a hypothesis, given a piece of evidence. The number that expresses the degree to which the evidence supports a particular hypothesis is represented by a function called the *basic probability assignment* or *bpa*; the resulting quantity of evidence is generally called the *measure of mass* or *mass of evidence*, and is denoted by $m(\psi)$, where ψ is any subset of θ. The sum of $m(\psi)$, for $\forall \psi$, must be equal to 1 (this is a rather important constraint). In our case, ψ can be one of the sets shown in Figure 9.5. If $m(\psi) = q$ and no other belief is assigned to the subsets of θ, then $m(\theta) = 1 - q$, that is, the remaining belief is assigned to the frame of discernment θ, rather than ψ's opponents. This point has already been emphasized in the previous section. An example is given below.

Suppose that a source of evidence indicates that a particular pixel of interest has 30% support for the belief that the pixel belongs to class $\{B\}$, 30% support for the belief that the pixel belonging to class $\{F\}$, and 40% support for the belief that the pixel belongs to class $\{P\}$. Note again that the sum of all $m(\psi)$ must equal 1. If one assumes that the source of evidence is fully reliable (i.e., the uncertainty is zero), then using the theory of evidence symbolism, one obtains

$$m(\{B, F, P\}) = <0.3, 0.3, 0.4>. \tag{9.21}$$

Suppose one is somewhat uncertain about the reliability of this data evidence and is only willing to commit oneself to label the pixel with, for example, 80% confidence. Then one should change the previous *bpa* by multiplying each element by 0.8, which leads to:

$$m(\{B, F, P, \theta\}) = <0.24, 0.24, 0.32, 0.2>. \qquad (9.22)$$

Note that the quantity of *bpa* expressed by θ is denoted by $m(\theta)$ and is expressed in this case by

$$1 - (0.24 + 0.24 + 0.32) = 0.2 \qquad (9.23)$$

which is the measure of uncertainty or ignorance; in our case 80% confidence leads to 20% uncertainty and this is reflected in the labeling process.

9.4.2 BELIEF FUNCTION AND BELIEF INTERVAL

In what follows, both the belief function (or support) and the plausibility of each labeling proposition within evidential reasoning are described. A belief function, denoted by *Bel*, for a hypothesis ψ is defined as the sum of the mass of evidence that is committed to the ψ and to its subsets (if any), while the plausibility of ψ, denoted by *Pl*, is defined as one minus the *Bel* committed to ψ's contradiction, denoted by $Bel(\sim\psi)$. It is more convenient to explain such definitions in terms of mathematical representation:

$$Bel(\psi) = \sum_{\omega \in \psi} m(\omega), \text{ and } Pl(\psi) = 1 - Bel(\sim \psi). \qquad (9.24)$$

For instance, using the example discussed earlier, the *Bel* for hypothesis $\{B, F\}$ can be represented by

$$Bel(\{B, F\}) = m(\{B\}) + m(\{F\}) + m(\{B, F\}) \qquad (9.25)$$

and it can be found that $Bel(\psi)$ is a measure of the total amount of belief in ψ including all ψ's subsets. When ψ is a singleton, $Bel(\psi) = m(\psi)$. For instance, $Bel(\{B\}) = m(\{B\})$. The plausibility for hypothesis $\{B, F\}$ can be represented by

$$Pl(\{B, F\}) = 1 - Bel(\sim\{B, F\}) = 1 - Bel(\{P\}). \qquad (9.26)$$

$Bel(\psi)$ can thus be interpreted as the minimum amount of evidence that a pixel is properly labeled with ψ, while $Pl(\psi)$ can be interpreted as the maximal extent to which the current evidence could allow one to believe ψ (Shafer, 1979). Note that $Pl(\psi)$ provides another choice for us to make a decision when there is no evidence, which exactly supports hypothesis ψ.

If a pixel belongs to ψ, then the probability $P(\psi)$ may lie somewhere between the interval

$$[Bel(\psi), Pl(\psi)] \text{ or } [Bel(\psi), 1 - Bel(\sim\psi)] \qquad (9.27)$$

with the range $Pl(\psi) - Bel(\psi)$, which in turn means that

$$Pl(\psi) \geq P(\psi) \geq Bel(\psi). \qquad (9.28)$$

That is, $Pl(\psi)$ and $Bel(\psi)$ provide the upper bound and lower bound of the probability of the subset. Recall from Equation (9.21) that the belief, plausibility, and interval for each class are

Belief:	$Bel(\{B\})$	$= 0.24$
	$Bel(\{F\})$	$= 0.24$
	$Bel(\{P\})$	$= 0.32$
Plausibility:	$Pl(\{B\})$	$= 1 - Bel(\{F\}) - Bel(\{P\}) = 0.44$
	$Pl(\{F\})$	$= 1 - Bel(\{B\}) - Bel(\{P\}) = 0.44$
	$Pl(\{P\})$	$= 1 - Bel(\{B\}) - Bel(\{F\}) = 0.52$
Interval:	$Pl(\{B\}) - Bel(\{B\}) = 0.2$	
	$Pl(\{F\}) - Bel(\{F\}) = 0.2$	
	$Pl(\{P\}) - Bel(\{P\}) = 0.2$	(9.29)

In Equation (9.29), the interval is equal to $m(\theta)$, and one may finally choose class P as the label of the pixel of interest because in Equation (9.29) each interval is the same, and the class *pasture* (i.e., subset $\{P\}$), has the highest belief and plausibility in comparison with the other two classes.

Now consider a more complicated instance. If one holds the following evidence for each hypothesis

$$m(\{B\}) = 0.19$$

$$m(\{F\}) = 0.19$$

$$m(\{B, F\}) = 0.1$$

$$m(\{P\}) = 0.32$$

$$m(\{\theta\}) = 0.2$$

then the belief, plausibility, and interval become

Belief:	$Bel(\{B\})$	$= 0.19$
	$Bel(\{F\})$	$= 0.19$
	$Bel(\{B, F\})$	$= 0.48$
	$Bel(\{P\})$	$= 0.32$

$$\textit{Plausibility:} \quad Pl(\{B\}) \qquad\qquad = 0.49$$

$$Pl(\{F\}) \qquad\qquad = 0.49$$

$$Pl(\{B, F\}) \qquad\quad = 0.68$$

$$Pl(\{P\}) \qquad\qquad = 0.52$$

$$\textit{Interval:} \quad Pl(\{B\}) - Bel(\{B\}) = 0.3$$

$$Pl(\{F\}) - Bel(\{F\}) = 0.3$$

$$Pl(\{B, F\}) - Bel(\{B, F\}) = 0.2$$

$$Pl(\{P\}) - Bel(\{P\}) = 0.2 \qquad\qquad (9.30)$$

In such circumstances, if one considers only single-class labeling, one may still choose $\{P\}$, because its belief and plausibility are the highest without considering the set $\{B, F\}$. However, it may be sensible to collect more evidence rather than to generate a label for the pixel using the single-class evidence alone. If other evidence is available, the evidence combination problem will arise. This issue is considered in Section 9.4.3.

To investigate further the relationship between belief and plausibility, one can speculate that plausibility is always equal to or greater than belief, that is

$$Pl(\psi) \geq Bel(\psi),$$

or equivalently,

$$Bel(\sim\psi) \geq Bel(\psi) \qquad\qquad (9.31)$$

which leads to

$$1 \geq Bel(\psi) + Bel(\sim\psi) \qquad\qquad (9.32)$$

Equation (9.32) is derived on the basis of the observation that both $Bel(\psi)$ and $Bel(\sim\psi)$ have no subset in common, and $Bel(\psi)$ and $Bel(\sim\psi)$ are each made by the sum of the *bpa* of its own subsets. If we let $\psi = P$, it follows that

$$1 - Bel(\psi) - Bel(\sim\psi) = 1 - Bel(\{P\}) - Bel(\{^\sim P\}) = 1 - Bel(\{P\}) - Bel(\{B, F\})$$

$$= 1 - m(P) - m(B) - m(F) - m(\{B, F\})$$

$$= m(\{B, P\}) + m(\{F, P\}) + m(\{B, F, P\}). \qquad\qquad (9.33)$$

Equation (9.33) specifies what the interval or uncertainty is, and it is then apparent that $1 - Bel(\psi) - Bel(\sim\psi)$ can be greater than 0, which confirms Equation (9.32).

9.4.3 EVIDENCE COMBINATION

In most cases, more than one set of evidence is available, and the decision to label a pixel is made based on the accumulation of all of the evidence. D-S theory acknowledges such a requirement and provides a formal proposal for multi-evidence management.

The aggregation of multiple belief functions is called *Dempster's orthogonal sum*, or *Dempster's rule of combination* (Shafer, 1979). Let Bel_a and Bel_b denote two belief functions, and let m_a and m_b be their corresponding *bpa*s. Dempster's orthogonal sum generates a new *bpa*, denoted by $m_a \oplus m_b$, which represents the result of combing m_a and m_b. The result of the accumulation of both belief functions, denoted by $Bel_a \oplus Bel_b$, is derived from $m_a \oplus m_b$. If $m(\psi)$ denotes the new aggregated *bpa*, the combination rule can be specified by

$$m(\psi) = K \sum_{\xi \cap \zeta = \psi} m_a(\xi) \times m_b(\zeta) \qquad (9.34)$$

where K is a normalizing constant, which is defined by

$$K^{-1} = 1 - \sum_{\xi \cap \zeta = \varnothing} m_a(\xi) \times m_b(\zeta). \qquad (9.35)$$

The accumulation of *bpa* $m_a : [m_a(\xi_1), m_a(\xi_2), ..., m_a(\theta)]$ and *bpa* $m_b : [m_b(\zeta_1), m_a(\zeta_2), ..., m_a(\theta)]$ is calculated by considering all products in the form of $m_a(\xi) \times m_b(\zeta)$ where ξ and ζ are individually varied over all subsets of Bel_a and Bel_b. Note that the resulting sum of $m_a(\xi) \times m_b(\zeta)$ is equal to one, as is required by the definition of a *bpa*.

Equations (9.34) and (9.35) illustrate that $m(\psi)$ is formed by the orthogonal summation of all products $m_a(\xi) \times m_b(\zeta)$, which have an intersection ψ. The normalizing constant K is formed by the reciprocal of the sum of all products $m_a(\xi) \times m_b(\zeta)$ for which there is no intersection. This normalizing constant ensures that no contribution is committed to the null set (i.e., $\xi \cap \zeta = \varnothing$). The evidence combination process is illustrated by two examples, which are described below.

In the first example, suppose that one observation supports $\{B, F\}$ to the degree of 0.7 (i.e., m_a), whereas another observation disconfirms $\{F\}$ to the degree of 0.6 (i.e., m_b) (note that the situation here is the same to confirm $\{B, P\}$ to the degree of 0.6). For illustrative purposes, it is convenient to use an intersection table with the values assigned by m_a and m_b along the rows and columns, respectively, and only nonzero values are taken into consideration. We define the entry (r, c) in the table as the intersection of the subsets in row r and column c. The result of the rule combination is shown in Table 9.1.

In the example, each subset appears only once in the intersection table and no null intersections occur. Hence $m_a \oplus m_b$ for each intersection is easily calculated as shown in each entry. Once the calculation of $m_a \oplus m_b$ is completed, the belief and plausibility of the combination $Bel_a \oplus Bel_b$ can be derived as

TABLE 9.1

Example of Dempster's Rule Combination

$m_a \rightarrow$ $m_{b\downarrow}$	$m_a(\{B, F\}) =$ (0.7)	$m_a(\theta) =$ (0.3)
$m_b(\{B, P\}) = (0.6)$	$\{B\} = (0.42)$	$\{B, P\} = (0.18)$
$m_b(\theta) = (0.4)$	$\{B, F\} = (0.28)$	$\{\theta\} = (0.12)$

TABLE 9.2

Extended Example of Dempster's Rule Combination

$m_d \rightarrow$ $m_{c\downarrow}$	$m_d(\{B\}) =$ (0.42)	$m_d(\{B, F\}) =$ (0.28)	$m_d(\{B, P\}) =$ (0.18)	$m_d(\theta) =$ (0.12)
$m_c(\{P\})(0.7)$	$\varnothing = 0.294$	$\varnothing = 0.196$	$\{P\} = 0.126$	$\{P\} = 0.084$
$m_c(\theta)(0.3)$	$\{B\} = 0.126$	$\{B, F\} = 0.084$	$\{B, P\} = 0.054$	$\theta = 0.036$

$$Bel(\{B\}) = m_a \oplus m_b(\{B\}) = 0.42$$

$$Bel(\{B, P\}) = m_a \oplus m_b(\{B\}) + m_a \oplus m_b(\{P\}) + m_a \oplus m_b(\{B, P\})$$

$$= 0.42 + 0 + 0.18 = 0.6$$

$$Bel(\{B, F\}) = m_a \oplus m_b(\{B\}) + m_a \oplus m_b(\{F\}) + m_a \oplus m_b(\{B, F\})$$

$$= 0.42 + 0 + 0.28 = 0.7$$

The *belief* for other subsets is zero. The plausibilities for subsets $\{B\}$, $\{B, F\}$, and $\{B, P\}$ are all equivalent to one, because

$$Bel(\sim\{B\}) = Bel(\sim\{B, F\}) = Bel(\sim\{B, P\}) = 0$$

The second example is an extension of the first. Suppose that there is a third observation that confirms $\{P\}$ to the degree of 0.7, denoted by m_c, and let $m_a \oplus m_b = m_d$. The orthogonal sum $m_c \oplus m_d$ is illustrated in Table 9.2. There are two null intersections (i.e., $\varnothing$). One contains the value 0.294, and the other 0.196. Therefore, the normalization constant K and the final evidence combination results are computed by

$$K^{-1} \qquad\qquad = 1 - (0.294 + 0.196) = 0.51$$

$$m_c \oplus m_d\,(\{B\}) \quad = 0.126/0.51 = 0.247$$

$$m_c \oplus m_d\,(\{P\}) \quad = (0.126 + 0.084)/0.51 = 0.412$$

$$m_c \oplus m_d\,(\{B, F\}) = 0.084/0.51 = 0.165$$

$$m_c \oplus m_d (\{B, P\}) = 0.054/0.51 = 0.106$$

$$m_c \oplus m_d (\theta) \quad\quad = 0.036/0.51 = 0.07$$

and for all other subsets, $m_c \oplus m_d = 0$

Both belief and plausibility can be derived in the similar way, as explained in example one, and so will not be duplicated here. Readers may also wish to combine *bpas* using the above examples but in different sequence. For example, one may compute the orthogonal sum first using evidence 1 and 3, and then combine to evidence 2. This exercise should show that the results of evidence combination are the same, regardless of the ordering.

9.4.4 DECISION RULES FOR EVIDENTIAL REASONING

Where a statistical classifier such as maximum likelihood is used, the label assigned to a given pixel is the one for which the class membership likelihood function is greatest. This is the most straightforward choice. However, in evidential reasoning, the decision rules are more complicated because the decision space involves two elements (i.e., belief and plausibility). According to Shafer (1979), there are three possible choices:

1. *Belief driven*: Label is chosen on the basis of maximal belief.
2. *Plausibility driven*: Label is chosen on the basis of maximal plausibility.
3. *Mean of the interval*: Label is chosen on the basis of the maximal average of belief and plausibility values.

These three choices may generate different results. If we look at the example given above, with three classes (bare soil $\{B\}$, forest $\{F\}$, and pasture $\{P\}$), it is possible to compute the belief and plausibility values for each.

$$Bel\{B\} = m(\{B\}), Pl\{B\} = 1 - m(\{F\}) - m(\{P\})$$

$$Bel\{F\} = m(\{F\}), Pl\{F\} = 1 - m(\{B\}) - m(\{P\})$$

$$Bel\{P\} = m(\{P\}), Pl\{P\} = 1 - m(\{B\}) - m(\{F\}) \quad\quad (9.36)$$

It is clear that if $m(\{B\}) > m(\{F\})$, i.e., $Bel\{B\} > Bel\{F\}$, then one can also conclude that $Pl\{B\} > Pl\{F\}$. In this situation, two of the decision rules (i.e., belief driven and plausibility driven) produce the same result. Thus, it shows that one still has just one choice to make the decision if one tends to perform the single-class labeling.

9.5 DEALING WITH SOURCE RELIABILITY

The methods described above show the promise for handling the multisource classification problem. The issue concerning the source weighting factors is not fully resolved, however. In practical circumstance, each available data source may not be fully certain and complete. Therefore, it is necessary to weight each of the sources so that the final classification reflects our knowledge of the reliability of each of the data sources.

The process for the determination of weighting factors is similar to that of computing reliability in consensus theory for managing different opinion sources (Winkler, 1968; McConway, 1980; French, 1985). The choice of source weighting factors will have a significant effect on the results of a multisource classification because the contribution of each source will be reduced or enhanced in proportion to the weight. It can be seen that the results of the classification process depends both on the information provided by the source and on the source weighting factor.

For example, if two sources, denoted by A and B respectively, are used for labeling a pixel of interest, source A may suggest that the pixel belongs to class a with probability 0.8 and to class b with probability 0.2. However, source B indicates that the pixel belongs to class b with probability 1.0 and assigns zero probability to the pixel belonging to class a. If both sources have been determined as fully reliable (i.e., their associated weights are unity), the result based on both statistical Bayesian or evidential reasoning approaches for source combination will label the pixel as class b. If we have determined that the weight of source B is 0.5, and that of source A is unity, then the pixel will be labeled as class a. Several possible methods for measuring the source weighting parameters are described below.

9.5.1 USING CLASSIFICATION ACCURACY

The derivation of data source weighting parameters from classification accuracy measurements is an instinctive approach. A data source should be assigned a higher weight if the resulting classification accuracy is high. However, if the classification derived from the data source is unsatisfactory, the data are considered to be relatively unreliable, and should be assigned a lower weight. For example, if the classification accuracy using data source A is 80%, and using data source B is only 50%, then the probability measure or evidence derived from data source A should be assigned a higher weight. This method is very easy to apply, but one should note that the 80 and 50% classification accuracies do not necessarily mean that the source weighting parameters are equal to 0.8 and 0.5, respectively since different methods can be used for analyzing classification accuracy (Chapter 2), and these accuracy analysis tools are likely to provide different measures. The reliability measure in terms of classification accuracy can thus only give us a rough guide.

9.5.2 USE OF CLASS SEPARABILITY

The second approach to the measurement of source weighting uses the concept of the separability of the information classes. One can assign a higher weight to a source that contributes more to the separability of the information classes, while sources that contribute little to interclass separability can be assigned low weights. The estimates of the weighting parameter values are therefore based on the contribution to the statistical separability of each data source. Several approaches to the estimation of separability, using distance, divergence, or separability functions, are proposed by Fukunaga (1972), Whitsitt and Landgrebe (1977), Swain and Davis (1978), and Richards (1986). Separability methods are discussed in Chapter 2. One problem may result from the employment of separability as a weighting parameter measure. The assumption underlying each approach is that the data distribution for each class is Gaussian, and consequently that interclass relationships are fully described in the covariance matrices. However, not all data sources are Gaussian in their distribution; for example, the frequency distribution of land elevation measurements is often skewed. In such cases, the use of separability measures may be misleading.

9.5.3 DATA INFORMATION CLASS CORRESPONDENCE MATRIX

Consider that there are a total of I information classes denoted by $\{\alpha_1, \alpha_2, ..., \alpha_I\}$, and that there are J data classes (e.g., J clusters) denoted by $\{\beta_1, \beta_2, ..., \beta_J\}$. The relationship between information classes α_i and data classes β_j can be represented using conditional probabilities, denoted by $P(\alpha_i/\beta_j)$, which can be used to construct a J by I correspondence matrix M, expressed as

$$M = \begin{bmatrix} P(\alpha_1/\beta_1) & P(\alpha_2/\beta_1) & \cdots & P(\alpha_{I-1}/\beta_1) & P(\alpha_I/\beta_1) \\ P(\alpha_1/\beta_2) & P(\alpha_2/\beta_2) & \cdots & P(\alpha_{I-1}/\beta_2) & P(\alpha_I/\beta_2) \\ \cdot & \cdot & \cdots & \cdot & \cdot \\ \cdot & \cdot & \cdots & \cdot & \cdot \\ \cdot & \cdot & \cdots & \cdot & \cdot \\ P(\alpha_1/\beta_J) & P(\alpha_2/\beta_J) & \cdots & P(\alpha_{I-1}/\beta_J) & P(\alpha_I/\beta_J) \end{bmatrix} \quad (9.37)$$

The objective of the correspondence matrix is to express the strength of the relationship between data classes and information classes. This kind of relationship is sometimes called *equivocation*. We will give a data source the higher weight if a data class strongly supports the information class. A data source should have a lower weight if there is no such strong link between some data classes and information classes. After constructing the correspondence matrix, a meaningful measure is needed to quantify the overall relationships.

In the case in which the linkage between the data classes and the information classes is high, there will be a unique conditional probability in each row of matrix M containing the value of 1, with the remaining entries taking the value 0. Conversely, if a data class is simultaneously linking to several information classes, then in the extreme case, each entry of matrix M will be equal (i.e., there will be no correspondence between the data classes and information classes). A method of expressing these relationships is the entropy measure of Shannon and Weaver (1963). For a particular data class β_j, entropy measure E is given by

$$E(\alpha/\beta_j) = -\sum_i P(\alpha_i/\beta_j) \times \ln P(\alpha_i/\beta_j) \quad (9.38)$$

and the overall measure for all observed data classes and information classes can be expressed as

$$E(\alpha/\beta) = -\sum_{i}\sum_{j} P(\alpha_i/\beta_j) \times \ln P(\alpha_i/\beta_j). \qquad (9.39)$$

It should be noted that as entropy is being used to quantify the relationship between data classes and information classes, the higher the value, the less weight should be allocated to the data source.

The methodology discussed above only provides a rough estimation of data source weighting parameters. One has to determine a mapping function to convert the measured results into the weighting parameters. It is not always easy to define such a mapping function, and the determination of the weighting parameter still relies on the analyst's *ad hoc* decisions, which cannot guarantee an optimal solution. Moreover, when the MRF-MAP estimate is attempted, the parameter selection issue will become more complicated because both weighting and potential parameters have to be determined simultaneously.

A straightforward approach to achieve optimal parameter assignments is to perform an exhaustive search of the total parameter space and then select the best solution. Searching for the best combination of weights can be time-consuming. For example, if there are 10 data sources, and if the weights are defined within the range [0, 1] with steps of 1/256, the total search space will be 256^{10}. A more efficient search tool is needed in order to reduce the computational cost. The genetic optimal search algorithm (Holland, 1975; Goldberg, 1989) appears to be a good tool for locating optimal solutions.

9.5.4 THE GENETIC ALGORITHM

Many practical search and optimization problems require the investigation of multiple maximums or minimums. Genetic algorithms (GA) have been increasingly used in solving such problems, and their general performance has been empirically shown to be robust (Holland, 1975; Goldberg, 1989). The central concept of GA is based on the mechanics of natural selection and natural genetics. GA operates by maintaining a set of trial structures, forming a population. Each trial structure is a finite-length string and generally is represented through special coding techniques (such as gray code or binary code). Once the trial structure has been determined by the user, the GA then starts a series of search processes within which, at each generation, the idea of the survival of the fittest structures in the current population is used to create a new population based on both structured and randomized information exchange between the present and the new population. The GA search process can be allowed to run for a prespecified number of iterations, or the search can be continued until the fitness function converges. The GA is said to have converged if approximately 95% of the structures forming the population contains the same value.

Why can the performance GA surpass that of more traditional search methodologies, such as calculus-based, enumerative schemes, and random search algorithms? As Goldberg (1989) notes, there are several significant differences between GA and other algorithms that contribute to GA's highly promising performance. The first difference is that GA uses multiple start points for performing the search process and deals with the special trial structures, which are codings of the original parameter set. The second reason is that GA uses an objective function (the measure of fitness of each trial structure) to lead its search process, and is not dependent on derivatives or auxiliary knowledge. Finally, the transition from one generation to the succeeding generation is based on probabilistic instead of deterministic rules.

The initial population used in a GA may be chosen at random or composed of heuristically selected initial points. A fitness function is used to evaluate each structure's performance. GA then constructs a new generation using three basic operators known as *reproduction, crossover,* and *mutation.* In the reproduction process, individuals are randomly selected from the present population. Although the selection is random, the expected number of times an individual is selected is associated with individual's performance. For example, if an individual's performance is three times the average performance of the population, this individual will be present three times in the new generation. As a result, individuals with low performance will be filtered out, and only high-performance individuals will survive.

In order to introduce new individuals into the search process in the new generation, GA uses the procedures of crossover and mutation. Crossover can be regarded as information exchange between two individuals. For instance, in a simple crossover process for two individuals (both represented by string of length λ), a cutting point ε within the string is randomly selected. Then two substrings, addressed from ε to λ, of both individuals are exchanged. This will form two new individuals. Figure 9.5 illustrates the process.

The mutation operation is carried out by randomly changing bits within the individual's coding from 1 to 0 or from 0 to 1. The main contribution of the mutation operation is to ensure that the search process by the genetic algorithm will not be limited by local maxima (or minima). Goldberg (1989) suggests that a mutation rate of 0.001 can achieve a relatively good result. This value is therefore used in the study reported here.

All GA operations discussed previously have the same purpose, which is to test different yet high-performance similarity templates or schema. A schema is a subset of strings with similarities at certain string positions (Holland, 1975). For example, suppose there are two individuals; individual A has a binary coded value 1010, and individual B contains the coding value 1111. One possible schema for these two individuals can therefore be 1*1*, where the symbol * means "don't care." One interesting and important result pointed out by Holland is that if P_i denotes the number of individuals in generation i, the number of schema being tested will be P_i^3. This result explains why a substantial search improvement can be achieved using a relatively small population size, and the concept is therefore termed *implicit parallelism.*

During the search process, the progress of GA survivals is determined by a user-defined function called the fitness function. The fitness function can be regarded as controlling GA by providing the ability to evaluate and hence select higher-quality individuals. Where a GA is used to determine data source weighting parameters, and if the aim of classification is to improve the overall classification accuracy, one can then simply use the overall classification accuracy as the fitness function.

9.6   EXPERIMENTAL RESULTS

Three classification approaches (the extension of Bayesian theory, Markov random fields, and evidential reasoning) are used in a classification experiment with the aim of evaluating their relative classification performance. Three data sources are used: (1) Landsat TM bands 1 through 5 and 7, (2) Shuttle Imaging Radar C (SIR-C) C-band HH and HV polarization images, and (3) SIR-C L-band HH and HV polarization images. Each image is 256 × 256 pixels in size. Examples of these data sources are shown in Figure 9.6. The study area is located in the Red Hills, Sudan, and the aim is to classify surface lithology into eight types, listed in Table 9.3, which also gives the number of training pixels selected for each class. Ground truth is available in the form of a paper map (Figure 9.7a).

For the six TM bands, the pixel values are used as inputs, while for the SIR-C radar images, textural features are extracted using the multiplicative autoregressive random field model (MAR) (Chapter 7). The window size for texture feature extraction is 5 × 5 pixels, and the neighbor support parameter for MAR is defined as {(1, 0), (−1, −1), (−1, 0)}. Three texture features, the mean value of neighborhood vector θ (Chapter 7, Equation [7.37]), covariance (Equation [7.38]), and mean (Equation [7.39]) for each radar image are used as inputs.

Class-conditional probabilities were generated for each data source using Equation (2.17), which assumes a Gaussian probability density function. The maximum likelihood method was used to classify each data source individually. The average producer's accuracy (Section 2.7) for the TM bands alone is 56.58%. Average producer's accuracy is 50.22% for the SIR-C C band HH and HV images, while the average

(a)

FIGURE 9.6   (a) Images of the test area (Red Sea Hills, Sudan). Landsat TM false color image. Landsat TM data are available from U.S. Geological Survey, EROS Data Center, Sioux Falls, SD, USA.

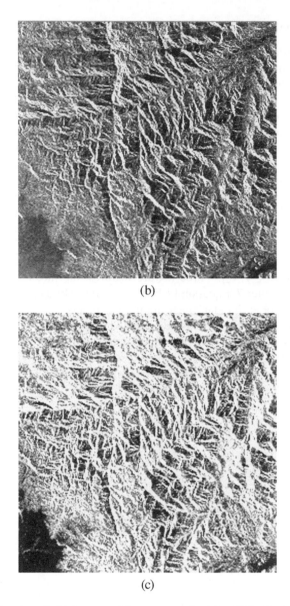

FIGURE 9.6 (continued) (b) Images of the test area (Red Sea Hills, Sudan). SIR-C C band HH polarization image. SIR-C data provided by NASA Jet Propulsion Laboratory, Pasadena, CA, USA. (c) Images of the test area (Red Sea Hills, Sudan). SIR-C L band HH polarization image. SIR-C data provided by NASA Jet Propulsion Laboratory, Pasadena, CA, USA.

producer's accuracy for SIR-C L band HH and HV images is only 36.75%. Clearly, the classification results based on each data source separately are unsatisfactory.

These class-conditional probabilities are first subjected to the extension of Bayesian theory for multisource consensus experiment. The prior probability in this

TABLE 9.3

Information Classes and Number of Ground Reference Pixels

Class	Information Class	Number of Pixels
1	Metarhyodacids (rhd)	429
2	Granodiorities	43496
3	Gneiss Basitic	5807
4	Metavolcanics	8157
5	Gneiss Acidic	1998
6	Pluvial Flood Plain	703
7	Quaternary Deposits	2529
8	Granites	385

stage of the experiment is abandoned. The genetic algorithm is used for detecting the optimal weighting parameters for each data source. Each string in the GA contains 21 bits in length, i.e., 7 bits for each weighting parameter, which is equivalent to dividing the weighting value within the range [0, 1] into $2^7 = 128$ choices. The fitness function is based on the average producer's accuracy. The initial population is set as 100 with a crossover rate of 0.6 and a mutation rate of 0.001. After 3000 search iterations, the average producer's accuracy is improved to 67.85% (see the confusion matrix in Table 9.6a), which is around 10% improvement. The corresponding weighting parameters are shown in Table 9.4, and the classification patterns are shown in Figure 9.7b.

In approaches using evidential reasoning, GA is applied in a similar way for detecting optimal weighting parameters. The probabilities derived from data sources 1 and 3 are first chosen to perform an orthogonal sum. After 2000 search iterations, the average producer's accuracy is only enhanced to 58.1%. The resulting evidence is then fused with the probabilities derived from data source 2. After 2000 iterations, the accuracy is enhanced to 65.63% (Table 9.6b), close to the accuracy level that the extension of Bayesian theory achieved. The corresponding weights and classification pattern is shown in Figure 9.8 and Figure 9.7c, respectively.

In this multisource classification exercise, which includes both contextual prior probabilities and Markov random fields, the GA is used for two purposes: first, finding optimal values for the weighting parameters, and second, for estimating the potential parameters. Only pairwise potential parameters β are used. For simplicity, we have also used the isotropic assumption (i.e., rotation invariant, single β is used), and both the fitness function and classification algorithm are coded using the iterated conditional mode (ICM) algorithm, which is described in Chapter 8. Each string in the GA contains 28 bits. The first 21 bits are used to determine the weighting parameters, while the last 7 bits are used for searching for an optimum value of the potential parameters β. The range of β is defined as [0, 3]. After 4000 iterations, a classification accuracy of 79% was achieved (Table 9.6c). This represents an improvement of close to 20% using the TM images alone, and an improvement of

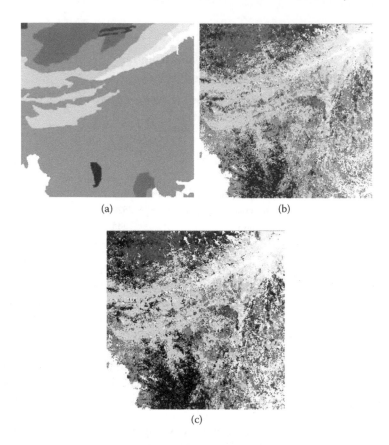

(a)   (b)

(c)

FIGURE 9.7 (a) Ground reference image. (b) Result of classification by using the extension of Baysian theory without including prior probability. (c) Classification using evidential reasoning.

more than 10% in comparison with the classification without contextual information. The corresponding weights are listed in Table 9.5, and the classified image is shown in Figure 9.7d. It is apparent that the patterns shown on the classified image are more patchlike, which gives a more realistic result.

TABLE 9.4

Optimal Consensus Weights Detected by GA for the Extension of Bayesian Theory

Source	TM	SIRC C Band HH, HV	SIRC L Band HH, HV
Weights	0.882	0.693	0.142

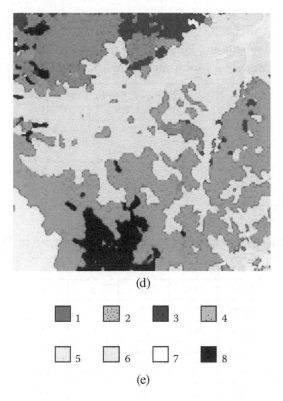

(d)

1 2 3 4

5 6 7 8

(e)

FIGURE 9.7 (continued) (d) Classification using a Markov random field model. (e) Key showing 8 classes listed in Table 9.3.

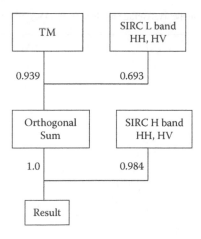

FIGURE 9.8 Evidence consensus sequence and its corresponding weighting parameters.

TABLE 9.5

Weights and Potential Parameter in MRF
Model for Multisource Classification

Source	TM	SIRC C Band HH, HV	SIRC L Band HH, HV
Weights	0.882	0.693	0.142
β	1.37		

TABLE 9.6a

Confusion Matrices Generated by the Extension of Bayesian Theory

No.	1	2	3	4	5	6	7	8
1	**280**	2548	1173	160	8	1	5	9
2	37	**14730**	700	1256	62	2	102	26
3	59	3464	**3056**	237	22	1	34	9
4	9	10432	340	**4960**	205	0	64	1
5	5	5912	107	1212	**1429**	104	70	1
6	8	850	39	70	269	**595**	78	0
7	0	1103	28	31	2	0	**2176**	0
8	31	4457	364	231	1	0	0	**339**

Note: The extension of Bayesian theory. Average producer's accuracy: 67.85%. Kappa coefficient: 0.262.

TABLE 9.6b
Confusion Matrices Generated by Evidential Reasoning

No.	1	2	3	4	5	6	7	8
1	**241**	2395	1114	143	11	0	5	7
2	19	**11934**	535	933	44	2	75	14
3	50	3436	**2916**	223	19	0	31	7
4	13	10666	385	**4860**	206	0	83	2
5	5	5996	139	1224	**1351**	89	49	1
6	43	1579	154	163	358	**612**	136	0
7	0	1348	39	66	0	0	**2150**	0
8	58	6142	525	545	9	0	0	**354**

Note: Evidential reasoning. Average producer's accuracy: 65.63%. Kappa coefficient: 0.233.

TABLE 9.6c
Confusion Matrices Generated by Iterative Conditional Mode Algorithms

No.	1	2	3	4	5	6	7	8
1	**409**	1247	784	25	0	0	0	0
2	2	**19433**	786	717	0	0	121	4
3	15	2743	**3916**	74	0	0	10	0
4	1	10985	111	**5983**	132	0	77	0
5	0	4697	5	1225	**1693**	118	65	0
6	1	446	0	25	172	**585**	105	0
7	0	782	0	0	1	0	**2151**	0
8	1	3163	205	108	0	0	0	**381**

Note: Iterated conditional mode algorithm. Average producer's accuracy: 79.09%. Kappa coefficient: 0.368.

Bibliography

Abe, S., and M. S. Lan. 1995. A method for fuzzy rules extraction directly from numerical data and its application to pattern classification. *IEEE Transactions on Fuzzy Systems* 3:18–28.

Achim, A., P. Tsakalides, and A. Bezerianos. 2003. SAR image denoising via Bayesian wavelet shrinkage based on heavy-tailed modelling. *IEEE Transactions on Geoscience and Remote Sensing* 41:1773–1784.

Agterberg, F. P., and Q. Cheng. 1999. Introduction to special issue on "Fractals and Multifractals." *Computers and Geosciences* 25:947–948.

Aiazzi, B., L. Alparone, S. Baronti, and A. Garzelli. 2002. Context-driven fusion of high spatial and spectral resolution images based on oversampled multi-resolution analysis. *IEEE Transactions on Geoscience and Remote Sensing* 40:2300–2312.

Aires, F., C. Prigent, and W. B. Rossow. 2004. Neural network uncertainty assessment using Bayesian statistics: A remote sensing application. *Neural Computation* 16:2415–2458.

Aiyer, S. V. B., M. Niranjan, and F. Fallside. 1990. A theoretical investigation into the performance of the Hopfield model. *IEEE Transactions on Neural Networks* 1:204–215.

Aizerman, M. A., E. M. Braverman, and L.I. Rozoner. 1964. Theoretical foundations of the potential function method in pattern recognition learning. *Automation and Remote Control* 25:821–837.

Alilat, F., S. Loumi, and B. Sansal. 2006. Modified fuzzy ARTMAP and supervised fuzzy ART: Comparative study with multispectral classification. *International Journal of Computer Science* 1:232–238.

Allen, T. R., and J. A. Kupfer. 2000. Application of spherical statistics to change vector analysis of Landsat data: Southern Appalachian spruce-fir forests. *Remote Sensing of Environment* 74:482–493.

Alt, M. 1990. *Exploring Hyperspace—A Non-mathematical Explanation of Multivariate Analysis*. London: McGraw-Hill.

Ardö, J., P. Pilesjö, and A. Skidmore. 1997. Neural networks, multi temporal TM data and topographic data to classify forest damage in the Czech Republic. *Canadian Journal of Remote Sensing* 23:217–229.

Asner, G. P., and K. B. Heidebrecht. 2002.Spectral unmixing of vegetation, soil and dry carbon cover in arid regions: Comparing multispectral and hyperspectral observations. *International Journal of Remote Sensing* 23:3939–3958.

Asner, G. P., and D. B. Lobell. 2000. A biophysical approach for automated SWIR unmixing of soils and vegetation. *Remote Sensing of Environment* 74:99–112.

Atkinson, P. M. 1991. Optimal ground-based sampling for remote sensing investigations: Estimating the regional mean. *International Journal of Remote Sensing* 12:559–567.

Atkinson, P. M. 1996. Optimal sampling strategies for raster-based geographical information systems. *Global Ecology and Biogeographical Letters* 5:271–280.

Atkinson, P. M., and P. Lewis. 2000. Geostatistical classification for remote sensing: An introduction. *Computers and Geosciences* 26:361–371.

Atkinson, P. M., and A. R. L. Tatnall. 1997. Introduction: Neural networks in remote sensing. *International Journal of Remote Sensing* 18:699–709.

Ball,, G. H., and D. J. Hall. 1967. A clustering technique for summarising multivariate data. *Behavioral Science* 12:153–155.

Baraldi, A., and F. Parmiggiani. 1995a. A refined Gamma MAP SAR speckle filter with improved geometrical adaptivity. *IEEE Transactions on Geoscience and Remote Sensing* 33:1245–1257.

Baraldi, A., and F. Parmiggiani. 1995b. An alternative form of the Lee filter for speckle suppression in SAR images. *Computer Vision, Graphics and Image Processing—Graphical Models and Image Processing* 57:75–78.

Baronti, S., R. Carla, S. Sigismondi, and L. Alparone. 1994. Principal component analysis for change detection on polarimetric multitemporal SAR data. In *Proceedings of the International Geoscience and Remote Sensing Symposium (IGARSS'94)*, 8–12 August, Pasadena, CA, 2152–2154. Piscataway, NJ: Institute of Electrical and Electronics Engineers (IEEE).

Bardossy, A., and L. Samaniego. 2002. Fuzzy rule-based classification of remotely sensed imagery. *IEEE Transactions on Geoscience and Remote Sensing* 40:362–374.

Barnett, J. A. 1981. Computational methods for a mathematical theory of evidence. In *Proceedings of the 7th International Conference on Artificial Intelligence*, Vancouver, British Columbia, 868–875. Los Altos, CA: William Kaufmann.

Barnsley, M. F. 1993. *Fractals everywhere*. London: Academic Press Limited.

Barnsley, M. F., and L. P. Hurd. 1993. *Fractal image compression*. Wellesley, MA: AK Peters Ltd.

Bastin, L. 1997. Comparison of fuzzy c-means classification, linear mixture modelling and MLC probabilities as tools for unmixing coarse pixels. *International Journal of Remote Sensing* 18:3629–3648.

Bavarian, B., and Z. Lo. 1991. On the rate of convergence in topology preserving neural networks. *Biological Cybernetics* 65:55–63.

Belousov, A. I., S. A. Verzakov, and J. Von Frese. 2002. A flexible classification approach with optimal generalisation performance: Support vector machines. *Chemometrics and Intelligent Laboratory Systems* 64:15–25.

Benediktsson, J. A., P. H. Swain, and O. K. Esroy. 1990. Neural network approaches versus statistical methods in classification of multisource remote sensing data. *IEEE Transactions on Geoscience and Remote Sensing* 28:540–552.

Benson, B. J., and M. D. Mackenzie. 1995. Effects of sensor spatial resolution on landscape structure parameters. *Landscape Ecology* 10:113–120.

Berry, B. J. L., and A. M. Baker. 1968. Geographical sampling. In *Spatial analysis*, ed. B. J. L. Berry and D. F. Marble. Englewood Cliffs, NJ: Prentice-Hall.

Bersini, H., M. Saerens, and L. G. Sotelino. 1994. Hopfield net generation, encoding and classification of temporal trajectories. *IEEE Transactions on Neural Networks* 5:945–953.

Besag, J. 1974. Spatial interaction and the statistical analysis of lattice systems. *Journal of the Royal Statistical Society* 36:192–236.

Besag, J. 1986. On the statistical analysis of dirty pictures. *Journal of the Royal Statistical Society* 48:259–302.

Bezdek, J. C. 1981. *Pattern recognition with fuzzy objective function algorithms*. New York: Plenum Press.

Bezdek, J. C., R. Ehrlich, and W. Full. 1984. FCM: The fuzzy c-means clustering algorithm. *Computers and Geosciences* 10:191–203.

Bian, L. 2003. Retrieving urban objects using a wavelet transform approach. *Photogrammetric Engineering and Remote Sensing* 69:133–141.

Bishop, C. M. 1996. *Neural networks for pattern recognition*. Oxford: Clarendon Press.

Blake, A., and A. Zisserman. 1987. *Visual reconstruction*. Cambridge, MA: MIT Press.

Blamire, P. A. 1996. The influence of relative sample size in training artificial neural networks. *International Journal of Remote Sensing* 17:223–230.

Blanz, V., B. Schölkopf, H. Bülthoff, C. Burges, V. N. Vapnik, and T. Vetter. 1996. Comparison of view–based object recognition algorithms using realistic 3D models. In *Proceedings of International Conference on Artificial Neural Networks—ICANN'96*, Berlin, 1112, 251–256.

Borak, J. S., and A. H. Strahler. 1999. Feature selection and land cover classification of a MODIS-like data set for semi-arid environment. *International Journal of Remote Sensing* 20:919–938.

Borel, C. C., and S. A. W. Gerstl. 1994. Nonlinear spectral mixing models for vegetative and soil surfaces. *Remote Sensing of Environment* 47:403–416.

Bork, E. W., N. E. West, and K. P. Price. 1999. Calibration of broad- and narrow-band spectral variables for rangeland cover component quantification. *International Journal of Remote Sensing* 20:3641–3662.

Born, M., and E. Wolf. 1980. *Principles of optics*. New York: Pergamon.

Bosdogianni, P., M. Petrou, and J. Kittler. 1997. Mixture models with higher order moments. *IEEE Transactions on Geoscience and Remote Sensing* 33:341–353.

Boser, B. E., I. M. Guyon, and V. N. Vapnik. 1992. A training algorithm for optimal margin classifiers. In *Proceedings of the 5th Annual Workshop on Computational Learning Theory*, 144–152. Pittsburgh, PA: ACM Press.

Breim, G. J., J. A. Benediktsson, and J. R. Sveinsson. 2002. Multiple classifiers applied to multisource remote sensing data. *IEEE Transactions on Geoscience and Remote Sensing* 40:2291–2299.

Breiman, L. 1996. Bagging predictors. *Machine Learning* 26:123–140.

Breiman, L. 2001. Random forests. *Machine Learning* 45:5–32.

Breiman, L., J. H. Friedman, R. A. Olsen, and C. J. Stone. 1984. *Classification and Regression Trees*. Belmont, CA: Wadsworth.

Brigham, E. O. 1974. *The fast Fourier transform*. Englewood Cliffs, NJ: Prentice-Hall.

Brodatz, P. 1966. *Texture: A photographic album for artists and designers*. New York: Dover Books.

Brodley, C., and M. A. Friedl. 1996. Improving automated land cover mapping by identifying and eliminating mislabelled observations from training data. *Proceedings of the International Geoscience and Remote Sensing Symposium (IGARSS'96)*, Lincoln, NE, May 27–31, 1996, 1382–1384. Piscataway, NJ: Institute of Electrical and Electronics Engineers (IEEE).

Brodley, C., and M. A. Friedl. 1999. Identifying mislabelled training data. *Journal of Artificial Intelligence Research* 11:131–167.

Brogaard, S., and R. Ólafsdóttir. 1997. *Ground-truth or Ground-lies? Environmental sampling for remote sensing application exemplified by vegetation cover data*. Lund Electronic Reports in Physical Geography, No. 1 (October). Lund, Sweden: University of Lund, Department of Physical Geography. (http://www.lub.lu.se/luft/rapporter/lerpg/1/1Abstract.htm).

Brown, D. E., V. Corruble, and C. L. Pittard. 1993. A comparison of decision tree classifiers with backpropagation neural networks for multimodal classification problems. *Pattern Recognition* 26:953–961.

Brown, L. G. 1992. A survey of image registration techniques. *ACM Computing Surveys* 24:325–376.

Bruzzone, J., R. Gossu, and G. Vernazza. 2004. Detection of land-cover transitions by combining multidate classifiers. *Pattern Recognition Letters* 25:1491–1500.

Burges, C. J. C. 1998. A tutorial on support vector machines for pattern recognition. *Data Mining and Knowledge Discovery* 2:121–167.

Burges, C. J. C., and B. Schölkopf. 1997. Improving the accuracy and speed of support vector learning machines. In *Advances in neural information processing systems*, ed. by M. Mozer, M. Jordan, and T. Petsche, 9, 375–381, Cambridge, MA: MIT Press.

Burrus, C. S., R. A. Gopinath, and H, Guo. 1998. *Introduction to wavelets and wavelet transforms: A primer*. Upper Saddle River, NJ: Prentice Hall.

Camacho-De Coca, F., F. J. Carcía-Haro, M. A. Gilabert, and J. Meliá. 2004. Vegetation cover seasonal changes assessment from TM imagery in a semi-arid landscape. *International Journal of Remote Sensing* 25:3451–3476.

Campbell, J. B. 1987. *Introduction to remote sensing*. London: Guilford Press.

Cannon, R. L., J. V. Dave, J. C. Bezdek, and M. M. Trivedi. 1986. Segmentation of a Thematic Mapper image using the fuzzy c-means clustering algorithm. *IEEE Transactions on Geoscience and Remote Sensing* GE-24:400–408.

Cappellini, V., and A. Chiuderi. 1994. Neural networks in remote sensing multisensor data processing. *Proceedings of the 14th EARSeL Symposium*, Sweden, ed. by J. Askne, 457–462. Rotterdam: Balkema.

Carpenter, G. A., M. N. Gjaja, S. Gopal, and C. E. Woodcock. 1997. ART neural networks for remote sensing: Vegetation classification from Landsat TM and terrain data. *IEEE Transactions on Geoscience and Remote Sensing* 33:308–325.

Carpenter, G. A., and S. Grossberg. 1987a. A massively parallel architecture for a self-organising neural pattern recognition machine. *Computer Vision, Graphics and Image Processing* 37:54–115.

Carpenter, G. A., and S. Grossberg. 1987b. ART2: Stable self-organisation of pattern recognition codes for analogue input patterns. *Applied Optics* 26:4919–4930.

Carpenter, G. A., S. Grossberg, N. Markuzon, J. H. Reynolds, and D. B. Rosen. 1992. Fuzzy ARTMAP: A neural network architecture for incremental supervised learning of analogue multidimensional maps. *IEEE Transactions on Neural Networks* 3:698–713.

Carpenter, G. A., S. Grossberg, and J. H. Reynolds. 1991. ARTMAP: Supervisor real-time learning and classification of non-stationary data by a self-organising neural network. *Neural Networks* 4:565–588.

Carpenter, G. A., S. Grossberg, and D. B. Rosen. 1991. Fuzzy ART: Fast stable learning and categorisation of analogue patterns by an Adaptive Resonance system. *Neural Networks* 4:759–771.

Carper, W. J., T. M. Lillesand, and R. W. Kiefer. 1990. The use of IHS transformations for merging SPOT panchromatic and multispectral image data. *Photogrammetric Engineering and Remote Sensing* 56:459–467.

Carr, J. R. 1996. Spectral and textural classification of single and multiple band images. *Computers and Geosciences* 22:849–865.

Carr, J. R. 1999. Classification of digital image texture using variograms. In *Advances in remote sensing and GIS analysis*, ed. P. M. Atkinson and N. J. Tate, 135–146. Chichester, U.K.: John Wiley and Sons.

Carr, J. R., and F. P. Miranda. 1998. The semivariogram in comparison to the co-occurrence matrix for classification of image texture. *IEEE Transactions on Geoscience and Remote Sensing* 36:1945–1952.

Cestnik, B., and I. Bratko. 1991. On estimating probabilities in tree pruning. In *Proceedings of European Working Session on Learning*, ed. Y. Kodrato, 138–150. New York: Springer Verlag.

Chandler, D. 1987. *Introduction to modern statistical mechanics*. Oxford: Oxford University Press.

Chang, C. C., and C. J. Lin. 2007. LIBSVM: A library for support vector machines, 2001. Software available from http://www.csie.ntu.tw/~cjlin/libsvm (accessed May 16, 2008).

Chang, P. C., Y. W. Wang, and C. Y. Tsai. 2005. Evolving neural network for printed circuit board sales. *Expert Systems with Applications* 29:83–92.

Chapelle, O., V. Vapnik, O. Bousquet, and S. Mukherjee. 2002. Choosing multiple parameters for support vector machines. *Machine Learning* 46:131–159.

Chavez, P. S., Jr. 1988. An improved dark-object subtraction technique for atmospheric scattering correction of multispectral data. *Remote Sensing of Environment* 24:459–479.

Chen, C. F., and J. M. Lee. 2001. The validity measurement of fuzzy c-means classification for remotely sensed images. In *Proceedings of the 22nd Asian Conference on Remote Sensing*, November 5–9, Singapore, 208–210.

Chen, J., P. Gong, C. He, R. Pu, and P. Shi. 2003. Land-use/land-cover change detection using improved change-vector analysis. *Photogrammetric Engineering and Remote Sensing* 69:369–379.

Chen, Y., and T. Huang. 2001. Hierarchical MRF model for model-based multi-object tracking. In *Proceedings of the IEEE International Conference on Image Processing (ICIP'01)*, October 2001, Thessaloniki, Greece, 385–388.

Chen, Y. H., L. Den, J. Li, X. Li, and P. J. Shi. 2006. A new wavelet-based image fusion method for remotely sensed data. *International Journal of Remote Sensing* 27:1465–1476.

Cheung, Y. M. 2005. On rival penalization controlled competitive learning for clustering with automatic cluster number selection. *IEEE Transactions on Knowledge and Data Engineering* 17:1583–1588.

Chica-Olmo, M., and F. Abarca-Hérnandez. 2000. Computing geostatistical image texture for remotely sensed data classification. *Computers and Geosciences* 26:373–383.

Civco, D. 1989 Topographic normalisation of Landsat Thematic Mapper imagery. *Photogrammetric Engineering and Remote Sensing* 55:1303–1309.

Clarke, K. C. 1986. Computation of the fractal dimension of topographic surfaces using the triangular prism surface area method. *Computers and Geosciences* 12:713–722.

Clausi, M., and M. Jernigan. 1998. A fast method to determine co-occurrence texture features. *IEEE Transactions on Geoscience and Remote Sensing* 36:298–300.

Coburn, C. A., and A. C. B. Roberts. 2004. A multiscale texture analysis procedure for improved forest stand classification. *International Journal of Remote Sensing* 25:4287–4308.

Cochran, W. G. 1977. *Sampling techniques*. New York: John Wiley and Sons.

Cohen, A., and J. Kovavcevic. 1996. Wavelets: The mathematical background. *Proceedings of the IEEE* 84:514–522.

Cohen, J. 1960. A coefficient of agreement for nominal scales. *Educational and Psychological Measurement* 20(1):37–46.

Colby, J. D. 1991. Topographic normalisation in rugged terrain. *Photogrammetric Engineering and Remote Sensing* 57:531–537.

Colditz, R. C., T. Wehrmann, M. Bachmann, K. Steinnocher, M. Schmidt, G. Strunz, and S. Dech. 2006. Influence of image fusion approaches on classification accuracy: A case study. *International Journal of Remote Sensing* 27(15):3311–3335.

Collins, J. B., and C. E. Woodcock. 1994. Change detection using the Gramm-Schmidt transformation applied to mapping forest mortality. *Remote Sensing of Environment* 50:267–279.

Collins, J. B., and C. E. Woodcock. 1996. An assessment of several linear change detection techniques for mapping forest mortality using multitemporal Landsat TM data. *Remote Sensing of Environment* 56:66–77.

Congalton, R. G. 1988. A comparison of sampling schemes used in generating error matrices for assessing the accuracy of maps generated from remotely sensed data. *Photogrammetric Engineering and Remote Sensing* 54:593–600.

Congalton, R. G. 1991. A review of assessing the accuracy of classifications of remotely sensed data. *Remote Sensing of Environment* 37:35–46.

Congalton, R. G., and K. Green. 1999. *Assessing the accuracy of remotely sensed data: Principles and practice*. New York: Lewis Publishers.

Congalton, R. G., and L. C. Plourde. 2000. Sampling methodology, sampling placement and other important factors in assessing the accuracy of remotely sensed forest maps. In *Proceedings of the Accuracy 2000 Conference*, ed. G. B. M Heuvelink and M. J. P. M. Lemmens, 117–124. Amsterdam: Delft University Press.

Coppin, P. R., and M. E. Bauer. 1994. Processing of multitemporal Landsat TM imagery to optimise extraction of forest cover change features. *IEEE Transactions on Geoscience and Remote Sensing* 32:918–927.

Coppin, P. R., and M. E. Bauer. 1995. The potential contribution of pixel-based canopy change information to stand-based forest management in the northern US. *Journal of Environmental Management* 44:69–82.

Coppin, P. R., E. Lambin, I. Jonckheere, K. Nackaerts, and B. Muys. 2004. Digital change detection methods in ecosystem monitoring: A review. *International Journal of Remote Sensing* 25:1565–1596.

Cortes, C., and V. Vapnik. 1995. Support vector networks. *Machine Learning* 20:273–297.

Cortijo, F. J., and N. Pérez de la Blanca. 1999. The performance of regularised discriminant analysis versus non-parametric classifiers applied to high dimensional image classification. *International Journal of Remote Sensing* 20:3345–3365.

Côté, S., and A. R. L. Tatnall. 1997. The Hopfield neural network as a tool for feature tracking and recognition from satellite sensor images. *International Journal of Remote Sensing* 18:871–885.

Courant, R., and D. Hilbert. 1953. *Methods of mathematical physics*, Volume 1. New York: Interscience Publishers.

Crammer, K., and Y. Singer. 2002. On the learnability and design of output codes for multi-class problems. *Machine Learning* 47:201–233.

Crapper, P. F. 1984. An estimate of the number of boundary cells in a mapped landscape coded to grid cells. *Photogrammetric Engineering and Remote Sensing* 50:1497–1503.

Craven, M. W. 1996. Extracting comprehensible models from trained neural networks. Ph.D. Thesis, University of Wisconsin, Madison, WI.

Crist, E. P., and R. C. Cicone. 1984a. A physically-based transformation of Thematic Mapper data—The TM Tasselled Cap. *IEEE Transactions on Geoscience and Remote Sensing* 22:256–263.

Crist, E. P., and R. C. Cicone. 1984b. Comparison of the dimensionality and features of simulated Landsat-4 MSS and TM data. *Remote Sensing of Environment* 14:235–246.

Cross, G. C., and A. K. Jain. 1983. Markov random field texture models. *IEEE Transactions on Pattern Analysis and Machine Intelligence* 5:25–39.

Csiszar, I., G. Guiman, P. Romanov, M. Leroy, and O. Hautecoeul. 2001. Using ADEOS/POLDER data to reduce angular variability of NOAA/AVHRR reflectances. *Remote Sensing of Environment* 76:399–409.

Curran, P. J. 1988. The semi-variogram in remote sensing: An introduction. *Remote Sensing of Environment* 24:493–507.

Curran, P. J., and P. M. Atkinson. 1998. Geostatistics and remote sensing. *Progress in Physical Geography* 22:61–78.

Cushnie, J. L. 1987. The interactive effect of spatial resolution and degree of internal variation within land cover types on classification accuracies. *International Journal of Remote Sensing* 8:15–29.

Daubechies, I. 1992. *Ten lectures on wavelets,* 2nd ed. Philadelphia, PA: Society for Industrial and Applied Mathematics (SIAM).

Davies, J. K. 1992. *Space Exploration.* New York: W & R Chambers.

De Bruin, S., and B. G. H. Gorte. 2000. Probabilistic image classification using geological map units applied to land-cover change detection. *International Journal of Remote Sensing* 21:2389–2402.

Decatur, S. E. 1989. Application of neural networks to terrain classification. In *Proceedings of International Joint Conference on Neural Networks.* Piscataway, NJ: Institute of Electrical and Electronics Engineers (IEEE), 1, 283–288.

De Colstoun, E. C. B., M. H. Story, C. Thompson, K. Commisso, T. G. Smith, and J. Irons. 2003. National park vegetation mapping using multitemporal Landsat 7 data and a decision tree classifier. *Remote Sensing of Environment* 85:316–327.

De Colstoun, E. C. B., and C. L. Walthall. 2006. Improving global scale land cover classifications with multi-directional POLDER data and a decision tree classifier. *Remote Sensing of Environment* 100:474–485.

De Jong, S. M., and P. A. Burrough. 1995. A fractal approach to the classification of Mediterranean vegetation types in remotely sensed images. *Photogrammetric Engineering and Remote Sensing* 61:1041–1053.

De Jong, S. M., and H. Van der Werff. 1998. The DAIS 7915 Peyne experiment: Using airborne imaging spectrometry for surveying minerals in vegetated and fragmented Mediterranean landscapes. In *Proceedings of the 1st EARSeL Workshop on Imaging Spectroscopy*, October 6–8, 1998, Remote Sensing Laboratories, University of Zurich, Switzerland, 341–347. Paris: EARSeL.

Dempster, A. P., N. M. Laird, and D. B. Bubin. 1977. Maximum likelihood from incomplete data via expectation-maximization (EM) algorithm. *Journal of the Royal Statistical Society*, Series B 39:1–38.

Dennison, P. E., and D. A. Roberts. 2003. Endmember selection for multiple spectral mixture analysis using endmember average RMSE. *Remote Sensing of Environment* 87:123–135.

Derin, H., and W. S. Cole. 1986. Segmentation of textured images using Gibbs random fields. *Computer Vision, Graphics and Image Processing* 35:72–98.

Derin, H., and H. Elliott. 1987. Modelling and segmentation of noisy and textured images using Gibbs random fields. *IEEE Transactions on Pattern Analysis and Machine Intelligence* 9:39–55.

Donoho, D. L. 1995. Denoising by soft-thresholding. *IEEE Transactions on. Information Theory* 41:613–627.

Dubes, R. C., and A. K. Jain. 1989. Random field models in image analysis. *Journal of Applied Statistics* 16:131–164.

Dubuc, B., J. F. Quiniou, C. Roques-Carmes, and S. W. Zucker. 1989. Evaluating the fractal dimension of profiles. *Physical Review* 39:1500–1512.

Duch, W., R. Adamczak, and K. Grabczewski. 2001. A new methodology of extraction, optimization and application of crisp and fuzzy logical rules. *IEEE Transactions on Neural Networks* 12:277–306.

Dymond, J. R., and J. D. Shepherd. 1999. Correction of the topographic effect in remote sensing. *IEEE Transactions on Geoscience and Remote Sensing* 37:2618–1619.

Eidenshink, J. C., and J. L. Faundeen. 1994. The 1 km AVHRR global land data set: First stages in implementation. *International Journal of Remote Sensing* 15:3443–3462.

Elachi, C. 1988. *Spaceborne Radar Remote Sensing: Applications and Techniques*. New York: IEEE Press.

Elmore, J. A., J. F. Mustard, S. J. Manning, and D. B. Lobell. 2000. Quantifying vegetation change in semiarid environments: Precision and accuracy of spectral mixture analysis and the normalized difference vegetation index. *Remote Sensing of Environment* 73:87–102.

El-Sheik, T. S., and A. G. Wacker. 1980. Effect of dimensionality and estimation on the performance of Gaussian classifiers. *Pattern Recognition* 12:115–126.

Erickson, W. K., and W. C. Lickens. 1984. An application of expert systems technology to remotely sensed image analysis. *IEEE Proceedings of Pecora IX Symposium*, October 2–4, 1984, Sioux Falls, SD, 258–276. Piscataway, NJ: Institute of Electrical and Electronics Engineers (IEEE), IEEE Catalogue Number CH2079 – 2/84/0000232.

Erwin, E., K. Obermayer, and K. Schulten. 1992a. Self-organizing maps: Stationary states, metastability and convergence rate. *Biological Cybernetics* 67:35–45.

Erwin, E., K. Obermayer, and K. Schulten. 1992b. Self-organizing maps: Ordering, convergence properties and energy functions. *Biological Cybernetics* 67:47–55.

Esposito, F., D. Malerba, and G. Semeraro. 1997. A comparative analysis of methods for pruning decision trees. *IEEE Transactions on Pattern Analysis and Machine Intelligence*
19:476–491.

Evans, D. L., T. G. Farr, J. J. Van Zyl, and H. A. Zebker. 1988. Radar polarimetry: Analysis tools
and application. *IEEE Transactions on Geoscience and Remote Sensing* 26:774–789.

Evans, F. 1998. An investigation into the use of maximum likelihood classifiers, decision
trees, neural networks and conditional probabilistic networks for mapping and pre-
dicting salinity. M.Sc. Thesis, Department of Computer Science, Curtin University of
Technology, Australia.

Feder, J. 1988. *Fractals*. New York: Plenum Press.

Fioravanti, S. 1994. Multifractals: Theory and application to image texture recognition. In
*Fractals in geoscience and remote sensing. Proceedings of a Joint JRC/EARSeL Expert
Meeting, Ispra, Italy, 14th–15th April*, ed. G. G. Wilkinson, I. Kanellopoulos, and
J. Megier, Image Understanding Research Series, 1, 152–175. Report EUR 16092,
Office for Official Publications of the European Communities, Luxembourg.

Fitzpatrick-Lins, K. 1981. Comparison of sampling procedures and data analysis for a land-use
and land-cover map. *Photogrammetric Engineering and Remote Sensing* 47(3):343–351.

Fletcher, R. 1987. *Practical methods of optimization*. 2nd ed. Chichester, U.K.: John Wiley
and Sons.

Foley, J. D., and A. Van Dam. 1982. *Fundamentals of interactive computer graphics*. Reading,
MA: Addison-Wesley.

Foody, G. M. 1992. On the compensation for chance agreement in image classification accuracy assessment. *Photogrammetric Engineering and Remote Sensing* 58:1459–1460.

Foody, G. M. 1996. Approaches for the production and evaluation of fuzzy land cover
classifications from remotely-sensed data. *International Journal of Remote Sensing*
17:1317–1340.

Foody, G. M. 1998. Sharpening fuzzy classification output to refine the representation of sub-
pixel land cover distribution. *International Journal of Remote Sensing* 19:2593–2599.

Foody, G. M. 1999a. The significance of border training patterns in classification by a feed-
forward neural network using back-propagation learning. *International Journal of
Remote Sensing* 20:3549–3562.

Foody, G. M. 1999b. Image classification with a neural network: From completely crisp
to fully-fuzzy situations. In *Advances in remote sensing and GIS analysis*, ed. P. M.
Atkinson and N. J. Tate, 17–37. Chichester, U.K.: John Wiley and Sons.

Foody, G. M. 2000a. Accuracy of thematic maps derived by remote sensing. In *Proceedings
of the Accuracy 2000 Conference*, ed. G. B. M. Heuvelink and M. J. P. M. Lemmens,
217–224. Amsterdam: Delft University Press.

Foody, G. M. 2000b. Estimation of sub-pixel land cover composition in the presence of
untrained classes. *Computers and Geosciences* 26:469–478.

Foody, G. M. 2002. Status of land cover classification accuracy assessment. *Remote Sensing
of Environment* 80:185–201.

Foody, G. M., and P. J. Curran, eds.. 1994. *Environmental remote sensing from regional to
global scales*. Chichester, U.K.: John Wiley and Sons.

Foody, G. M., R. M. Lucas, P. J. Curran, and M. Honzak. 1997. Non-linear mixture modelling without end-members using an artificial neural network. *International Journal of
Remote Sensing* 18:937–953.

Foody, G. M., and A. Mathur. 2004. Toward intelligent training of supervised image classifications: Directing training data acquisition for SVM classification. *Remote Sensing of
Environment* 93:107–117.

Foody, G. M., and A. Mathur. 2006. The use of small training sets containing mixed pixels for
accurate hard image classification: Training on mixed spectral responses for classification by a SVM. *Remote Sensing of Environment* 103:179–189.

Foody, G. M., A. Mathur, C. Sanchez-Hernandez, and D. S. Boyd. 2006. Training set size requirements for the classification of a specific class. *Remote Sensing of Environment* 104:1–14.

Foody, G. M., M. B. McCullagh, and W. B. Yates. 1995. The effect of training set size and composition on artificial neural net classification. *International Journal of Remote Sensing* 16:1707–1723.

Franklin, S. E., and D. R. Peddle. 1990. Classification of SPOT HRV imagery and texture features. *International Journal of Remote Sensing* 11:551–556.

Franklin, S. E., D. R. Peddle, and J. A. Dechka. 2002. Evidential reasoning with Landsat TM, DEM and GIS data for landcover classification in support of grizzly bear habitat mapping. *International Journal of Remote Sensing* 23:4633–4652.

Frankot, R. T., and R. Chellappa. 1987. Lognormal random-field models and their applications to radar image synthesis. *IEEE Transactions on Geoscience and Remote Sensing* 25:195–207.

French, S. 1985. Group consensus probability distributions: A critical survey. In *Bayesian statistics 2*, ed. J. M. Bernardo, M. H. DeGroot, D. V. Lindley, and A. F. M. Smith, 183–202. New York: North Holland.

Freund, Y., and R. E. Schapire. 1999. A short introduction to boosting. *Journal of Japanese Society for Artificial Intelligence* 14:771–780.

Friedl, M. A., and C. E. Brodley. 1997. Decision tree classification of land cover from remotely sensed data. *Remote Sensing of Environment* 61:399–409.

Friedl, M. A., C. E. Brodley, and A. H. Strahler. 1999. Maximizing land cover classification accuracies produced by decision trees at continental to global scales. *IEEE Transactions on Geoscience and Remote Sensing* 37:969–977.

Friedman, J. H. 1994. An overview of predictive learning and function approximation. In *From statistics to neural networks*, ed. V. Cherkassky, J. Friedman, and H. Wechsler, 1–55. Proceedings of NATO/ISI Workshop. Berlin: Springer-Verlag.

Friedman, L. 1981. Extended plausible inference. In *Proceedings of the Seventh International Joint Conference on Artificial Intelligence*, Vancouver, BC, 487–495.

Frizzelle, B. G., and A. Moody. 2001. Mapping continuous distributions of land cover— a comparison of maximum likelihood estimation and artificial neural networks. *Photogrammetric Engineering and Remote Sensing*, 67693–67705.

Frost, V. S., J. A. Stiles, K. S. Shanmugan, and J. C. Holtzman. 1982. A model for radar images and its application to adaptive digital filtering of multiplicative Noise. *IEEE Transactions on Pattern Analysis and Machine Learning* 4:157–166.

Fukuda, S., and H. Hirosawa. 1998. Suppression of speckle in synthetic aperture radar images using wavelet. *International Journal of Remote Sensing* 19:507–519.

Fukuda, S., and H. Hirosawa. 1999. Smoothing effect of wavelet-based speckle filtering: The Haar basis case. *IEEE Transactions on Geoscience and Remote Sensing* 37:1168–1172.

Fukunaga, K. 1972. *Introduction to statistical pattern recognition*. New York: Academic Press.

Fung, T. 1990. An assessment of TM imagery for land cover change detection. *IEEE Transactions on Geoscience and Remote Sensing* 28:681–684.

Fung, T., and E. Le Drew. 1987. Application of principal components analysis to change detection. *Photogrammetric Engineering and Remote Sensing* 53:1649–1658.

Furrer, R., A. Barsch, C. Olbert, and M. Schaale. 1994. Multispectral imaging of land surface. *Geojournal* 32:7–16.

Gagnon, L., and A. Jouan. 1997. Speckle filtering of SAR images—a comparative study between complex-wavelet based and standard filters. *Proceedings of the SPIE* 3169:80–91.

Gahegan, M., and G. West. 1998. The classification of complex geographic datasets: An operational comparison of artificial neural networks and decision tree classifiers. In *Proceedings of the 3rd International Conference on GeoComputation* University of Bristol, United Kingdom, 17–19 September 1998, http://www.geocomputation.org/1998/61/gc_61.htm. Accessed 19 December 2008.

Garrigues, S., D. Allard, F. Baret, and M. Wesis. 2006. Quantifying spatial heterogeneity at the landscape scale using variogram models. *Remote Sensing of Environment* 103:81–96.

Garson, G. D. 1998. *Neural networks: An introductory guide for social scientists*. London: Sage Publications.

Garvey, T. D., J. D. Lowrance, and M. A. Fischler. 1979. An inference technique for integrating knowledge from disparate sources. In *Proceedings 7th International Conference on Artificial Intelligence*, Vancouver, British Columbia, 319–325.

Gellman, D. I., S. F. Biggar, M. C. Dinguirard, P. J. Henry, M. S. Moran, K. T. Thome, and P. N. Slater. 1993. Review of SPOT-1 and -2 calibration at White Sands from launch to present. *Proceedings of SPIE* 1938:118–125.

Geman, D., and B. Gidas. 1991. *Image analysis and computer vision*, chapter 2, 9–36. New York: Academic Press.

Geman S., and D. Geman. 1984. Stochastic relaxation Gibbs distributions and the Baysian restoration of the image. *IEEE Transactions on Pattern Analysis and Machine Intelligence* 6:721–741.

Gens, R., and J. L. Van Genderen. 1996. SAR interferometry—issues, techniques and applications. *International Journal of Remote Sensing* 17:1803–1835.

Gercek, D. 2004. Improvement of image classification with the integration of topographical data. In *Proceedings of 20th ISPRS Congress,* Istanbul, Turkey, 12–23 July 2004. 35-B8, 53–58.

Ghosh, J., K. Tumer, S. Beck, and L. Deser. 1995. Integration of neural classifiers for passive sonar signals. In *DSP theory and applications*, ed. C. T. Leondes, 301–338. New York: Academic Press.

Gillespie, A. R., A. B. Kahle, and R. E. Walker. 1986. Colour enhancement of highly correlated images. I. Decorrelation and HSI contrast stretches. *Remote Sensing of Environment* 20:209–235.

Ginevan, M. E. 1979. Testing land-use map accuracy: Another look. *Photogrammetric Engineering and Remote Sensing* 45:1371–1377.

Goldberg, D. E. 1989. *Genetic algorithms in search, optimization and machine learning*. Reading, MA: Addison-Wesley.

Goldon, J., and E. H. Shortliffe. 1985. Method for managing evidential reasoning in a hierarchical hypothesis space. *Artificial Intelligence* 26:323–357.

Goldstein, R. M., H. Engelhardt, B. Kamb, and R. M. Frolich. 1993. Satellite radar interferometry for monitoring ice sheet motion: Application to an Antarctic ice stream. *Science* 262:1525–1530.

Gong, P., and P. J. Howarth. 1990. An assessment of some factors influencing multispectral land cover classification. *Photogrammetric Engineering and Remote Sensing* 36:597–603.

González-Audícana, M., X. Otazu, O. Fors, and A. Seco. 2005. Comparison between Mallat's and the "a trous" discrete wavelet transform based algorithms for the fusion of multispectral and panchromatic images. *International Journal of Remote Sensing* 26(3):595–614.

Gonzato, G. 1998. A practical implementation of the box counting algorithm. *Computers and Geosciences* 24:95–100.

Gorte, B., and A. Stein. 1998. Bayesian classification and class area estimation of satellite images using stratification. *IEEE Transactions on Geoscience and Remote Sensing* 36:803–812.

Goth, I. and Geva, A.B. 1989. Unsupervised optimal fuzzy clustering. *IEEE Proceedings* on *Pattern Analysis and Machine Intelligence*, 11, 773–781.

Gouvernet, J., J. Ayme, S. Sanchez, E. Mattei, and F. Giraud. 1980. Diagnosis assistance in medical genetics based on belief functions and a tree structured thesaurus: A conversational mode realization. In *Proceedings of MEDINFO 80*, Tokyo, Japan, 798.

Goyal, S. K., M. S. Seyfried, and P. E. O'Neill. 1999. Correction of surface roughness and topographic effects on airborne SAR in mountainous rangeland areas. *Remote Sensing of Environment* 67:124–136.

Grasso, D. N. 1993. Applications of the IHS colour transformation for 1:24,000 scale geologic mapping: A low cost SPOT alternative. *Photogrammetric Engineering and Remote Sensing* 59:73–80.

Green, A. A., M. Berman, P. Switzer, and M. D. Craig. 1988. A transformation for ordering multispectral data in terms of image quality with implications for noise removal. *IEEE Transactions on Geoscience and Remote Sensing* 26:65–74.

Grossberg, S. 1976. Adaptive pattern classification and universal recoding, II: Feedback, expection, olfaction and illusions. *Biological Cybernetics* 23:187–202.

Gunn, S. R. 1997. Support vector machines for classification and regression. In *Technical Report ISIS-1-98*, Image Speech and Intelligent Systems Research Group (ISIS). Southampton, U.K.: University of Southampton.

Gurelli, M. I., and L. Onural. 1994. On a parameter estimation method for Gibbs-Markov random fields. *IEEE Transactions on Pattern Analysis and Machine Intelligence* 4:824–830.

Gurney, C. M. 1980. Threshold selection for line detection algorithms. *IEEE Transactions on Geoscience and Remote Sensing* 18:204–211.

Gurney, C. M. 1981. The use of contextual information to improve land cover classification of digital remotely sensed data. *International Journal of Remote Sensing* 2:379–388.

Guyon, I., J. Weston, S. Barnhill, and V. Vapnik. 2002. Gene selection for cancer classification using support vector machines. *Machine Learning* 46:389–422.

Haar, A. 1910. Zur Theorie der orthogonalen Funktionensysteme, *Mathematische Annalen* 69:331–371.

Halmos, P. R. 1967. *A Hilbert space problem book*. Princeton, NJ: D. Van Nostrand Company, Inc.

Hansen, L. K., and P. Salomon. 1990. Neural network ensembles. *IEEE Transactions on Pattern Recognition and Machine Intelligence* 12:993–1001.

Hansen, M. C., R. S. Defries, J. R. G. Townshend, M. Carroll, C. Dimicelli, and R. A. Sohlberg. 2003. Global percent tree cover at a spatial resolution of 500 metres: First results of the MODIS vegetation continuous fields algorithm. *Earth Interactions* 7:1–15.

Hansen, M. C., R. S. Defries, J. R. G. Townshend, and R. Sohlberg. 2000. Global land cover classification at 1km resolution using a classification tree approach. *International Journal of Remote Sensing* 21:1331–1364.

Hansen, M. C., R. Dubayah, and R. S. Defries. 1996. Classification trees: An alternative to traditional land cover classifiers. *International Journal of Remote Sensing* 17:1075–1081.

Haralick, R. M., and K. S. Fu. 1983. Pattern recognition and classification, In *Manual of Remote Sensing, Vol. I*, Second edition, ed. R. N. Colwell, D. S. Simonett and F. T. Ulaby, 881–884. Falls Church, VA: American Society of Photogrammetry.

Haralick, R. M., M. Shanmugam, and I. Dinstein. 1973. Texture feature for image classification. *IEEE Transactions on Systems, Man and Cybernetics* 3:610–621.

Harrell, F. E. 2001. *Regression modeling strategies: With applications to linear models, logistic regression, and survival analysis*. New York: Springer.

Harsanyi, J. C., and C. I. Chang. 1994. Hyperspectral image classification and dimensionality reduction: An orthogonal subspace projection approach. *IEEE Transactions on Geoscience and Remote Sensing* 32:779–785.

Haslett, J. 1985. Maximum likelihood discriminant analysis on the plane using a Markovian model of spatial context. *Pattern Recognition* 18:287–296.

Hassibi, B., and D. G. Stork. 1993. Second order derivatives for network pruning. In *Advances in neural information processing 5*, ed. S. J. Hansen, J. D. Cowan, and C. L. Giles, 164–171. San Mateo, CA: Morgan Kaufmann.

Hay, A. M. 1979. Sampling designs to test land-use map accuracy. *Photogrammetric Engineering and Remote Sensing* 45:529–533.

Hayes, D. J., and S. A. Sader. 2001. Comparison of change-detection techniques for monitoring tropical vegetation clearing and vegetation regrowth in a time series. *Photogrammetric Engineering and Remote Sensing* 67:1067–1075.

Haykin, S. 1999. *Neural networks: A comprehensive foundation.* Englewood Cliffs, NJ: Prentice-Hall.

Hebb, D. O. 1949. *The organisation of behaviour.* New York: John Wiley and Sons.

Hein, A. 2003. *Processing of SAR data: Fundamentals, signal processing, interferometry.* Berlin: Springer-Verlag.

Hellman, M. 2002. SAR polarimetry tutorial. http://epsilon.nought.de/. (Accessed 15 May 2008.)

Hentchel, H., and I. Procaccia. 1983. The infinite number of generalized dimensions of fractals and strange attractors. *Physica* 8D:435–444.

Hewitson, B. C., and R. G. Crane (eds.) 1994. *Neural nets: Applications in geography.* London: Kluwer Academic Publishers.

Hisdal, E. 1994. Interpretative versus prescriptive fuzzy set theory. *IEEE Transactions on Fuzzy Systems* 2:22–26.

Holben, B. N., and D. Kimes. 1986. Directional reflectance response in AVHRR red and near-infrared bands for three cover types and varying atmospheric conditions. *Remote Sensing of Environment* 19:213–226.

Holland, J. H. 1975. *Adaptation in natural and artificial systems.* Ann Arbor: University of Michigan Press.

Hong, D. H., and C. Hwang. 2003. Support vector fuzzy regression machines. *Fuzzy Sets and Systems* 138:271–281.

Hopfield, J. J. 1982. Neural networks and physical systems with emergent collective computational abilities. *Proceedings of the National Academy of Sciences of the USA* 79:2554–2558.

Hopfield, J. J. 1984. Neurones with graded response have collective computational properties like those of two-state neurones. *Proceedings of the National Academy of Sciences of the USA* 81:3088–3092.

Hopfield, J. J., and D. W. Tank. 1985. Neural computation of decisions in optimisation problems. *Biological Cybernetics* 52:141–152.

Hord, R. M., and W. Brooner. 1976. Land use map accuracy criteria. *Photogrammetric Engineering and Remote Sensing* 42:671–677.

Hostert, P., A. Röder, and J. Hill. 2003. Coupling spectral unmixing and trend analysis for monitoring of long-term vegetation dynamics. *Remote Sensing of Environment* 87:183–197.

Hseih, P. F., and D. Landgrebe. 1998. *Classification of high dimensional data.* Report TR-ECE 98-4, School of Electrical and Computer Engineering. W. Lafayette, IN: Purdue University.

Hsu, C. W., and C. J. Lin. 2002. A comparison of methods for multi-class support vector machines. *IEEE Transactions on Neural Networks* 13:415–425.

Hu, B., W. Lucht, A. H. Strahler, C. B. Schaaf, and M. Smith. 2000. Surface albedo and angle-corrected NDVI from AVHRR observations of South America. *Remote Sensing of Environment* 71:119–132.

Huang, C., L. S. Davis, and J. R. G. Townshend. 2002. An assessment of support vector machines for land cover classification. *International Journal of Remote Sensing* 23:725–749.

Huang, C., and J. R. G. Townshend. 2003. A stepwise regression for nonlinear approximation: Applications to estimating subpixel land cover. *International Journal of Remote Sensing* 24:75–90.

Huang, C., B. Wylie, L. Yang, C. Homer, and G. Zylstra. 2002. Derivation of a tasselled cap transformation based on Landsat 7 at-satellite reflectance. *International Journal of Remote Sensing* 23:1741–1748.

Huang, S. C., and Y. F. Huang. 1991. Bounds on the number of hidden neurons in multilayer perceptron. *IEEE Transactions on Neural Networks* 2:47–55.

Huber, P. 1981. *Robust statistics*. New York: John Wiley and Sons.

Hughes, G. F. 1968. On the mean accuracy of statistical pattern recognizers. *IEEE Transactions on Information Theory* 14:55–63.

Hur, A. B., D. Horn, H. T. Siegelmann, and V. Vapnik. 2001. Support vector clustering. *Journal of Machine Learning Research* 2:125–137.

Hutchinson, C. F. 1982. Techniques for combining Landsat and ancillary data for digital classification improvement. *Photogrammetric Engineering and Remote Sensing* 48:123–130.

Ichoku, C., A. Karnieli, Y. Arkin, J. Chorowicz, T. Fleury, and J. P. Rudant. 1998. Exploring the utility potential of SAR interferometric coherence images. *International Journal of Remote Sensing* 19:1147–1160.

Ince, F. 1987. Maximum likelihood classification, optimal or problematic: A comparison with nearest neighbour classification. *International Journal of Remote Sensing* 8:1829–1838.

Ishibuchi, H., and T. Nakashima. 2001. Effect of rule weights in fuzzy rule-based classification systems. *IEEE Transactions on Fuzzy Systems* 9:506–515.

Ishibuchi, H., T. Nakashima, and T. Morisawa. 1999. Voting in fuzzy rule-based systems for pattern classification problems. *Fuzzy Sets and Systems* 103:223–238.

Ishibuchi, H., K. Nozaki, and H. Tanaka. 1992. Distributed representation of fuzzy rules and its application to pattern classification. *Fuzzy Sets and Systems* 52:21–32.

Ishibuchi, H., and T. Yamamoto. 2002. Comparison of heuristic rule weight specification methods. *Proceedings of the 2002 IEEE International Conference on Fuzzy Systems*, Honolulu, USA, 12–17 May 2002, 908–913.

Ishibuchi, H., and T. Yamamoto. 2004. Fuzzy rule selection by multi-objective genetic local search algorithms and rule evaluation measures in data mining. *Fuzzy Sets and Systems* 141:59–88.

Ising, E. 1925. Beitrag zur Theorie des Ferromagnetismus. *Zeitschrift Physik* 31:253–258.

Jaakkola, T. S., and D. Haussler. 1999. Probabilistic kernel regression models. In *Proceedings of the 1999 Conference M on AI and Statistics*.

Jackson, Q., and D. Landgrebe. 2001. An adaptive classifier design for high dimensional data analysis with a limited training data set. *IEEE Transactions on Geoscience and Remote Sensing* 39:2664–2679.

Jackson, R. D. 1983. Spectral indices in *n*-space. *Remote Sensing of Environment* 13:409–421.

Jacobsen, A., K. B. Heidebreacht, and A. A. Neilsen. 1998. Monitoring grasslands using convex geometry and partial unmixing—a case study. *1st EARSeL Workshop on Imaging Spectroscopy*, 6–8 October 1998, Remote Sensing Laboratories, University of Zurich, Switzerland, 309–316. Paris: EARSeL.

Jansen, L. L. F., M. N. Jarsma, and E. T. M. Linden. 1990. Integrating topographic data with remote sensing for land-cover classification. *Photogrammetric Engineering and Remote Sensing* 56:1503–1506.

Jansen, L. L. F., and F. J. M. Van de Wel. 1994. Accuracy assessment of satellite derived land cover data: A review. *Photogrammetric Engineering and Remote Sensing* 48:595–604.

Jaynes, E. 1982. On the rationale of maximum-entropy methods. *Proceedings of IEEE* 70:939–952.

Jensen, J. R. 1983. Urban change detection mapping using Landsat digital data. *The American Cartographer* 81:127–147.

Jensen, J. R. 1986. *Introductory digital image processing: A remote sensing perspective.* Englewood Cliffs, NJ: Prentice-Hall.

Jhung, Y., and P. H. Swain. 1996. Bayesian contextual classification based on modified M-estimates and Markov random fields. *IEEE Transactions on Geoscience and Remote Sensing* 34:67–75.

Ji, C., R. R. Snapp, and D. Psaltis. 1990. Generalising smoothness constraints from discrete samples. *Neural Computation* 2:188–197.

Jin, Y. Q., and C. Liu. 1997. Biomass retrieval from high-dimensional active/passive remote sensing data by using an artificial neural network. *International Journal of Remote Sensing* 18:971–979.

Joachims, T. 1999. Transductive inference for text categorization using support vector machines. In *Proceedings of ICML-99, 16th International Conference on Machine Learning*, San Francisco, CA, Morgan Kaufmann, 200–209.

Johnson, R. D., and E. S. Kasischke. 1998. Change vector analysis: A technique for the multi-spectral monitoring of land cover and condition. *International Journal of Remote Sensing* 19:411–426.

Jones, K. 1998. A comparison of algorithms used to compute hill slope as a property of the DEM. *Computers and Geosciences* 24:315–323.

Ju, J., E. D. Kolaczyk, and S. Gopal. 2003. Gaussian mixture discriminant analysis and sub-pixel land cover characterization in remote sensing. *Remote Sensing of Environment* 84:550–560.

Kalluri, S. N. V., C. Huang, S. Mathieu, J. Townshend, K. Yang, and R. Chellapa. 1997. A comparison of mixture modelling algorithms and their applicability to MODIS data. *Proceedings of the IEEE International Geoscience and Remote Sensing Symposium (IGARSS'97)*, Singapore, August 3–8, 1997. Piscataway, NJ: IEEE, 1, 171–173.

Kanellopoulos, I., A. Varfis, G. G. Wilkinson, and J. Mégier. 1992. Land-cover discrimination in SPOT HRV imagery using an artificial neural network: A 20-class experiment. *International Journal of Remote Sensing* 13:917–924.

Kanellopoulos, I., and G. G. Wilkinson. 1997. Strategies and best practice for neural network image classification. *International Journal of Remote Sensing* 18:711–725.

Kanellopoulos, I., G. G. Wilkinson, F. Roli, and J. Austin (eds.). 1997. *Neurocomputation in remote sensing data analysis.* Berlin: Springer-Verlag.

Kangas, J. A., T. K. Kohonen, and J. T. Laaksonen. 1990. Variants of self-organising maps. *IEEE Transactions on Neural Networks* 1:93–99.

Karen, C., K. C. Seto, and W. Liu. 2003. Comparing ARTMAP neural network with maximum likelihood for detecting urban change: The effect of class resolution. *Photogrammetric Engineering and Remote Sensing* 69:981–990.

Karnin, E. D. (1990) A simple procedure for pruning back-propagation trained neural networks. *IEEE Transactions on Neural Networks* 1:239–242.

Karpouzli, E., and T. Malthus. 2003. The empirical line method for the atmospheric correction of IKONOS imagery. *International Journal of Remote Sensing* 24:1143–1150.

Kashyap, R. L., and R. Chellappa. 1983. Estimation and choice of neighbors in spatial-interaction models of images. *IEEE Transactions on Information Theory* 29:60–72.

Kass, G. V. 1980. An exploratory technique for investigating large quantities of categorical data. *Applied Statistics* 29:119–127.

Kauth, R. J., and G. Thomas. 1976. The tasselled cap—a graphic description of the spectral-temporal development of agricultural crops as seen by Landsat. In *Proceedings of the Symposium on Machine Processing of Remotely-Sensed Data 1976*, 4 B-41–4 B-51. West Lafayette, IN: Purdue University.

Kavzoglu, T. 2001. An investigation of the design and use of feed-forward artificial neural networks in the classification of remotely sensed images. Ph.D. thesis, School of Geography, The University of Nottingham.

Kavzoglu, T., and P. M. Mather. 1999. Pruning artificial neural networks: An example using land cover classification of multi-sensor images. *International Journal of Remote Sensing* 20:2787–2803.

Kavzoglu, T., and P. M. Mather. 2002. The role of feature selection in artificial neural network applications. *International Journal of Remote Sensing* 23:2919–2937.

Keller, J. M., and S. Chen. 1989. Texture description and segmentation through fractal geometry. *Computer Vision, Graphics and Image Processing* 45:150–166.

Keuchel, J., S. Naumann, M. Heiler, and A. Siegmun. 2003. Automatic land cover analysis for Tenerife by supervised classification using remotely sensed data. *Remote Sensing of Environment* 86:530–541.

Kim, B., and D. A. Landgrebe. 1991. Hierarchical classifier design in high-dimensional, numerous class cases. *IEEE Transactions on Geoscience and Remote Sensing* 29:518–528.

Kim, D. W., K. H. Lee, and D. Lee. 2004. On cluster validity index for estimation of the optimal number of fuzzy clusters. *Pattern Recognition* 37:2009–2025.

Kim, H., and P. H. Swain. 1995. Evidential reasoning approach to multisource-data classification in remote sensing. *IEEE Transactions on Systems, Man and Cybernetics* 25:1257–1265.

Kim, Y., and J. Van Zyl. 2000. Overview of polarimetric interferometry. In *Proceedings of IEEE Aerospace Conference*, 3, 231 –236, Big Sky, MT, March 18–25.

Kindermann, R., and J. L. Snell. 1980. *Markov random fields and their applications*. Providence, RI: American Mathematical Society.

Knerr, S., L. Personnaz, and G. Dreyfus. 1990. Single-layer learning revisited: A stepwise procedure for building and training neural network. In *Neurocomputing: Algorithms, architectures and applications*, ed. J. Fogelman, NATO ASI, 41–50. Berlin: Springer-Verlag,.

Kneubuehler, M., M. Shaepman, D. Schlaepfer, and C. Itten. 1998. Endmember selection procedures for partial spectral unmixing of DAIS 7915 imaging spectrometer data in highly vegetated areas. In *1st EARSeL Workshop on Imaging Spectroscopy*, 6–8 October 1998, Remote Sensing Laboratories, University of Zurich, Switzerland, 255–261. Paris: EARSeL.

Koch, C. 1988. Computing motion in the presence of discontinuities: Algorithm and analog networks. In *Neural computers*, ed. R. Eckmiller and C. D. Malsburg, Volume F41, NATO ASI Series, 101–110. Berlin: Springer-Verlag.

Kohavi, R., and G. H. John. 1997. Wrappers for feature subset selection. *Artificial Intelligence Journal* 97:273–324.

Kohonen, T. 1982. Self-organized formation of topologically correct feature maps. *Biological Cybernetics* 43:59–69.

Kohonen, T. 1988a. An introduction to neural computing. *Neural Networks* 1:3–16.

Kohonen, T. 1988b. The "neural" phonetic typewriter. *Computer* 21:11–22.

Kohonen, T. 1989. *Self-organization and associative memory*. New York: John Wiley and Sons.

Kolmogorov, A. N., and S. V. Fomin. 1970. *Introductory real analysis*. Englewood Cliffs, NJ: Prentice-Hall, Inc.

Kosko, B. 1992. *Neural networks and fuzzy systems*. Englewood Cliffs, NJ: Prentice Hall.

Kowalik, W. S., and S. E. Marsh. 1982. A relation between Landsat Digital Numbers, surface reflectance and the cosine of the solar zenith angle. *International Journal of Remote Sensing* 12:39–55.

Kraaijveld, M. A., J. Mao, and A. K. Jain. 1995. A nonlinear projection method based on Kohonen's topology preserving maps. *IEEE Transactions on Neural Networks* 6:548–558.

Kriegel, H. P., R. Kroger, A. Pryakhin, and M. Schubert. 2004. Using support vector machines for classifying large sets of multi-represented objects. In *Proceedings of the International Conference on Data Mining (SDM'04)*, 102–114.

Krishnapuram, R., and J. M. Keller. 1993. A possibilitic approach to clustering. *IEEE Transactions on Fuzzy Systems* 1:98–110.

Krishnapuram, R., and J. M. Keller. 1996. The possibilitic c-means algorithm: Insights and recommendations. *IEEE Transactions on Fuzzy Systems* 4:385–393.

Kropatsch, W. G., and D. Strobl. 1990. The generation of SAR layover and shadow maps from digital elevation models. *IEEE Transactions on Geoscience and Remote Sensing* 28:98–107.

Kuan, D. T., A. A. Sawchuk, T. C. Strand, and P. Chavel. 1987. Adaptive restoration of images with speckle. *IEEE Transactions on Acoustics, Speech and Signal Processing* 35:373–383.

Kuemmerle, T., A. Röder, and J. Hill. 2006. Separating grassland and shrub vegetation by multidate pixel-adaptive spectral mixture analysis. *International Journal of Remote Sensing* 27:3251–3271.

Kulkarni, A. V., and N. K. Laveen. 1976. An optimisation approach to hierarchical classifier design. In *Proceedings of the Third International Joint Conference on Pattern Recognition*, IEEE Catalogue No. 76CH1140-3C, 459–466. Piscataway, NJ: IEEE.

Kurzynski, M. W. 1983. The optimal strategy of a tree classifier. *Pattern Recognition* 16:81–87.

Kwarteng, A. Y., and P. S. Chavez. 1998. Change detection study of Kuwait City and environs using multitemporal Landsat Thematic Mapper data. *International Journal of Remote Sensing* 19:1651–1662.

Kwok, R. J., J. C. Curlander, and S. Pang. 1987. Rectification of terrain induced distortions in radar imagery. *Photogrammetric Engineering and Remote Sensing* 53:507–513.

Lam, H. K., and F. H. F. Leung. 2007. Fuzzy controller with stability and performance rules for nonlinear systems. *Fuzzy Sets and Systems* 158:147–163.

Lambin, E. F., and A. H. Strahler. 1994. Change vector analysis in multitemporal space: A tool to categorise land-cover change processes using high-temporal-resolution satellite data. *Remote Sensing of Environment* 48:231–244.

Lanari, R., G. Fornaro, D. Riccio, M. Migliaccio, K. P. Papathanassiou, J. R. Moreira, M. Schwäbisch, L. Dutra, G. Puglisi, G. Franceschetti, and M. Coltelli. 1996. Generation of digital elevation models by using SIR-C/X-SAR multifrequency two-pass interferometry: The Etna case study. *IEEE Transactions on Geoscience and Remote Sensing* 34(5):1097–1114.

Landgrebe, D. 1998. Information extraction principles and methods for multispectral and hyperspectral image data. In *Information processing for remote sensing*, ed. C. H. Chen. River Edge, NJ: World Scientific Publishing.

Law, K. H., and J. Nichol. 2004. Topographic correction for differential illumination effects on IKONOS satellite imagery. In *Proceedings of 20th ISPRS Congress*, Istanbul, Turkey, 12–23 July 2004. 35-B3, 641–646.

Le Cun, Y., J. S. Decker, and S. A. Solla. 1999. Optimal brain damage. In *Advances in neural information processing 5*, ed. D. S. Touresky, 598–605. San Mateo, CA: Morgan Kaufmann.

Le Hegarat-Mascle, S., R. Seltz, L. Hubert-Moy, S. Corgne, and N. Stach. 2006. Performance of change detection using remotely sensed data and evidential fusion: Comparison of three cases of application. *International Journal of Remote Sensing* 27(16):3515–3532.

Lee, B. W., and B. J. Sheu. 1991. Modified Hopfield neural networks for retrieving the optimal solution. *IEEE Transactions on Neural Networks* 2:137–142.

Lee, J., R. C. Weger, S. K. Sengupta, and R. M. Welch. 1990. A neural network approach to cloud classification. *IEEE Transactions on Geoscience and Remote Sensing* 28:846–855.

Lee, J. B., A. S. Woodyatt, and M. Berman. 1990. Enhancement of high spectral resolution remote sensing data by noise-adjusted principal components transform. *IEEE Transactions on Geoscience and Remote Sensing* 28:295–304.

Lee, J. S. 1980. Digital image enhancement and noise filtering by use of local statistics. *IEEE Transactions on Pattern Analysis and Machine Intelligence* 2:165–168.

Lee, J. S. 1981. Speckle suppression and analysis for synthetic aperture radar images. *Optical Engineering* 15:380–389.

Lee, J. S. 1986. Speckle suppression and analysis for synthetic aperture radar. *Optical Engineering* 25:636–643.

Lee, S., and R. G. Lathrop. 2005. Sub-pixel estimation of urban land cover components with linear mixture model analysis and Landsat Thematic Mapper imagery. *International Journal of Remote Sensing* 26:4885–4905.

Lee, S., and R. G. Lathrop. 2006. Subpixel analysis of Landsat ETM⁺ using self-organizing map (SOM) neural network for urban land cover characterisation. *IEEE Transactions on Geoscience and Remote Sensing* 44(6):1642–1654.

Lee, T., and J. A. Richards. 1985. A low-cost classifier for multitemporal applications. *International Journal of Remote Sensing* 6:1405–1417.

Lee, T., J. A. Richards, and P. H. Swain. 1987. Probabilistic and evidential approaches for multisource data analysis. *IEEE Transactions on Geoscience and Remote Sensing* 25:283–293.

Lewis, H. G., S. Côté, and A. R. L. Tatnall. 1997. Determination of spatial and temporal characteristics as an aid to neural network cloud classification. *International Journal of Remote Sensing* 18:899–915.

Li, L., S. L. Ustin, and M. Lay. 2005. Application of multiple endmember spectral mixture analysis (MESMA) to AVIRIS imagery for coastal salt marsh mapping: A case study in China Camp, CA, USA. *International Journal of Remote Sensing* 26:5193–5207.

Li, S. Z. 1990. Invariant surface segmentation through energy minimization with discontinuities. *International Journal of Computer Vision* 5:161–194.

Li, S. Z. 1995a. Markov random field modelling in computer vision. In *Computer science workbench*, ed. T. L. Kunii. Berlin: Springer-Verlag.

Li, S. Z. 1995b. On discontinuity-adaptive smoothness priors in computer vision. *IEEE Transactions on Pattern Analysis and Machine Intelligence* 17:576–586.

Lillesand, T. M., and R. W. Keifer. 2000. *Remote sensing and image interpretation*. New York: John Wiley and Sons.

Lim, C., and M. Kafatos. 2002. Frequency analysis of natural vegetation distribution using NDVI/AVHRR data from 1981 to 2000 for North America: Correlations with SOI. *International Journal of Remote Sensing* 23:3347–3383.

Liu, D., M. Kelly, and P. Gong. 2006. A spatial-temporal approach to monitoring forest disease spread using multi-temporal high spatial resolution imagery. *Remote Sensing of Environment* 101:167–180.

Liu, W., K. C. Seto, E. Y. Wu, S. Gopal, and C. E. Woodcock. 2004. ART-MAP: A neural network approach to subpixel classification. *IEEE Transactions on Geoscience and Remote Sensing* 42:1976–1983.

Liu, W., and E. Y. Wu. 2005. Comparison of non-linear mixture models: Sub-pixel classification. *Remote Sensing of Environment* 94(2):145–154.

Liu, Y., S. Nishiyama, and Y. Tomohisa. 2004. Analysis of four change detection algorithms in bi-temporal space with a case study. *International of Remote Sensing* 25:2121–2139.

Liu, Z., A. Liu, C. Wang, and S. Qi. 2004. Statistical ratio rank-ordered differences filter for radar speckle removal. *Proceedings of the SPIE* 5238:524–531.

Lloyd, C. D., S. Berberoglu, P. J. Curran, and P. M. Atkinson. 2004. A comparison of texture measures for the per-field classification of Mediterranean land cover. *International Journal of Remote Sensing* 25:3943–3965.

Loh, W. Y., and Y. S. Shih. 1997. Split selection methods for classification trees. *Statistica Sinica* 7:815–840.

Lopes, A., E. Nezry, and R. Touzi. 1990. Maximum a posteriori speckle filtering and first order texture models in SAR images. In *Proceedings of the International Geoscience and Remote Sensing Symposium (IGARSS'90)*, University of Maryland, College Park, MD, 20–24 May 1990, 3, 2409–2412. Piscataway, NJ: IEEE.

Lopez-Martinez, C., and X. Fabregas. 2003. Polarimetric SAR speckle model. *IEEE Transactions on Geosciences and Remote Sensing* 41:2232–2242.

Lu, C. S., P. C. Chung, and C. F. Chen. 1997. Unsupervised texture segmentation via wavelet transform. *Pattern Recognition* 30:729–742.

Lu, D., P. Mausel, M. Batistella, and E. Moran. 2005. Land-cover binary change detection methods for use in the moist tropical region of the Amazon: A comparative study. *International Journal of Remote Sensing* 26(1):101–114.

Lu, D., P. Mausel, E. Brondizio, and E. Moran. 2004. Change detection techniques. *International Journal of Remote Sensing* 25:2365–2407.

Luntz, A., and V. Brailovsky. 1969. On estimation of characters obtained in statistical procedure of recognition. *Technicheskaya Kibernetica*, 3 (in Russian).

Lyon, J., D. Yuan, R. Lunetta, and C. Elvidge. 1998. A change detection experiment using vegetation indices. *Photogrammetric Engineering and Remote Sensing* 64:143–150.

Mackay, G., M. D. Steven, and J. A. Clarke. 1994. A pseudo-5S code for atmospheric correction of ATSR-2 visible and near-IR land-surface data. In *Proceeding of the 6th International Symposium—Physical Measurements and Signatures in Remote Sensing*, Val d'Isère, France, 17–21st January, 103–110.

Macleod, R. B., and R. G. Congalton. 1998. A quantitative comparison of change-detection algorithms for monitoring eelgrass from remotely sensed data. *Photogrammetric Engineering and Remote Sensing* 64:207–216.

Magnussen, S., P. Boudewyn, and M. Wulder. 2004. Contextual classification of Landsat TM images to forest inventory cover types. *International Journal of Remote Sensing* 25:2421–2440.

Mali, K., and S. Mitra. 2005. Symbolic classification, clustering and fuzzy radial basis function network. *Fuzzy Sets and Systems* 152:553–564.

Malila, W. 1980. Change vector analysis an approach for detection forest changes with Landsat. In *Proceedings of the 6th Annual Symposium on Machine Processing of Remotely Sensed Data*, 3–6 June 1980, 326–335. West Lafayette, IN: Purdue University Press.

Mallat, S. G. 1989. A theory for multi-resolution signal decomposition: The wavelet representation. *IEEE Transactions on Pattern Analysis and Machine Intelligence* 11, 674–693.

Mallat, S., and S. Zhong. 1992. Characterization of signals from multiscale edges. *IEEE Transactions on Pattern Analysis and Machine Intelligence* 8:679–698.

Mandelbrot, B. B. 1977. *Fractals: Form, chance and dimension*. San Francisco. CA: Freeman.

Mandelbrot, B. B. 1982. *The fractal geometry of nature*. San Francisco, CA: Freeman.

Mannan, B., J. Rot, and A. K. Ray. 1998. Fuzzy ARTMAP supervised classification of multi-spectral remotely-sensed images. *International Journal of Remote Sensing* 19:767–774.

Marceau, D. J., and G. J. Hay. 1999. Remote sensing contributions to the scale issue. *Canadian Journal of Remote Sensing* 25:357–366.

Marceau, D. J., P. J. Howarth, and D.J. Gratton. 1994. Remote sensing and the measurement of geographical entities in a forested environment. Part I: The scale and spatial aggregation problem. *Remote Sensing of Environment* 49:93–104.

Marroquin, J. L., J. S. Mitter, and T. Poggio. 1987. Probabilistic solution of ill-posed problems in computational vision. *Journal of the American Statistical Association* 82:76–89.

Marsh, S. E., P. Switzer, W. S. Kowalik, and R. J. Lyon. 1980. Resolving the percentage of component terrains within single resolution elements. *Photogrammetric Engineering and Remote Sensing* 46:1079–1086.

Martin, R. S., and J. H. Wilkinson. 1971. Reduction of the symmetric eigenproblem $Ax = \lambda Bx$ and related problems to standard form. In *Handbook for automatic computation, volume II: Linear algebra*, ed. J. H. Wilkinson and C. Reinch, 303–314. Berlin: Springer-Verlag,.

Maselli, F., C. Conese, T. De Filippis, and S. Norcini. 1995. Estimation of forest parameters through fuzzy classification of TM data. *IEEE Transactions on Geoscience and Remote Sensing* 33:77–84.

Mason, D. C., D. G. Corr, A. Cross, D. C. Hogg, D. H. Lawrence, M. Petrou, and A. M. Taylor. 1988. The use of digital map data in the segmentation and classification of remotely-sensed images. *International Journal of Geographical Information Systems* 2:195–215.

Massonnet, D., and T. Rabaute. 1993. Radar interferometry: Limits and potential. *IEEE Transactions on Geoscience and Remote Sensing* 31:45–464.

Massonnet, D., M. Rossi, C. Carmona, F. Adragna, G. Peltzer, K. Feigi, and T. Rabaute. 1993. The displacement field of the Landers earthquake mapped by radar interferometry. *Nature* 364:138–142.

Mather, P. M. 1976. *Computational Methods of Multivariate Analysis in Physical Geography.* Chichester: John Wiley and Sons.

Mather, P. M. 1985. A computationally-efficient maximum-likelihood classifier employing prior probabilities for remotely-sensed data. *International Journal of Remote Sensing* 6:369–376.

Mather, P. M. 1987. Pre-processing of training data for multispectral image classification. In *13th Annual Conference of the Remote Sensing Society*, Nottingham, England, 111–120.

Mather, P. M. 1992. Remote sensing and the detection of change. In *Land use change: The cause and consequences. ITE Symposium Proceedings No. 27,* ed. M. C. Whitby, 60–68.

Mather, P. M. 1999. Land cover classification revisited. In *Advances in remote sensing and GIS analysis*, ed. P. M. Atkinson and N. J. Tate, 7–16. Chichester: John Wiley and Sons.

Mather, P. M. 2004. *Computer processing of remotely-sensed images: An introduction*, 3rd edition. Chichester: John Wiley and Sons.

Mather, P. M., B. Tso, and M. Koch. 1998. An evaluation of combining spectral and textural information for lithological classification. *International Journal of Remote Sensing* 19:587–604.

Mathieu-Marni, S., P. Leymarie, and M. Berthod. 1996. Removing ambiguities in a multi-spectral image classification. *International Journal of Remote Sensing* 17:1493–1504.

Mausel, P. W., W. J. Kramber, and J. K. Lee. 1990. Optimum band selection for supervised classification of multispectral data. *Photogrammetric Engineering and Remote Sensing* 56:55–60.

McConway, K. J. 1980. The combination of experts' opinions in probability assessment: Some theoretical considerations, Ph.D. thesis, University College London.

McCulloch, W. S., and W. Pitts. 1943. A logical calculus of the ideas imminent in nervous activity. *Bulletin of Mathematical Biophysics* 5:115–133.

Mégier, J., W. Mehl, and R. Ruppelt. 1984. Per-field classification and application to SPOT simulated SAR and combined SAR-MSS data. In *18th International Symposium on Remote Sensing of Environment*. Ann Arbor, MI: Environmental Research Institute of Michigan, 1011–1017.

Mehrotra, K. G., C. K. Mohan, and S. Ranka. 1991. Bounds on the number of samples needed for neural learning. *IEEE Transactions on Neural Networks* 2:548–558.

Metropolis, N., A. W. Rosenbluth, M. N. Rosenbluth, A. H. Teller, and E. Teller. 1953. Equations of state calculations by fast computational machine. *Journal of Chemical Physics* 21:1087–1091.

Metternicht, G. I., and A. Fermont. 1998. Estimating erosion surface features by linear mixture modelling. *Remote Sensing of Environment* 12:1887–1903.

Meygret, A. 2005. Absolute calibration: From SPOT1 to SPOT5. *Proceedings of SPIE* 5882:335–346.

Michalek, J. L., T. Wagner, J. J. Luczkovich, and R. W. Stoffle. 1993. Multispectral change vector analysis for monitoring coastal marine environments. *Photogrammetric Engineering and Remote Sensing* 59:381–384.

Michie, D., D. J. Spiegelhalter, and C. C. Taylor. 1994. *Machine learning, neural and statistical classification*. Chichester, U.K.: Ellis Horwood.

Miller, A. J. 1990. *Subset selection in regression*. New York: Chapman & Hall.

Miller, J. W. V., J. B. Farison, and Y. Shin. 1992. Spatially invariant image sequences. *IEEE Transactions on Image Processing* 1:148–161.

Milne, A. K. 1988. Change detection analysis using Landsat imagery: A review of methodology. *Proceedings of the IGARSS'88 Symposium*, Edinburgh, Scotland, ESA SP-284, 541–544.

Minnaert, M. 1941. The reciprocity principle in Lunar photometry. *Astrophysics Journal* 93:403–410.

Moran, M. S., R. Bryanta, K. Thomeb, W. Nia, and Y. Nouvellona. 2001. A refined empirical line approach for reflectance factor retrieval from Landsat-5 TM and Landsat-7 ETM+. *Remote Sensing of Environment* 78:71–82.

Moussouris, J. 1974. Gibbs and Markov systems with constraints. *Journal of Statistical Physics* 10:11–33.

Muchoney, D., J. Borak, H. Chi, M. Friedl, S. Gopal, J. Hodges, N. Morrow, and A. H. Strahler. 2000. Application of MODIS global supervised classification model to vegetation and land cover mapping in Central America. *International Journal of Remote Sensing* 21:1115–1138.

Muchoney, D. M., and B. N. Haack. 1994. Change detection for monitoring forest defoliation. *Photogrammetric Engineering and Remote Sensing* 60:1243–1251.

Murphy, S. K., F. Kasif, and S. Salzberg. 1994. A system for induction of oblique decision trees. *Journal of Artificial Intelligence Research* 2:1–32.

Mustard, J. F. 1993. Relationships of soil, grass and bedrock over the Kameah Serpentine Melange through spectral mixture analysis of AVIRIS data. *Remote Sensing of Environment* 44:293–308.

Nackaerts, K., K. Caesen, B. Muys, and P. Coppin. 2005. Comparative performance of a modified change vector analysis in forest change detection. *International Journal of Remote Sensing* 26:839–852.

Nelson, R. F. 1983. Detecting forest canopy change due to insect activity using Landsat MSS. *Photogrammetric Engineering and Remote Sensing* 49:1303–1314.

Nemmour, H., and Y. Chibani. 2006. Fuzzy neural network architecture for change detection in remotely sensed imagery. *International Journal of Remote Sensing* 27:705–717.

Nezry, E., A. Lopes, and R. Touzi. 1991. Detection of structure and textural features for SAR images filtering. *Proceedings of the International Geoscience and Remote Sensing Symposium (IGARSS'91)*, Espoo, Finland, June 03–6, 1991, 3, 2169–2172. Piscataway, NJ: IEEE.

Nielsen, A. A. 1994. Analysis of regularly and irregularly sampled spatial, multivariate and multi-temporal data. Ph.D. thesis, Institute of Mathematical Modelling, Technical University of Denmark, Lyngsby, Denmark.

Nielsen, A. A. 1998. Linear mixture modelling and partial unmixing in multi- and hyper-spectral data. In *EARSeL Workshop on Imaging Spectroscopy, University of Zurich, 6–8 October 1988*, 165–172. Paris: EARSeL.

Neilsen, A. A., K. Conradsen, and J. J. Simpson. 1998. Multivariate Alteration Detection (MAD) and MAF post-processing in multispectral, bi-temporal image data: New approaches to change detection studies. *Remote Sensing of Environment* 64:1–19.

Nikhil, R.P., and J. C. Bezdek. 1995. On cluster validity for the fuzzy c-means model. *IEEE Transactions on Fuzzy Systems* 3(3):370–379.

Nishii, R., and S. Eguchi. 2005. Supervised image classification by contextual AdaBoost based on posteriors in neighbourhoods. *IEEE Transactions on Geoscience and Remote Sensing* 43:2547–2554.

Núñez, J., X. Otazu, O. Fors, A. Prades, V. Palá, and R. Arbiol. 1999. Multiresolution-based image fusion with additive wavelet decomposition. *IEEE Transactions on Geoscience and Remote Sensing* 37:1204–1211.

Olsen, S. I. 1993. Estimation of noise in images: An evaluation. *Graphical Models and Image Processing* 55:319–323.

Opper, M., and O. Winter. 2000. Gaussian processes and SVM: Mean field and leave-one-out. In *Advances in Large Margin Classifiers*, ed.A. J. Smola, P. L. Bartlett, B. Schölkopf, and D. Schuurmans, 311–326. Cambridge, MA: MIT Press.

Ouaidrari, H., and E. F. Vermote. 2001. Operational atmospheric correction of Landsat TM data. *Remote Sensing of Environment* 70:4–15.

Ouma, Y. O., T. G. Ngigi, and R. Tateishi. 2006. On the optimization and selection of wavelet texture for feature extraction from high-resolution satellite imagery with application towards urban-tree delineation. *International Journal of Remote Sensing* 27:73–104.

Pal, M. 2005. Random forest classifiers for remote sensing classification. *International Journal of Remote Sensing* 26:217–222.

Pal, M. 2006. Support vector machine-based feature selection for land cover classification: A case study with DAIS hyperspectral data. *International Journal of Remote Sensing* 27:2877–2894.

Pal, M., and P. M. Mather. 2003. An assessment of the effectiveness of decision tree methods for land cover classification. *Remote Sensing of Environment* 86:554–565.

Pal, M., and P. M. Mather. 2005. Support vector machines for classification in remote sensing. *International Journal of Remote Sensing* 26:1007–1011.

Pal, M., and P. M. Mather. 2006. Some issues in the classification of DAIS hyperspectral data. *International Journal of Remote Sensing* 27:2895–2916.

Pal, N. R., and J. C. Bezdek. 1995. On cluster validity for fuzzy c-means model. *IEEE Transactions on Fuzzy Systems* 3:370–379.

Pal, N. R., and J. C. Bezdek. 1997. Correction to "On cluster validity for fuzzy c-means model." *IEEE Transactions on Fuzzy Systems* 5:152–153.

Pal, N. R., J. C. Bezdek, and C. K. Tsao. 1993. Generalised clustering networks and Kohonen's self-organising scheme. *IEEE Transactions on Neural Networks* 4:549–557.

Pal, N. R., A. Laha, and J. Das. 2005. Designing fuzzy rule based classifier using self-organizing feature map for analysis of multispectral satellite images. *International Journal of Remote Sensing* 26:2219–2240.

Pal, S. K., and S. Mitra. 1992. Multilayer perceptron, fuzzy sets and classification. *IEEE Transactions on Neural Networks* 3:683–697.

Paola, J. D., and R. A. Schowengerdt. 1995. A review and analysis of backpropagation neural networks for classification of remotely-sensed multi-spectral imagery. *International Journal of Remote Sensing* 16:3033–3058.

Papoulis, A. 1991. *Probability random variables and stochastic processes*. New York: John Wiley and Sons.

Parisi, G. 1988. *Statistical field theory*. Reading MA: Addison-Wesley.

Parisi, G., and U. Frish. 1985. A multifractal model of intermittency. In *Turbulence and predictability in geophysical fluid dynamics and climate dynamics*, ed. M. Ghil, 84–89. New York: North Holland.

Park, C. H., H. Park, and P. Pardalos. 2004. A comparative study of linear and nonlinear feature extraction methods. In *Proceedings of the Fourth IEEE International Conference on Data Mining (ICDM'04)*, 1–4 November, Brighton, U.K., 495–498.

Pecknold, S., S. Lovejoy, D. Scherter, and C. Hooge. 1997. Multifractals and resolution dependence of remotely sensed data: GSI to GIS. In *Scale in remote sensing and GIS*, ed. D. A. Quattrochi and M. F. Goodchild, 361–394. New York: Lewis Publishers.

Peddle, D. R., and A. M. Smith. 2005. Spectral mixture analysis of agricultural crops: Endmember validation biophysical estimation in potato plots. *International Journal of Remote Sensing* 26(22):4959–4979.

Pedrycz, W. 1989. *Fuzzy control and fuzzy systems*. New York: John Wiley and Sons.

Peleg, S., J. Naor, R. Hartley, and D. Avnir. 1984. Multiple resolution texture analysis and classification. *IEEE Transactions on Pattern Analysis and Machine Intelligence* 6:1984, 518–523.

Pena, J. M., J. A. Lozano, and P. Larranaga. 1999. An empirical comparison of four initialization methods for the k-means algorithm. *Pattern Recognition Letters* 20:1027–1040.

Pentland, A. P. 1984. Fractal based description of natural scenes. *IEEE Transactions on Pattern Analysis and Machine Intelligence* 6:661–674.

Pickard, D. K. 1980. Unilateral Markov fields. *Advance in Applied Probability*, 12:655–671.

Pierce, L. E., F. T. Ulaby, K. Sarabandi, and M. C. Dobson. 1994. Knowledge-based classification of polarimetric SAR images. *IEEE Transactions on Geoscience and Remote Sensing* 33(5):1081–1086.

Pizurica, A., W. Philips, I. Lemahieu, and M. Acheroy. 2001. Despeckling SAR images using wavelets and a new class of adaptive shrinkage estimators. In *Proceedings of IEEE International Conference on Image Processing*, Thessaloniki, Greece, 7–10 October 2001, 233–236.

Platt, J. C., N. Cristianini, and Shawe-Taylor. 2000. Large margin DAGs for multiclass classification. In *Advances in Neural Information Processing Systems*, 547–553. Cambridge, MA: MIT Press.

Prasad, N., S. Saran, S. P. S. Kushwaha, and P. S. Roy. 2001. Evaluation of various image fusion techniques and imaging scales for forest features interpretation. *Research Communication* 81:1218–1224.

Preiss, M., D. Gray, and N. Stacy. 2003. A change detection statistic for repeat pass interferometric SAR. In *Proceedings of IEEE International Conference on Acoustics, Speech and Signal Processing (ICASSP 2003)*, April 6 – 10 2003, Hong Kong. 5, 241–244.

Puissant, A. J. Hirsch, and C. Weber. 2005. The utility of texture analysis to improve per-pixel classification for high to very high spatial resolution imagery. *International Journal of Remote Sensing* 26:733–745.

Puyou-Lacassies, P. H., G. Flouzat, M. Gay, and C. Vignolles. 1994. Validation of the use of multiple linear regression as a tool for unmixing coarse spectral resolution images. *Remote Sensing of Environment* 49:155–166.

Qing-Yun, N, S., and K. S. Fu. 1983. A method for design of binary tree classifiers. *Pattern Recognition* 16:593–603.

Quattrochi, D. A., and M. A. Goodchild, eds. 1997. *Scale in remote sensing and GIS*. New York: Lewis Publishers.

Quinlan, J. R. 1979. Discovering rules by induction from large collections of examples. In *Expert systems in the micro-electronic age*, ed. D. Michie, 168–201. Edinburgh, Scotland: Edinburgh University Press.

Quinlan, J. R. 1986. Induction of decision trees. *Machine Learning* 1:81–106.

Quinlan, J. R. 1993. *C4.5: Algorithm for machine learning*. San Mateo: Morgan Kaufmann.

Quinlan, J. R. 1996. Bagging, boosting and C4.5. In *Thirteenth National Conference on Artificial Intelligence*, 725–730. Portland, OR: American Association for Artificial Intelligence.

Quintano, C., A. Fernández-Mando, O. Fernández-Manso, and Y. E. Shimabukuro. 2006. Mapping burned areas in Mediterranean countries using spectral mixture analysis from a uni-temporal perspective. *International Journal of Remote Sensing* 27:645–662.

Radeloff, V., D. Mladenoff, and M. Boyce. 1999. Detecting jack pine budworm defoliation using spectral mixture analysis: Separating effects from determinants. *Remote Sensing of Environment* 69:156–169.

Ramstein, G., and M. Raffy. 1989. Analysis of the structure of radiometric remotely-sensed images. *International Journal of Remote Sensing* 10:1049–1073.

Rashed, T., J. R. Weeks, D. Stow, and D. Fugate. 2005. Measuring temporal compositions of urban morphology through spectral mixture analysis: Toward a soft approach to change analysis in crowded cities. *International Journal of Remote Sensing* 26(4):699–718.

Raudys, S., and V. Pikelis. 1980. On dimensionality, sample size, classification error and complexity of classification algorithms in pattern recognition. *IEEE Transactions on Pattern Analysis and Machine Intelligence* 2:242–252.

Ray, T. W., and B. C. Murray. 1996. Non-linear mixing in desert vegetation. *Remote Sensing of Environment* 55:59–64.

Reed, R. 1993. Pruning algorithms—a survey. *IEEE Transactions on Neural Networks* 4:740–747.

Reeves, S. J. 1992. A cross-validation framework for solving image restoration problems. *Journal of Visual Communication and Image Representation* 3:433–445.

Reillt, D. L., L. N. Cooper, and C. Elbaum. 1982. A neural model for category learning. *Biological Cybernetics* 45:35–41.

Riano, D., E. Chuvieco, J. Salas, and I. Aguado. 2003. Assessment of different topographic corrections in Landsat-TM data for mapping vegetation types. *IEEE Transactions on Geoscience and Remote Sensing* 41:1056–1061.

Ricchetti, E. 2000. Multispectral satellite image and ancillary data integration for geological classification. *Photogrammetric Engineering and Remote Sensing* 66:429–436.

Richards, J. A. 1986. *Remote sensing digital image analysis: An introduction*. Berlin: Springer-Verlag.

Riegler, G., and W. Mauser. 1998. Geometric and radiometric terrain correction of ERS SAR data for applications in hydrologic modelling. In *Proceedings of the International Geoscience and Remote Sensing Symposium (IGARSS'98)*, Seattle, WA, July 6–10, 1998, 2603–2605. Piscataway, NJ: IEEE.

Ripley, B. D. 1996. *Pattern recognition and neural networks*. Cambridge: Cambridge University Press.

Ritter, H., and K. Schulten. 1988. Convergence properties of Kohonen's topology conserving maps: Fluctuations, stability and dimension selection. *Biological Cybernetics* 60:59–71.

Roberts, D. A., M. Gardner, R. Church, S. L. Ustin, G. Scheer, and R. O. Green. 1998. Mapping Chaparral in the Santa Monica Mountains using multiple endmember spectral mixture models. *Remote Sensing of Environment* 65:267–279.

Robinson, G. D., H. N. Gross, and J. R. Schott. 2000. Evaluation of two applications of spectral unmixing models to image fusion. *Remote Sensing of Environment* 71:272–281.

Roli, F., G. Giacinto, and G. Vernazza. 1997. Comparison and combination of statistical and neural net algorithms for remote sensing image classification. In *Neurocomputation in remote sensing data analysis,* ed. I. Kanellopoulos, G. G. Wilkinson, G. G. Roli, and J. Austin, 117–124. Berlin: Springer-Verlag.

Rosenblatt, R. 1959. *Principles of neurodynamics*. New York, Spartan Books.

Rosenfield, G. 1981. Analysis of variance of thematic mapping experiment data. *Photogrammetric Engineering and Remote Sensing* 47:1685–1692.

Rosenfield, G. H., K. Fitzpatrick-Lins, and H. S. Ling. 1982. Sampling for thematic map accuracy testing. *Photogrammetric Engineering and Remote Sensing* 48:131–137.

Rosin, P. L., and J. Hervás. 2005. Remote sensing image thresholding methods for determining landslide activity. *International Journal of Remote Sensing* 26(6):1075–1092.

Rumelhart, D. E., G. E. Hinton, and R. J. Williams. 1986a. Learning internal representation by error propagation. In *Parallel distributed processing: Explorations in the microstructures of cognition*, 318–362. Cambridge, MA: MIT Press.

Rumelhart, D. E., G. E. Hinton, and R. J. Williams. 1986b. Learning representations by back-propagating errors. *Nature* 323:533–536.

Sá, A. C. L., J. M. C. Pereira, M. J. P. Vasconcelos, J. M. N. Silva, N. Ribeiro, and A. Awasse. 2003. Assessing the feasibility of sub-pixel burned area mapping in Miombo woodlands of northern Mozambique using MODIS imagery. *International Journal of Remote Sensing* 24(8):1783–1796.

Sabol, D. E., A. R. Gillespie, J. B. Adams, M. O. Smith, and C. J. Tucker. 2002. Structural stage in Pacific Northwest forests estimated using simple mixing models of multispectral images. *Remote Sensing of Environment* 80:1–16.

Safavian, S. R., and D. Landgrebe. 1991. A survey of decision tree classifier methodology. *IEEE Transactions on Systems, Man and Cybernetics* 21:660–674.

Saguib, S. S., C. A. Bouman, and K. Sauer. 1998. ML parameter estimation for Markov random fields with applications to Bayesian tomography. *IEEE Transactions on Image Processing* 7:1029–1044.

Salford Systems. 2008. CART overview. http://www.Salford-systems.com/cart.php. (Accessed 20 December 2008.)

Sali, E., and H. Wolfson. 1992. Texture classification in aerial photographs and satellite data. *International Journal of Remote Sensing* 13:3395–3408.

Sarkar, N., and B. B. Chaudhuri. 1994. An efficient differential box-counting approach to compute the fractal dimension of images. *IEEE Transactions on Systems, Man and Cybernetics* 24:115–120.

Sarle, W. S. 2000. *Neural network frequently asked questions,* ftp://ftp.sas.com/pub/neural/FAQ.html (accessed 15 May 2008).

Schaale, M., and R. Furrer. 1995. Land surface classification by neural networks. *International Journal of Remote Sensing* 16:3003–3031.

Schaffer, C. 1993. Selecting a classification method by cross validation. *Machine Learning* 13:135–143.

Schapire, R. E., and Y. Singer. 1998. BoosTexter: A system for multiclass multi-label text categorization. *Machine Learning* 39:135–168

Schepers, J. 2004. Integrating remote sensing and ancillary information into management systems. In *Remote sensing for agriculture and the environment*, ed. S. Stamadiadis, J. M. Lynch, and J. S. Schepers, 254–259. Larissa Greece: Peripheral Editions.

Schistad, A. H., A. K. Jain, and T. Taxt. 1994. Multisource classification of remotely sensed data: Fusion of Landsat TM and SAR images. *IEEE Transactions on Geoscience and Remote Sensing* 32:768–778.

Schistad, A. H., T. Taxt, and A. K. Jain. 1996. A Markov random field model for classification of multisource satellite imagery. *IEEE Transactions on Geoscience and Remote Sensing* 34:100–113.

Schouten, T. E., and M. S. Klein Gebbinck. 1997. A neural net approach to spectral mixture analysis. In *Neurocomputation in remote sensing data analysis*, ed. I. Kanellopoulos, G. G. Wilkinson, F. Roli, and J. Austin, 79–85. Berlin: Springer-Verlag.

Schowengerdt, R. A. 1983. *Techniques for image processing and classification in remote sensing*. New York: Academic Press.

Schowengerdt, R. A. 1996. Soft classification and spectral-spatial mixing. In *Soft computing in remote sensing data analysis*, ed. E. Binaghi, P. A. Brivio, and A. Rampini, 1–6. Singapore: World Scientific.

Schowengerdt, R. A. 1998. *Remote sensing: Models and methods for image processing.* New York: Academic Press.

Schröder, M., M. Walessa, H. Rehrauer, K. Seidel, and M. Datcu. 2000. Gibbs random field models: A toolbox for spatial information extraction. *Computers and Geosciences* 26:423–432.

Serneels, S., M. Said, and E. F. Lambin. 2001. Land-cover changes around a major East African wildlife reserve: The Mara ecosystem. *International Journal of Remote Sensing* 22:3397–3420.

Settle, J. 1996. On the relationship between spectral unmixing and subspace projection. *IEEE Transaction on Geoscience and Remote Sensing* 34:1045–1046.

Settle, J. J., and N. A. Drake. 1993. Linear mixing and the estimation of ground cover proportions. *International Journal of Remote Sensing* 14:1159–1177.

Shafer, G. 1979. *A mathematical theory of evidence.* Princeton, NJ: Princeton University Press.

Shafer, G. 1987. Belief functions and possibility measures. In *Analysis of fuzzy information. Vol. 1: Mathematics and logic*, ed. J. C. Bezdek, 51–84. Boca Raton, FL: CRC Press.

Shafer, G., and R. Logan. 1987. Implementing Dempster's rule for hierarchical evidence. *Artificial Intelligence* 33:271–298.

Shannon, C. E. 1948. A mathematical theory of communication. *Bell System Technical Journal* 27:379–423, 623–656.

Shannon, C. E., and W. Weaver. 1963. *The mathematical theory of communication.* Chicago: University of Illinois Press.

Sheng, Y. 1996. *Wavelet transform. The transforms and applications handbook*, ed. A. D. Poularikas, 747–827. Boca Raton, FL: CRC Press.

Sheng, Y., and Z. G. Xia. 1996. A comprehensive evaluation of filters for radar speckle suppression. *Proceedings of the International Geoscience and Remote Sensing Symposium (IGARSS'96)*, Lincoln, Nebraska, May 27–31, 1996, 1559–1561. Piscataway, NJ: IEEE.

Shölkopf, B., K. K. Sung, C. Burges, F. Girosi, P. Niyogi, T. Poggio, and V. Vapnik. 1997. Comparing support vector machines with Gaussian kernels to radial basis function classifiers. *IEEE Transactions on Signal Processing* 45:2758–2765.

Shoshany, M., and T. Svoray. 2002. Multidate adaptive unmixing and its application to analysis of ecosystem transition along a climatic gradient. *Remote Sensing of Environment* 82:5–20.

Siegel, A. F. 1982. Robust regression using repeated medians. *Biometrika* 69:242–244.

Simard, M., G. De Grandi, S. Saatchi, and P. Mayaux. 2002. Mapping tropical coastal vegetation using JERS-1 and ERS-1 radar data with a decision tree classifier. *International Journal of Remote Sensing* 23:1461–1474.

Simoncelli, E. P. 1999. Bayesian denoising of visual images in the wavelet domain. In *Bayesian inference in wavelet based models*, ed. P. Muller and B. Vidakovic, 291–308. New York: Springer-Verlag.

Singh, A. 1984. Some clarifications about the pairwise divergence method in remote sensing. *International Journal of Remote Sensing* 5:623–627.

Singh, A. 1986. Change detection in the tropical forest environment of North-eastern India using Landsat. In *Remote sensing and tropical land management*, 237–254. New York: John Wiley and Sons.

Singh, A. 1989. Digital change detection techniques using remotely-sensed data. *International Journal of Remote Sensing* 10:989–1003.

Skidmore, A. K. 1989. A comparison of techniques for calculating gradient and aspect from a gridded digital elevation model. *International Journal of Geographical Information Systems* 3:323–334.

Skidmore, A. K., B. J. Turner, W. Brinkhof, and E. Knowles. 1997. Performance of a neural network: Mapping forests using GIS and remote sensing data. *Photogrammetric
Engineering & Remote Sensing* 63(5):501–514.

Small, C. 2001. Estimation of urban vegetation abundance by spectral mixture analysis. *International Journal of Remote Sensing* 22:1305–1334.

Smith, G. M., and E. J. Milton. 1999. The use of the empirical line method to calibrate remotely
sensed data to reflectance. *International Journal of Remote Sensing* 20:2653–2662.

Smith, J. A., T. L. Lin, and K. Ranson. 1980. The Lambertian assumption and Landsat data.
Photogrammetric Engineering and Remote Sensing 46:1183–1189.

Smits, P. C. 2002. Multiple classifier systems for supervised remote sensing image classification based on dynamic classifier selection. *IEEE Transactions on Geoscience and
Remote Sensing* 40:801–813.

Solaiman, B., M. C. Mouchot, and A. Hillion. 1996. Contextual dynamic neural networks
learning in multispectral images classification. In *Proceedings of the International
Geoscience and Remote Sensing Symposium (IGARSS'96)*, Lincoln, Nebraska, May
27–31, 1996, 523–525. Piscataway, NJ: IEEE.

Songh, C. W., C. E. Woodstock, K. C. Seto, M. P. Lenney, and S. A. Macomber. 2001. Classification and change detection using Landsat TM data: When and how to correct atmospheric effects. *Remote Sensing of Environment* 75:230–244.

SPSS. 2006. http://www.spss.com (accessed 20 December 2007).

Srinivasan, A., and J. A. Richards. 1990. Knowledge-based techniques for multi-source classification. *International Journal of Remote Sensing* 11:505–525.

Starck, J.-L., F. Murtagh, and A. Bijaoui. 1998. *Image processing and data analysis*. Cambridge:
Cambridge University Press.

Steele, B. M. 2000. Combining multiple classifiers-an application using spatial and remotely
sensed information for land cover type mapping. *Remote Sensing of Environment*
74:545–556.

Stehman, S. V. 1992. Comparison of systematic and random sampling for estimating the accuracy of maps generated from remotely sensed data. *Photogrammetric Engineering and
Remote Sensing* 58:1343–1350.

Stehman, S. V., and R. L. Czaplewinski. 1998. Design and analysis for thematic map accuracy
assessment—fundamental principles. *Remote Sensing of Environment* 64:331–344.

Strahler, A. H. 1980. The use of prior probabilities in maximum likelihood classification of
remotely sensed data. *Remote Sensing of Environment* 10:135–163.

Strahler, A. H. 1981. Stratification of natural vegetation for forest and rangeland inventory
using Landsat digital imagery and collateral data. *International Journal of Remote
Sensing* 2:15–41.

Strahler, A. H., and N. A. Bryant. 1978. Improving forest cover classification accuracy from
Landsat by incorporating topographic information. In *Proceeding 12th International
Symposium on Remote Sensing of the Environment*, Ann Arbor, MI, 927–942.

Strand, E., A. M. S. Smith, S. C. Bunting, L. A. Verling, D. B. Hann, and P. E. Gessler. 2006.
Wavelet estimation of plant spatial patterns in multi-temporal aerial photography. *International Journal of Remote Sensing* 27:2049–2054

Strat, T. M. 1984. Continuous belief functions for evidential reasoning. In *Proceedings of
the Fourth National Conference on Artificial Intelligence*, 6–10, August, Austin, TX,
308–313.

Sun, W., G. Xu, P. Gong, and S. Liang. 2006. Fractal analysis of remotely sensed images: A review
of methods and applications. *International Journal of Remote Sensing* 27:4963–4990.

Swain, P. H., and S. M. Davis. 1978. *Remote sensing: The quantitative approach*. New
York: McGraw-Hill.

Swain, P. H., and H. Hauska. 1977. The decision tree classifier: Design and potential. *IEEE Transactions on Geoscience and Remote Sensing* 15:142–147.

Switzer, P., and A. A. Green. 1984. *Min/Max autocorrelation factors for multivariate spatial imagery*. Technical Report No. 6, Department of Statistics, Stanford University, Stanford, CA.

Szeliski, R. 1989. *Bayesian modeling of uncertainty in low-level vision*. Rotterdam: Kluwer.

Taha, I. A., and J. Ghosh. 1999. Symbolic interpretation of artificial neural networks. *IEEE Transactions on Knowledge and Data Engineering* 11:448–463.

Takagi, H., and L. Hayashi. 1991. NN-driven fuzzy reasoning. *International Journal of Approximate Reasoning*, 5:191–212.

Takagi, H., and M. Sugeno. 1985. Fuzzy identification of systems and its application to modelling and control. *IEEE Transactions on Systems, Man and Cybernetics* 15:116–132.

Tanré, D., C. Deroo, P. Duhaut, M. Herman, J. J. Morcette, J. Perbos, and P. Y. Deschamps. 1986. Simulation of the satellite signal in the solar spectrum (5S), *Technical Report*, Laboratoire d'Optique Atmospherique, Universite des Sciences et Techniques de Lille, 59655 Villeneuve d'ascq Cedex, France.

Tanré, D., C. Deroo, P. Duhaut, M. Herman, J. J. Morcette, J. Perbos, and P. Y. Deschamps. 1990. Description of a computer code to simulate the satellite signal in the solar spectrum: The 5S code. *International Journal of Remote Sensing* 11:659–668.

Tatem, A. J., H. G. Lewis, P. M. Atkinson, M. S. Nixon. 2002. Multiple-class land-cover mapping at the sub-pixel scale using a Hopfield neural network. *International Journal of Applied Earth Observation and Geoinformation* 3:184–190.

Teggi, S., R. Cecchi, and F. Serafina. 2003. TM and IRS-1C-PAN data fusion multiresolution decomposition methods based on the "a trous" algorithm. *International Journal of Remote Sensing* 24(6):1287–1301.

Teillet, P. M., J. L. Barker, B. L. Markham, R. R. Irish, G. Fedosejevs, and J. C. Storey. 2001. Radiometric cross-calibration of the Landsat-7 ETM+ and Landsat-5 TM sensors based on tandem data sets. *Remote Sensing of Environment* 78:39–54.

Terzopoulos, D. 1984. Multiresolution computation of visible-surface representation. Ph.D. Thesis, Department of Electrical Engineering and Computer Science, MIT, Cambridge, MA.

Therrien, C. W. 1989. *Decision, estimations and classification: An introduction to pattern recognition and related topics*. New York: John Wiley and Sons.

Theseira, M. A., G. Thomas, J. C. Taylor, F. Gemmell, and J. Varjo. 2003. Sensitivity of mixture modelling to end member selection. *International Journal of Remote Sensing* 24:1559–1575.

Thome, K, P. Slater, S. Biggar, B. Markham, and J. Barker. 1997. Radiometric calibration of Landsat. *Photogrammetric Engineering and Remote Sensing* 63:853–858

Thornton, M. W., P. M. Atkinson, and D. A. Holland. 2006. Sub-pixel mapping of rural land cover objects form fine spatial resolution satellite sensor imagery using super-resolution pixel-swapping. *International Journal of Remote Sensing* 27:473–491.

Tian, Y., P. Guo, and M. R. Lyu. 2005. Comparative studies on feature extraction methods for multispectral remote sensing image classification. *Proceedings of IEEE International Conference on Systems, Man and Cybernetics*, 10–12 October 2005, Waikoloa, Hawaii, 2, 1275–1279.

Tikhonov, A. N., and V. A. Arsenin. 1977. *Solutions of ill-posed problems*. Washington: Winston and Sons.

Touzi, R., A. Lopes, and P. Bousquet. 1988. A statistical and geometrical edge detector for SAR images. *IEEE Transactions on Geoscience and Remote Sensing* 26:764–773.

Towell, G., and J. Shavlik. 1994. Knowledge-based artificial neural networks. *Artificial Intelligence* 70:119–165.

Townshend, J. R. G., ed. 1981. *Terrain analysis and remote sensing*. London: George Allen & Unwin.

Townshend, J. R. G., C. Huang, S. N. V. Kalluri, R. S. Defries, S. Liang, and K. Yang. 2000. Beware of per-pixel characteristics of land cover. *International Journal of Remote Sensing* 21:839–843.

Townshend, J. R. G., C. O. Justice, C. Gurney, and J. McManus. 1992. The impact of misregistration on change detection. *IEEE Transactions on Geoscience and Remote Sensing* 30:1054–1060.

Tso, B. 1997. Investigation of alternative strategies for incorporating spectral, textural and contextual information in remote sensing image classification. Ph.D. Thesis, School of Geography, The University of Nottingham, Nottingham, England, U.K.

Tso, B., and P. M. Mather. 1999. Classification of multisource remote sensing imagery using a Genetic Algorithm and Markov Random Fields. *IEEE Transactions on Geoscience and Remote Sensing* 37:1255–1260.

Tso, B., and R. C. Olsen. 2005a. Combining spectral and spatial information into hidden Markov models for unsupervised image classification. *International Journal of Remote Sensing* 26:2113–2133.

Tso, B., and R. C. Olsen. 2005b. A contextual classification scheme based on MRF model with improved parameter estimation and multiscale fuzzy line process. *Remote Sensing of Environment* 97:127–136.

Tukey, J. W. 1977. *Exploratory data analysis*. Reading, MA: Addison-Wesley.

Tumer, K., and J. Ghosh. 1995. *Theoretical foundations of linear and order statistics combiners for neural pattern classifiers*, Technical Report TR-95-02-98, The Computer and Vision Research Center, The University of Texas at Austin.

Tupin, F., H. Maitre, J. F. Mangin, J. M. Nicolas, and E. Pechersky. 1998. Detection of linear features in SAR images: Application to road network extraction. *IEEE Transactions on Geoscience and Remote Sensing* 36:434–454.

Turket, P. D. 1995. Cost-sensitive classification: Empirical evaluation of a hybrid genetic decision tree induction algorithm. *Journal of Artificial Intelligence Research* 2:369–409.

Ulaby, F. T., A. K. Fung, and R. K. Moore. 1982. *Microwave remote sensing* (two volumes). Reading, MA: Addison-Wesley.

Ulaby, F. T., A. K. Fung, and R. K. Moore. 1986a. *Microwave remote sensing*, Vol. 3. Dedham, MA: Artech House.

Ulaby, F. T., F. Kouyate, B. Brisco, and L. Williams. 1986b. Texture information in SAR images. *IEEE Transactions on Geoscience and Remote Sensing* 24:235–245.

Ulfarsson, M. O., J. A. Benediktsson, and J. R. Sveinsson. 2003. Data fusion and feature extraction in the wavelet domain. *International Journal of Remote Sensing* 24:3933–3945.

Ustin, S. L., D. Dipietro, K. Olmstead, E. Underwood, and G. J. Scheer. 2002. Hyperspectral remote sensing for invasive species detection and mapping. In *Proceedings of the International Geoscience and Remote Sensing Symposium (IGARSS'02)*, 24–28 June, Toronto, Canada, 1658–1660. Piscataway, NJ: IEEE.

Van de Hulst, H. C. 1981. *Light scattering by small particles*. New York: Dover,

Van der Meer, F. 1995. Spectral unmixing of Landsat Thematic Mapper data. *International Journal of Remote Sensing* 16:3189–3194.

Van der Meer, F. 1999a. Iterative spectral unmixing. *International Journal of Remote Sensing* 20:3431–3436.

Van der Meer, F. 1999b. Geostatistical approaches for image classification and assessment of uncertainty in geologic processing. In *Advances in remote sensing and GIS analysis*, ed. P. M. Atkinson and N. J. Tate, 147–166. Chichester: John Wiley and Sons.

Van der Meer, F., and S. M. De Jong. 2000. Improving the results of spectral unmixing of Landsat TM imagery by enhancing the orthogonality of end-members. *International Journal of Remote Sensing* 21:2781–2797.

Van der Meer, F., K. Scholte, S. De Jong, and M. Dorrestijn. 1998. Scaling to MERIS resolution: Mapping accuracy and spatial variability. In *EARSeL Workshop on Imaging Spectroscopy,* University of Zurich, 6–8 October 1988, 147–153. Paris: EARSeL.

Van Genderen, J. L., and B. F. Lock. 1977. Testing land use map accuracy. *Photogrammetric Engineering and Remote Sensing* 43:1135–1137.

Van Zyl, J.J., B. D. Chapman, P. Dubois, and J. C. Shi. 1993. The effect of topography on SAR calibration. *IEEE Transactions on Geoscience and Remote Sensing.* 31:1036–1043.

Van Zyl, J. J., H. A. Zebker, and C. Elachi. 1987. Image radar polarization signatures: Theory and observation, *Radio Science* 22:529–543.

Vapnik, V. 1979. *Estimation of dependences based on empirical data* [in Russian]. Nauka, Moscow. (English translation: Springer-Verlag, New York).

Vapnik, V. 1995. *The nature of statistical learning theory.* New York: Springer-Verlag.

Vapnik, V. 1998. *Statistical learning theory.* New York: John Wiley.

Vapnik, V., and O. Chapelle. 2000. Bounds on error expectation for support vector machines. *Neural computation* 12:2013–2036.

Verbyla, D. L., and S. H. Boles. 2000. Bias in land cover change estimates due to misregistration. *International Journal of Remote Sensing* 21:3553–3560.

Vermote, E. F., D. Tanré, J. L. Deuze, M. Herman, and J. J. Morcette. 1997. Second simulation of the satellite signal in the solar spectrum, 6S: An overview. *IEEE Transactions on Geoscience and Remote Sensing* 35:675–686.

Vieira, C. A. O. 2000. Accuracy of remote sensing classification of agricultural crops: A comparative study. Ph.D. thesis, School of Geography, The University of Nottingham, Nottingham, UK.

Vieira, C. A. O., and P. M. Mather. 1999. Assessing the accuracy of classifications using remotely sensed data. In *Proceedings of the 4th International Airborne Remote Sensing Conference/21st Canadian Symposium on Remote Sensing*, Ottawa, Canada, 21–24 June 1999, 2, 823–830.

Voss, R. 1986. Random fractals: Characterization and measurement. In *Scaling phenomena in disordered systems*, ed. R. Pynn and A. Skjeltorp, 37–48. New York: Plenum Press.

Wahbe, G. 1990. *Spline models for observation data.* Philadelphia, PA: Society for Industrial and Applied Mathematics.

Wald, L. 1999. Some terms of reference in data fusion. *IEEE Transactions on Geosciences and Remote Sensing* 37(3):1190–1193

Wald, L. 2000. Quality of high resolution synthesized images: Is there a simple criterion. International Conference on Fusion of Earth Data, *France*, Nice, France, 99–105.

Wald, L. 2002. *Data fusion.* Paris: Ecole des Mines de Paris.

Wald, L., T. Ranchin, and M. Mangolini. 1997. Fusion of satellite images of different spatial resolutions: Assessing the quality of resulting images. *Photogrammetric Engineering and Remote Sensing* 63:691–699.

Walpole, R. E. 1982. *Introduction to statistics.* New York: Macmillan.

Wang, F. 1990a. Fuzzy supervised classification of remote sensing images. *IEEE Transactions on Geoscience and Remote Sensing* 28:194–201.

Wang, F. 1990b. Improving remote sensing image analysis through fuzzy information representation. *Photogrammetric Engineering and Remote Sensing* 56:1163–1169.

Wang, G., G. Gertner, and A. B. Anderson. 2005. Sampling design and uncertainty based on spatial variability of spectral variables for mapping vegetation cover. *International Journal of Remote Sensing* 26:3255–3274.

Wang, G. J., Y. D. Cheng, and K. Chang. 2005. A new neural fuzzy control system with heuristic learning. *International Journal of Fuzzy Systems* 7:158–168.

Wang, L., and D. C. He. 1990. A new statistical approach for texture analysis. *Photogrammetric Engineering and Remote Sensing* 56:61–66.

Wang, X., and H. Wang. 2004. Markov random field modeled range image segmentation. *Pattern Recognition Letters* 25(3):367–375.

Wang, Y., and D. Dong. 1997. Retrieving forest stand parameters from SAR backscatter data using a neural network trained by a canopy backscatter model. *International Journal of Remote Sensing* 18:981–989.

Wang, Z., and A. C. Bovik. 2002. A universal image quality index. *IEEE Signal Processing Letters* 9:81–84

Warner, T. 2005. Hyperspherical direction cosine change vector analysis. *International Journal of Remote Sensing,* 26:1201–1215.

Wei, J., and I. Gertner. 2003. MRF-MAP-MFT visual object segmentation based on motion boundary field. *Pattern Recognition Letters* 24(16):3125–3139.

Welch, R. M., S. K. Sengupta, A. K. Goroch, P. Rabindra, N. Rangaraj, and M. S. Navar. 1992. Polar cloud and surface classification using AVHRR imagery: An intercomparison of methods. *Journal of Applied Meteorology* 31:405–420.

Wessman, C. A., C. A. Bateman, and T. A. Benning. 1997. Detecting fire and grazing patterns in tallgrass prairie using spectral mixture analysis. *Ecological Applications* 7:493–511.

Whitsitt, S. J., and D. A. Landgrebe. 1977. *Error estimation and separability measures in feature selection for multiclass pattern recognition,* LARS Publication 082377, Laboratory for Application of Remote Sensing. W. Lafayette, IN: Purdue University.

Widrow, B., and M. E. Hoff. 1960. Adaptive switching circuits. In *IRE WESCON Convention Record, August,* Part 4, 96–104.

Wilkinson, G. G., F. Fierens, and I. Kanellopoulis. 1995. Integration of neural and statistical approaches in spatial data classification. *Geographical Systems* 2:1–20.

Wilkinson, G. G., and J. Mégier. 1990. Evidential reasoning in a pixel classification hierarchy—a potential method for integrating image classifiers and expert system rules based on geographic context. *International Journal of Remote Sensing* 11:1963–1968.

Winkler, R. L. 1968. The consensus of subjective probability distributions. *Management Science* 15:B61–B75.

Witten, I. H., and E. Frank. 2005. *Data Mining: Practical Machine Learning Tools and Techniques.* San Francisco, CA: Morgan Kaufmann.

Won, C. S., and H. Derin. 1992. Unsupervised segmentation of noisy and textured images using Markov random fields. *Computer Vision, Graphics and Image Processing (CVGIP): Graphics Models and Image Processing* 54:308–328.

Woodcock, C. E., and A. H. Strahler. 1983. Characterising spatial patterns in remotely sensed data. In *Proceedings of the 17th International Symposium on Remote Sensing of Environment,* 839–852. Ann Arbor, MI: University of Michigan.

Woodcock, C. E., and A. H. Strahler. 1987. The factor of scale in remote sensing. *Remote Sensing of Environment* 21:311–332.

Woodcock, C. E., and A. H. Strahler, and D. B. Jupp. 1988a. The use of variogram in remote sensing: I. Scene models and simulated images. *Remote Sensing of Environment* 25:323–348.

Woodcock, C. E., and A. H. Strahler, and D. B. Jupp. 1988b. The use of variogram in remote sensing: II. Real digital images *Remote Sensing of Environment* 25:349–379.

Wu, Y., and H. Maitre. 1992. Smoothing speckled synthetic aperture radar image by using maximum homogeneous region filters. *Optical Engineering* 31:1785–1792.

Xu, M., P. Watanachaturaporn, P. K. Varshney, and M. K. Arora. 2005. Decision tree regression for soft classification of remote sensing data. *Remote Sensing of Environment* 97:322–336.

Yang, X., and C. P. Lo. 2000. Relative radiometric normalization performance for change detection from multi-data satellite images. *Photogrammetric Engineering and Remote Sensing* 66:967–980.

Yokoya, N., K. Yamamoto, and N. Funakubo. 1989. Fractal-based analysis and interpolation of 3D natural surface shapes and their application to terrain modelling. *Computer Vision, Graphics and Image Processing* 46:284–302.

Zadeh, L. A. 1965. Fuzzy sets. *Information Control* 8:338–353.

Zebker, H.A., and J. J. Van Zyl. 1991. Imaging radar polarimetry: A review. *Proceedings of the IEEE* 79:1583–1606.

Zebker, H.A., J. J. Van Zyl, S. L. Durden, and L. Norikane. 1991. Calibrated imaging radar polarimetry: Technique, examples and applications. *IEEE Transactions on Geoscience and Remote Sensing* 29:942–961.

Zebker, H. A., J. J. Van Zyl, and D. N. Held. 1987. Imaging radar polarimetry from wave synthesis. *Journal of Geophysical Research* 92(B1):683–701.

Zebker, H.A., C. L. Werner, P. A. Rosen, and S. Hensley. 1994. Accuracy of topographic maps derived from ERS-1 interferometric radar. *IEEE Transactions on Geoscience and Remote Sensing* 32:823–836.

Zhang, C., S. E. Franklin, and M. A. Wulder. 2004. Geostatistical and texture analysis of airborne-acquired images used in forest classification. *International Journal of Remote Sensing* 25:859–865.

Zheng, J. S., and Y. W. Leung. 2004. Improved possibilistic c-means clustering algorithms. *IEEE Transactions on Fuzzy Systems* 12:209–217.

Zhong, J., and R. Wang. 2006. Multi-temporal remote sensing change detection based on independent component analysis. *International Journal of Remote Sensing* 27(10): 2055–2061.

Zhu, C., and X. Yang. 1998. Study of remote sensing image texture analysis and classification using wavelets. *International Journal of Remote Sensing* 13:3167–3187.

Zhu, H. 2005. Linear spectral unmixing assisted by probability guided and minimum residual exhaustive search for subpixel classification. *International Journal of Remote Sensing* 26(24):5585–5601.

Zhu, L., and R. Tateishi. 2006. Fusion of multisensor multitemporal satellite data for land cover mapping. *International Journal of Remote Sensing* 27(5):903–918.

Zitová, B., and J. Flusser. 2003. Image registration methods: A survey. *Image and Vision Computing* 21:977–1000.

Index

T - #0091 - 101024 - C0 - 234/156/20 [22] - CB - 9781420090727 - Gloss Lamination